BEYOND THE IONOSPHERE

FIFTY YEARS OF SATELLITE COMMUNICATION

October 1995 marked the fiftieth anniversary of Arthur C. Clarke's famous article in *Wireless World* proposing the use of satellites placed in geosynchronous orbit for worldwide communications relay. The article proved prophetic, for it heralded the modern era of telecommunications. Beginning in the early 1960s, several series of satellites were launched into Earth's orbit; collectively they transformed the latter twentieth century, creating a global village of instantaneous communications.

Previously, the ionosphere had defined the limits of radio communication; today, by going beyond the ionosphere, broadband telecommunication has entered a new age. This book describes the first attempts to go beyond the ionosphere, including both the earliest uses of the Moon as a passive, natural relay satellite and Project Echo, the massive inflated satellite off which Earth stations bounced radio signals, as well as contemporary communications via active-repeater artificial moons in orbit about the Earth. It analyzes both American and foreign satellite communications, the histories of several satellite communications companies, the roles of government agencies, and the contribution of research laboratories.

The book is a collection of papers originally presented during an international symposium held in Washington, D.C., at the time of the fiftieth anniversary of Clarke's 1945 article. Contributions from historians and other scholars from throughout the world present a stimulating analysis of one of the most important global technologies at work today—and how it originated and evolved.

Andrew J. Butrica has edited a comprehensive and illuminating collection of articles on this important subject. It is vital reading for anyone seeking to understand the history of global telecommunications in the twentieth century.

Andrew J. Butrica received his Ph.D. in the history of science and technology at Iowa State University. He is a research historian in Franklin Park, New Jersey, specializing in the history of science. In 1990 Praeger Publishers issued his *Out of Thin Air: A History of Air Products and Chemicals, Inc., 1940–1990*; in 1996 his book *To See the Unseen: A History of Planetary Radar Astronomy* was published in the NASA History Series.

About the Cover: "Endeavour and Intelsat," acrylic by Jim Alford. A nine-day mission (7–16 May 1992) accomplished the capture of the Intelsat VI satellite and the subsequent mating of the satellite to a booster and its eventual redeployment. In this painting, the floating architectural components symbolize the union of the Space Shuttle Endeavour with the communications satellite Intelsat. Courtesy of the NASA Art Program, no. 92-HC-716.

BEYOND THE IONOSPHERE:

Fifty Years of Satellite Communication

NASA SP-4217

BEYOND THE IONOSPHERE:

Fifty Years of Satellite Communication

Andrew J. Butrica, Editor

The NASA History Series

National Aeronautics and Space Administration
NASA History Office
Washington, D.C. 20546   1997

Library of Congress Cataloguing-in-Publication Data

Beyond the Ionosphere: Fifty Years of Satellite Communication/Andrew J. Butrica, editor.
p. cm.—(NASA historical series) (NASA SP: 4217)
Includes bibliographical references (p. 289) and index.
1. Artificial satellites in telecommunication. I. Butrica, Andrew J. II. Series. III. Series: NASA SP; 4217.
TK5104.B48   1997
621.382'38—dc21   97-10856
                                                                                      CIP

To the pioneering dreamers and doers of satellite communications: Arthur C. Clarke, who envisioned the geosynchronous orbit and a global satellite television network; John Robinson Pierce, for transforming the visions of his science fiction into a science fact, Echo; Col. John H. DeWitt, Jr., and Zoltán Bay, for achieving the first successful space communications links; and James H. Trexler, who turned the Moon into humanity's first communications satellite.

Contents

Europe

Part III: The Unfolding of the World System

Geography, Politics, and Culture

Applications

Appendices

Acknowledgments

This volume originated several years ago out of my growing interest in communications satellites and their beginnings—an interest that had been sparked by my research into the history of radar astronomy. The two, I discovered, had much in common. I then met a pair of George Washington University graduate students who were preparing, or thinking of preparing, doctoral theses on the history of satellite communications. We decided to do something—initially undefined, though perhaps a conference session—on the subject sometime in the near future.

Like the proverbial "best laid plans of mice and men," this plan went astray, but in a good direction—namely, an entire conference on the development of satellite communications. That conference would not have taken place, however, were it not for the generosity of Roger D. Launius, NASA Chief Historian, and the NASA History Office. Dr. Launius made available the NASA Headquarters auditorium and the History Office staff, as well as a small stipend to cover limited expenses. Without his continuing support, the conference—and this volume—would not have come into existence. It was, moreover, a pleasure to work with Dr. Launius and with the entire History Office staff. In particular, the diligent service of Nadine J. Andreassen was above and beyond the call of duty.

I also want to thank those who served as session moderators during the conference: Sylvia K. Kraemer, senior policy analyst in the Office of Policy and Plans at the National Aeronautics and Space Administration, and former NASA Chief Historian; John Krige, formerly of the European Space Agency History Project, European University Institute, Florence, and currently director, Center for Research in the History of Science and Technology, Cité des Sciences et de l'Industrie, Paris; Pamela E. Mack, Professor of History, Clemson University; and Joseph N. Tatarewicz, Professor of History, University of Maryland, Baltimore County. Thanks, too, go to John V. Evans, Vice President and Director of Comsat Laboratories, and Michael Onufry for arranging the tour of Comsat Labs. Finally, appreciation is deserved by those who helped in the editing and composition of this volume: Janie Penn, Michael Crnkovic, Patricia Lutkenhouse Talbert, Jonathan L. Friedman, and Stanley Artis.

Andrew J. Butrica, Editor

Introduction

The year 1995 marked a number of anniversaries in the development of satellite communications:

- The thirty-fifth anniversary of the launch of Echo 1, the first passive communications balloon
- The thirtieth anniversary of the April 1965 launch of Early Bird (Intelsat I), Comsat's first satellite, which effectively began global satellite communications
- The twenty-fifth anniversary of the launch of NATOSAT, the first satellite stationed over the Atlantic Ocean to carry military traffic between the United States and its NATO allies
- Most notably, the fiftieth anniversary of Arthur C. Clarke's article published in *Wireless World*, in which he proposed the use of geosynchronously orbiting satellites for communications relay sites[1]

Satellite communications are at the very heart of the notion of a "global village" and constitute a continually growing, multibillion-dollar, nearly ubiquitous civil and military enterprise deserving recognition. Much fanfare accompanied the first satellite television broadcasts. Yet, as the technology has grown increasingly pervasive, satellites have became an almost invisible part of the cultural landscape. Simultaneously, satellite communication has become a tremendous international commercial success, currently worth around $15 billion dollars per year; it is on the verge of expanding spectacularly in the near future, perhaps to $30 billion per year by the end of the decade. Despite the expanding network of fiber optic cables, approximately 60 percent of all overseas communications pass via satellites. More than 200 countries and territories rely on nearly 200 satellites for defense, direct broadcast, navigational, and mobile communications, not to mention data collection and faxing, via domestic, regional, and global links.

Despite the commercial success and ubiquity of satellite communications, far too little attention has been paid to its development. For the most part, scholars have focused on politics and policy studies of the period roughly from 1958 to the mid-1970s (customarily centering their discussions on the passage of the Communications Satellite Act of 1962 and Intelsat negotiations), neglecting economic and technological questions and slighting the earlier work of the 1940s and 1950s. Recently, however, some scholars have attempted to address these overlooked areas of research. I saw a need to bring the results of their research to light, while in the process stimulating others to take up research in satellite communications history. Roger Launius, NASA's Chief Historian, and the NASA History Office staff graciously offered to help organize a symposium on the development of satellite communications titled "Beyond the Ionosphere: The Development of Satellite Communications." Through their efforts and hard work, the symposium took place at NASA Headquarters, 17–18 October 1995, and a tour of Comsat's research laboratories followed, thanks to John V. Evans, Vice President and Director of Comsat Laboratories. In addition, George Washington University hosted a one-day celebration of government-industry cooperation prior to the NASA History Office symposium.

---

1.   Arthur C. Clarke, "Extra-Terrestrial Relays: Can Rocket Stations Give World-Wide Radio Coverage?," *Wireless World* 51 (October 1945): 305–08. Serial issues are identified only when each issue is paginated separately; otherwise, only the volume number and year of publication are provided.

That symposium served as a forum for presenting not only the research results of scholars, but also the experiences of practitioners. Indeed, one of the motives for organizing the symposium was to create a vehicle that would facilitate fruitful interaction among scholars and practitioners. A few papers not presented during the symposium have been included in this book to provide additional temporal, geographical, and thematic coverage of the symposium's vast subject—the development of satellite communications. While complete coverage of such a subject is impossible at present, these "proceedings" attempt to present a broad, but systematic survey of the evolution of satellite communications. It is hoped that this diffusion of scholarly research and practical experience will both fill gaps in the literature and provide a framework for future studies. Finally, this work will have achieved its goal if it stimulates others to take up research in satellite communications history.

Figure 1

Arthur C. Clarke during a visit to the Goddard Space Flight Center in 1976. His article on "Extra-Terrestrial Relays," published in 1945, proposed a series of geostationary satellites. Today, the geostationary orbit is known also as the Clarke orbit. (Courtesy of NASA, photo no. 76-007273)

The contributions to this volume demonstrate, if nothing else, the dramatic temporal and geographic breadth and thematic richness of satellite communications history. The narrative, which has not a single strand, but many, reaches back nearly a half century to the first attempts to communicate via natural and artificial satellites; it extends from the United States and its northern and southern neighbors to the countries of Western and Eastern Europe, to India, Australia, and Asia, and to the rest of the globe. This book, then, is organized along temporal and geographical lines.

The Three Stages of Satellite Communications Development

The temporal ordering of the development of satellite communications is somewhat problematic. As the readings make clear, satellite communications developed along several evolutionary lines that variously intersected and diverged. In addition, there is the difficulty of periodizing something whose existence was only dreamed 50 years ago. This work posits three stages of satellite communications development as a suggested framework for future research and discussion, but also to provide the diverse contributions of this volume with a rational organization.

The first stage of development, extending from the 1940s into the early 1960s, was distinguished by experiments with passive artificial and natural satellites. Long before the launch of Sputnik, investigators in the United States and Europe attempted to establish long-distance communications using the Moon as a passive relay satellite, while others sought to create an artificial ionosphere (Project Needles) or to use meteor ionization trails (meteor burst communications). Military and business funding and the communications needs of both shaped these experiments.

Preceding these experiments was, of course, several decades of experience with radio communications, dating back to the pioneering work of Guglielmo Marconi and others in Europe and the United States. To achieve transmission distances beyond the horizon, long-distance radio systems ricocheted signals off an ionized portion of the atmosphere called the ionosphere. The ionosphere retained this communications role until the advent of satellite communications, initially in the guise of lunar relay experiments. Thus, the development of satellite communications can be thought of as a prolonged attempt to achieve long-distance communications by going "beyond the ionosphere."

The second stage of satellite communications development began in 1958, not with Sputnik, but with the launch of the first communications satellite (SCORE) and the first teletype relay by satellite (Courier 1B). Project Echo followed, launched in 1960, then came Telstar (equipped with an active repeater) and Relay (the first satellite to transmit television worldwide) in 1962, Syncom 2 (the first geosynchronous communications satellite) in 1963, the first operational commercial communications satellite (Intelsat I, alias "Early Bird") in 1965, and in 1966 the first operational military communications satellite (IDSCS). By 1966, then, the era of satellite communications was well on its way.

The placement of these satellites in orbit by the United States alone signaled that country's dominance of the field and precipitated a series of highly political negotiations on both sides of the Atlantic Ocean that eventually led to the creation in 1964 of Intelsat (International Telecommunications Satellite organization), an international framework for the growth of satellite communications. Comsat, a corporation created by congressional passage of the Communications Satellite Act of 1962, became the key organizational instrument through which the United States influenced the making of decisions and the letting of contracts within Intelsat. This second stage of development, then, saw the creation of satellite communications institutions and the establishment of management at the international level, while a single country, the United States, dominated satellite communications technology and services.

Although that domination continued throughout the 1960s and 1970s, more and more countries acquired access to communications satellites, especially in Europe, and challenged the U.S. monopoly. The nature of satellite communications has undergone deep changes since the late 1970s to the present, the period of the third stage of satellite communications development. During this period, the U.S. lead in satellite communications technologies and services established during the 1960s and 1970s waned, while Europe and Japan invested heavily in satellite communications research and development in the hopes of harvesting economic benefits and began to pose a major technological and economic challenge to the United States. Although the United States retained a leading position in the marketplace, it lost ground in the technologies and systems that held the key to future communications markets.

More than ever before, the geopolitics of satellite communications came into its own during the third stage of development, as the small club of countries with satellite access grew into a global public enterprise, embracing first the countries of North America and Western Europe, then outward into Eastern Europe, Asia, Africa, and South America. As satellite geographic coverage increased, the types of services offered multiplied. Fixed (as opposed to mobile) satellite communications services reached maturity during the 1960s and 1970s, while mobile and broadcast services underwent explosive growth during the third stage of satellite communications development. Growth in fixed satellite services slowed to an annual rate of about 10 percent, while broadcast and mobile communications services thrived (growing more than 20 percent annually). The International

Figure 2

The telephone shown here will provide voice, fax, and data services via the Iridium satellite communications system, as well as through terrestrial cellular telephone networks (availability and compatibility permitting), beginning in 1998. (Courtesy of John Windolph)

Maritime Satellite Organization (Inmarsat) introduced maritime service; it then branched out into aircraft and mobile land services. By the mid-1990s, the fastest growing field was personal communications services via satellite using hand-held transceivers similar to those used in cellular radio.

Put in geopolitical terms, though, what characterized the third stage of satellite communications development was the creation of a true international satellite system—a system that was international not only because satellite coverage was global, but because of the increasing number of countries with satellite access. Indeed, the global satellite system was built country by country. The nation-state was a fundamental engine of growth, serving as both the initiator and chief consumer of satellite communications services in a given geographical and political territory. This circumstance was to be expected; in most countries of the world, the state traditionally was the sole provider and major user of communications systems. Those systems embraced everything from postal services (including financial services) and telecommunications (telegraphy, telephony, and telex) to roads, bridges, and waterways (communications understood in the broadest sense of the word).

Because the global satellite system was built largely at the state level, to understand the character of the third stage of satellite communications development, we must consider each country (each telecommunications system provider) as a separate case. The results would mirror the complexity and heterogeneity of factors that influenced the development of satellite communications across the planet. Political and cultural factors likely would dominate these country-focused case studies, as the contributions to this volume affirm, although certainly technology and economics would play roles as well.

The Organization of This Book

The three overlapping stages of satellite communications development outlined above provide the three-part framework for the organization of the papers contained in this book. Part I, "Passive Origins," treats the first stage of satellite communications development, extending from the 1940s into the early 1960s, when passive artificial and natural satellites funded by the military and private enterprise established the field. Part II, "Creating the Global, Regional, and National Systems," addresses events that constituted the second stage of development. Early in this stage, which stretched from the 1960s into the 1970s, satellite systems began to make their appearance in the United States, while domestic and international efforts sought to bring order to this new, but chaotic, field in the form of Comsat and Intelsat.

While the first two parts of this book involve the United States and Western Europe, Part III, "The Unfolding of the World System," explores the development of satellite communications in the remainder of the world, with a strong emphasis on Asia. Thus, while the positing of three stages of satellite communications development serves as a temporal framework, the course taken in Part III is less determined by temporal limits than by geographical expansion. The political and cultural landscape of each country takes the lead in shaping that country's accession to satellite communications capability.

Politics and culture utilize satellite communications within the national space in one of two ways: to create internal political and cultural cohesion or to create cohesion among nation-states. Countries occupying large land masses, such as the United States, Canada, the Soviet Union, China, or the Indian subcontinent, inaugurated domestic satellite services to foster national cohesion, and so did the smaller states of Asia, but for different reasons. On the other hand, Western Europe initiated satellite programs in the name of regional integration, and the Soviet Union created Intersputnik, its own clone of Intelsat, to interconnect its client states (including Cuba). Moreover, to bolster colonial ties in the postcolonial era, Britain and France hoped to use satellites to communicate with their former colonies.

These are but some examples of the geopolitical motivations of states in acquiring satellite communications capability. However motivated the states were, though, the question of how the satellite communications capability was utilized must be addressed as well. Therefore, Part III includes a section on satellite applications in education and medicine, in mobile communications and navigation, and in corporate business strategies.

The three temporal stages of satellite communications development, and the focus in Part III on geographical expansion linked to factors of politics, culture, and national space, represent only a tentative framework for studying the development of satellite communications. Indeed, many themes not alluded to in this analytical schema merit study. Several of these themes will become apparent after a brief review of the papers contained in this volume.

Passive Origins

To understand the circumstances leading to the emergence of global satellite communications, we first must examine the technological, economic, and political world of cable telegraphy and telephony. That is the raison d'être for Daniel R. Headrick's study of the rivalry between radio and cable. Until the advent of wireless radio a century ago, all electrical communications traveled by cable. Cable and radio cohabited peacefully for a couple of decades until the arrival of shortwave radio, which was faster and cheaper than conventional long-wave radio. Headrick sees a parallel between the 1920s' rivalry of shortwave and cable and today's rivalry between satellites and fiber optic cable. Also from the era discussed by Headrick came the international carriers of record, such as American Telephone and Telegraph (AT&T), International Telephone and Telegraph (ITT), and Western Union International, which later played a key role in shaping U.S. satellite communications.

Although natural (meteors) and artificial (Project Needles) objects served as early passive relay satellites, the era of space communications actually began with the efforts on both sides of the Atlantic Ocean to use the Moon as a passive communications satellite. The Massachusetts Institute of Technology's (MIT's) Lincoln Laboratory carried out Project Needles, formerly known as Project West Ford, on behalf of the Department of Defense

(DoD). This project involved launching nearly 500 million hair-like copper wires into orbit in 1963, thereby forming a belt of dipole antennas. Lincoln Laboratory then used this artificial ionosphere to send messages between Camp Parks, California, and Westford, Massachusetts. British radio astronomers, including Sir Martin Ryle and Sir Bernard Lovell, as well as optical astronomers, objected fervently to Project Needles, and the Council of the Royal Astronomical Society formally protested to the U.S. President's Science Advisor.[2]

Meteor burst communications outlived Project West Ford and has endured to the present. The purpose of meteor burst communications is to obtain secure, point-to-point radio connections. It originated in the United States in radio propagation work at Stanford University carried out initially during the early 1950s. Von R. Eshleman, an electrical engineering graduate student, laid out in his dissertation a general theory of detecting meteor ionization trails and its application in actual experiments. After graduation, he developed this method of communicating in collaboration with his Stanford colleagues and with funding from the Air Force.[3] The Stanford research had nontrivial consequences. Eshleman's dissertation has continued to provide the theoretical foundation of modern meteor burst communications—a communications mode that promises to function even after a nuclear holocaust has rendered useless all normal wireless communications. The pioneering work at Stanford, as well as at the National Bureau of Standards and the Air Force's Cambridge Research Laboratories (without leaving out Jodrell Bank in Britain), received new attention in the 1980s when the Space Defense Initiative ("Star Wars") revitalized interest in using meteor ionization trails for jam-proof communications. Nonmilitary applications of meteor burst communications also have arisen in recent years.[4]

Concurrent with the research on meteor burst communications and Project West Ford, civilian and military investigators on both sides of the Atlantic Ocean attempted to use the Moon as a passive communications satellite. These lunar communications efforts succeeded. As a result, they constitute the first passive satellite communications link— a place in history usually accorded to the Echo balloon experiments. Those efforts are the subject of papers by David K. van Keuren and Jon Agar.

---

2.   Overhage to Ferguson, 26 June 1963, 1/24/AC 134, MIT Archives; Overhage and Radford, "The Lincoln Laboratory West Ford Program: An Historical Perspective," *Proceedings of the IEEE* 52 (1964): 452–54; "Project West Ford Releases and Reports," folder, Lincoln Laboratory Library Archives. Much of Volume 52 of the *Proceedings of the IEEE* addresses exclusively Project West Ford. On the antagonism of radio astronomers to Project Needles, see Bernard Lovell, *Astronomer by Chance* (New York: Basic Books, 1990), pp. 331–34; Martin Ryle and Bernard Lovell, "Interference to Radio Astronomy from Belts of Orbiting Dipoles (Needles)," *Quarterly Journal of the Royal Astronomical Society* 3 (1962): 100–08; D.E. Blackwell and R. Wilson, "Interference to Optical Astronomy from Belts of Orbiting Dipoles (Needles)," *ibid.*, pp. 109–17; Hermann Bondi, "The West Ford Project," *ibid.*, p. 99.

3.   Von R. Eshleman, interview with Andrew Butrica, 9 May 1994, Stanford University, JPL Archives; Von R. Eshleman, "The Mechanism of Radio Reflections from Meteoric Ionization," Ph.D. diss., Stanford University, 1952; Von R. Eshleman, *The Mechanism of Radio Reflections from Meteoric Ionization*, Technical Report No. 49 (Stanford, CA: Stanford Electronics Research Laboratory, 15 July 1952), pp. ii–iii, 3; Laurence A. Manning, "Meteoric Radio Echoes," *Transactions of the Institute of Radio Engineers* 2 (1954): 82–90; Laurence A. Manning and Von R. Eshleman, "Meteors in the Ionosphere," *Proceedings of the Institute of Radio Engineers* 47 (1959): 186–199.

4.   Robert Desourdis, telephone conversation, 22 September 1994; Donald Spector, telephone conversation, 22 September 1994; Donald L. Schilling, ed., *Meteor Burst Communications: Theory and Practice* (New York: John Wiley and Sons, 1993); Jacob Z. Schanker, *Meteor Burst Communications* (Boston: Artech House, 1990). On the civilian use of meteor burst communications, see Henry S. Santeford, *Meteor Burst Communication System: Alaska Winter Field Test Program* (Silver Spring, MD: U.S. Department of Commerce, National Oceanic and Atmospheric Administration, National Weather Service, Office of Hydrology, 1976).

The Moon communications relay had its origins in radar experiments conducted in 1946 by a team of investigators at the U.S. Army Signal Corps's Evans Signal Laboratory, near Belmar, New Jersey, working under the laboratory's director, John H. DeWitt, Jr. They successfully detected radar waves transmitted to the Moon on 10 January 1946. As early as 1940, DeWitt had failed to bounce radio signals off the Moon, to study the Earth's atmosphere, using the transmitter of Nashville's radio station WSM. He wrote in his notebook: "It occurred to me that it might be possible to reflect ultrashort waves from the Moon. If this could be done it would open up wide possibilities for the study of the upper atmosphere. So far as I know, no one has ever sent waves off the earth and measured their return through the entire atmosphere of the earth."[5] Later, in 1946, a Hungarian physicist and director of the research laboratory of the United Incandescent Lamps and Electric Company (Tungsram), Zoltán Bay, succeeded in bouncing radar waves off the Moon. Other experimenters had preceded DeWitt and Bay, but they had failed to detect lunar radar echoes.[6] The tentative, but successful trailblazing efforts of DeWitt and Bay opened up new vistas in ionospheric and communications research using radio echoes reflected off the Moon.

Private enterprise showed no lack of interest in developing a lunar communications relay. Experiments conducted at ITT's Federal Telecommunications Laboratories, Inc., in New York City, shortly after World War II, attempted to use the Moon as a passive relay for radio telephone communications between New York and Paris. The lunar relay would allow ITT to compete with AT&T, which held a monopoly on transatlantic cable traffic. What the Federal Telecommunications Laboratories imagined, however, the Collins Radio Company in Cedar Rapids, Iowa, and the National Bureau of Standards's Central Radio Propagation Laboratory in Sterling, Virginia, accomplished on 8 November 1951, when a slowly hand-keyed telegraph message was sent over the Iowa-Virginia circuit several times. The message was the same sent by Samuel Morse over the first U.S. public telegraph line: "What hath God wrought?"[7]

Meanwhile, though, the first use of the Moon as a relay in a communications circuit had been achieved only a few days earlier by military researchers at the Naval Research Laboratory (NRL). The efforts of the NRL to use the Moon as a passive communications satellite are the subject of David van Keuren's contribution. For the Navy, secure, reliable long-distance communications were a tactical necessity, especially during the Cold War of

5. DeWitt notebook, 21 May 1940, and DeWitt biographical sketch, HL Diana 46 (04), Historical Archives, U.S. Army Communications-Electronics Command, Ft. Monmouth, NJ.

6. Among those were Thomas Gold, Von R. Eshleman, and A.C. Bernard Lovell. Gold, a retired Cornell University professor of astronomy, claims to have proposed a lunar radar experiment to the British Admiralty during World War II; Eshleman, a Stanford University professor of electrical engineering, unsuccessfully attempted a lunar radar experiment aboard the *U.S.S. Missouri* in 1946, while returning from the war; and Lovell proposed a lunar bounce experiment in a paper of May 1946. Thomas Gold, interview with Andrew Butrica, Ithaca, NY, 14 December 1993; Von R. Eshleman, interview with Butrica, Caltech, 9 May 1994; and Bernard Lovell, "Astronomer by Chance," manuscript, February 1988, p. 183, personal papers, Sir Bernard Lovell.

7. D.D. Grieg, S. Metzger, and R. Waer, "Considerations of Moon-Relay Communication," *Proceedings of the IRE* 36 (May 1948): 652–63; "Via the Moon: Relay Station to Transoceanic Communication," *Newsweek* 27 (11 February 1946): 64; Peter G. Sulzer, G. Franklin Montgomery, and Irvin H. Gerks, "A U-H-F Moon Relay," *Proceedings of the IRE* 40 (1952): 361. A few years later, three amateur radio operators, "hams" who enjoyed detecting long-distance transmissions (DXing), succeeded in bouncing 144-megahertz radio waves off the Moon on 23 and 27 January 1953. E.P.T., "Lunar DX on 144 Mc!," *QST* 37 (1953): 11–12, 116. Their success sparked an ongoing ham interest in lunar DXing, which continues today in the form of contests to detect lunar radio echoes.

the 1950s. Moreover, ionospheric storms had shown the vulnerability of radio transmissions to natural jamming. The Navy's project called Communication Moon Relay (also known as "Operation Moon Bounce") sought to exploit the Moon as a high-tech communications option in the years before the launch of the first artificial satellite. Through his discussion of Operation Moon Bounce, moreover, van Keuren shows the close linkages between classified and unclassified research and development programs that existed within U.S. military laboratories during the Cold War.

Jon Agar takes up the lunar communications relay program undertaken on the other side of the Atlantic Ocean by Sir Bernard Lovell and his colleagues at the Nuffield Radio Astronomy Laboratories at Jodrell Bank. Unlike the work at the NRL, the Jodrell Bank experiments received underwriting from private enterprise—namely, Pye Telecommunications Ltd., a British electronics firm based in Cambridge. Thus, this early stage of satellite communications development not only witnessed activity on both sides of the Atlantic Ocean, but interest and funding came from both military and business sources. Agar also points out the role of demonstration—that is, public displays of scientific spectacle such as the bouncing of signals off the Moon—in funding the construction and operation of the giant Jodrell Bank radio telescope, as well as in supporting Jodrell Bank research in general.

Creating the Global, Regional, and National Systems

The United States and Canada

In light of the NRL, Jodrell Bank, and other lunar relay tests, as well as the West Ford and meteor burst experiments, Project Echo can be understood as a turning point, rather than a starting point, in the development of satellite communications. Similar to its predecessors, Echo was a passive communications relay; however, unlike them, it was an artificial satellite. Lunar echo and meteor burst techniques used natural satellites. Although Project Needles created an artificial ionosphere, it was not placed in orbit until 1963, three years after the launch of the first Echo balloon. As a passive relay, then, Echo continued past practice, but as an artificial satellite, it symbolized the future of space communications. Part II of this volume begins with two papers on this transitional satellite communications program.

In the first paper, Donald C. Elder, drawing on research conducted for a lengthier and more detailed study,[8] reviews the origins of Project Echo. Echo received inheritance from the military interest in satellite communications begun with the first lunar passive relay experiments. Initial funding for Echo came from DoD's research and development agency, the Advanced Research Projects Agency (ARPA), as well as NASA's predecessor, the National Advisory Committee for Aeronautics (NACA). The Pentagon was interested in space surveillance, such as taking photographs from space, and doing so involved the principle of overflight: who owned the space above each country?

Elder also argues that Echo originated with a prophetic science fiction article written by John Robinson Pierce, the director of research at AT&T's Bell Telephone Laboratories. Although Arthur C. Clarke first published the notion of space communications via geo-

---

8. Donald C. Elder, *Out From Behind the Eight-Ball: A History of Project Echo*, AAS History Series, Vol. 16 (San Diego: American Astronautical Society, 1995).

Figure 3

John Robinson Pierce, the former director of research at AT&T's Bell Telephone Laboratories. In a piece of science fiction, Pierce predicted in 1952 the potential benefits of satellite communications. A few years later, he helped realize the first artificial communications satellite, Echo. (Courtesy of NASA)

synchronous satellites, seven years later Pierce published in 1952 a story in *Astounding Science Fiction* in which he discussed the potential benefits of satellite communications. Echo came into existence, though, only after the Eisenhower administration gave its approval in 1955 to the launch of a satellite as part of the 1957–1958 International Geophysical Year (IGY). Thus, Echo ultimately involved collaboration by AT&T, NASA (as NACA's successor), and the IGY coordinators. Elder also recounts the technical difficulties encountered in constructing a balloon appropriate for the Echo experiments—namely, one that would not fall apart before being placed in orbit.

Also significant for the future development of satellite communications is that Project Echo saw the intertwining of private enterprise and the nation's space agency. The relationship between NASA and business was to color the future development of satellite communications. The association of NASA—through the Jet Propulsion Laboratory (JPL)—and AT&T's Bell Telephone Laboratories is the substance of Craig B. Waff's contribution. Continuing the Echo story, Waff focuses on the problem of target acquisition, which is a prerequisite for all satellite communications. Indeed, as the geosynchronous orbit reaches saturation, and as Iridium and other satellite communications systems create a growing demand for low and medium orbital slots, the problem of acquiring satellite targets, first encountered on Project Echo, takes on new relevance. Recalling the lunar communications tests of the 1950s, JPL and Bell personnel conducted a range of tests, including a lunar bounce experiment, to prepare for Project Echo. Taking part in Project Echo also served JPL's new research direction, for it was a key milestone in the creation of the Deep Space Network, NASA's worldwide space communications network.

That NASA would play a central role in the development of satellite communications was determined by its monopoly on civilian launchers. Daniel R. Glover reviews NASA's experimental communications satellite program from the agency's founding to the present.

NASA's launcher monopoly assured its place on Project Echo, as well as AT&T's Telstar, a system of communications satellites to be placed in polar and equatorial orbits. Telstar, similar to Echo, involved NASA and AT&T going forward in tandem in space communications. For launch, tracking, and telemetry services on Telstar, AT&T paid NASA $6 million. For its part, AT&T hoped to use Telstar to extend into space its historical monopoly on American wire and transatlantic cable traffic.

After reviewing NASA's roles in Project Relay and Syncom, Glover discusses NASA's generation of experimental communication satellites, known as the Applications Technology Satellite (ATS) series. NASA's search for ATS series funding highlighted the ongoing relationship between the space agency and private enterprise. This time, though, the business was not AT&T, but the Communications Satellite Corporation (better known as Comsat). Congress objected to NASA's continuation of Syncom (built by Hughes) out of fear that the space agency was developing technology for the benefit of a single private company, namely Comsat. As Glover explains, NASA responded by broadening the project's objectives to include meteorology and other scientific experiments and renaming it the ATS series. Years later, in 1973, the ATS series came to a halt, when Congress canceled it as a budget reduction measure, so the commercial satellite communications industry, not NASA, would have to support its own research and development. When NASA resumed its experimental communications satellite program with the launch of the Advanced Communications Technology Satellite (ACTS) in September 1993, the space agency made it available to industry, universities, and other government agencies to conduct experiments. ACTS emerged only after a long and tortuous debate carried on throughout the 1980s by Congress and the White House over whether NASA should develop technology for the U.S. satellite communications industry.

In the background of that debate, as well as throughout the history of satellite communications in the United States, was the role of the military. After NASA and private enterprise, the military constitutes a third strand in the development of U.S. satellite communications. As we have seen, DoD supported space communications research and development as early as the 1940s and 1950s. Later, the Pentagon influenced NASA to include in its ATS series technology for gravity-gradient stabilization (on ATS-2, ATS-4, and ATS-5) and for medium altitude orbits (ATS-2). These are only two examples drawn from a long history of development outlined by David N. Spires and Rick W. Sturdevant in their paper. Focusing on Air Force satellite communications from the 1960s to the present, these two writers show that the Air Force, as the chief provider of military launch vehicles, supporting infrastructure, and communications satellites, confronted a variety of interrelated technical, political, and institutional problems. Although Air Force engineers often surpassed their commercial counterparts in the design of communications satellites, Spires and Sturdevant point out that the special endurance requirements of military communications satellites drove their costs upward even as commercial costs dropped. The high cost of military communications satellites persuaded Congress to pressure the Pentagon to cut costs and to consider using commercial satellite systems.

Although the military often is portrayed as a unified entity, Spires and Sturdevant demonstrate its organizational disunity, at least for the case of satellite communications. Within DoD, moreover, the fractured, complicated system of satellite communications management has impeded the integration of military satellite communications planning and activities across the three services, as well as the transition of new technology from the research and development laboratory to the operational satellite.

A key military satellite communications research facility was MIT's Lincoln Laboratory, which oversaw Project West Ford. By 1963, when Project West Ford was launched, Lincoln Laboratory had established a reputation as a major defense research laboratory and as a vital center for state-of-the-art electronics and computer research. Heir to MIT's Radiation Laboratory, which had been at the heart of U.S. radar research and development during World War II, Lincoln Laboratory was underwritten jointly by the three armed services. The Air Force provided most of the funding, though. The laboratory designed and developed what became known as SAGE (Semi-Automatic Ground Environment), a digital, integrated computerized North American network of air defense. SAGE involved a diversity of applied research in digital computing and data processing, long-range radar, and digital communications. Lincoln Laboratory also worked on the Distant Early Warning (DEW) Line, a network of radome-enclosed radars intended to search for incoming enemy aircraft, and its successor, the Ballistic Missile Early Warning System (BMEWS).

William W. Ward and Franklin W. Floyd, drawing on their personal experiences, describe the military satellite communications development work that took place at Lincoln Laboratory under their direction—primarily Project Needles and the Lincoln Experimental Satellite (LES) series. In particular, they focus on the development and testing of communications satellite hardware and electronics. Scholarly studies of satellite communications have tended to concentrate on politics and policy studies, in contrast to the conspicuous emphasis on technology offered by Ward and Floyd.

Through the Lincoln Laboratory's research and development work described by Ward and Floyd, as well as the efforts outlined by Spires and Sturdevant, the military sought to create a system of space communications. The military is only one thread of the story of the development of U.S. satellite communications, however. The second thread is the independent, though at times intertwining, development of satellite communications by NASA. Yet a third thread is woven through this strange narrative fabric: private enterprise. Businesses interacted with the military side of satellite communications, as well as with NASA, but as seen in the case of lunar relay communications, private enterprise took an active interest in and funded tests of space communications technologies as early as the 1950s.

The importance of private enterprise in the development of satellite communications is the core of David J. Whalen's offering. Conventional wisdom, Whalen argues, holds that the government developed satellite communications technology because industry was either unwilling or unable to face the high costs and high risks of research and development. Before the launch of Echo in 1960, private industry—notably AT&T and ITT—invested substantial amounts in satellite communications research and development and expected to reap a profit. AT&T's participation in Project Echo was a further expression of its commitment to and competence in satellite communications. AT&T—not a government laboratory—had made that possible.

The government curtailed AT&T's involvement in the development of satellite communications, according to Whalen, largely out of fear that AT&T would extend its terrestrial telecommunications monopoly into space. The Kennedy administration, while not against private industry, stood firmly against AT&T enjoying a monopoly of space communications. In some ways, this moment marked the beginning of the end of AT&T's monopoly. As Whalen so convincingly argues, the problem was not an industry unwilling or unable to face the costs and risks of developing satellite communications systems, but the efforts of the government (including NASA) to restrain AT&T from extending its monopoly into space.

Figure 4

The U.S.S. Kingsport *carrying the first shipboard satellite communications station into the Pacific in 1964. Previously, this satellite station had served in the Atlantic and the Mediterranean for military communications via Syncom.* (Courtesy of NASA)

NASA checked the AT&T monopoly by awarding contracts to RCA and Hughes—not AT&T—to build Relay and Syncom, respectively. Congress and the White House reaffirmed their belief in private enterprise by establishing a private corporation (Comsat) through the Communications Satellite Act signed by President Kennedy in August 1962. Nonetheless, the intent behind the creation of Comsat was to ensure that any existing telecommunications company could not establish a monopoly on space communications, although the new law did grant Comsat a monopoly on satellite communications.

Comsat's monopoly did not come to an end until the June 1972 decision by the Federal Communications Commission (FCC) that opened up domestic satellite communications. Subsequently, Western Union launched WESTAR in 1973, and ComSat General launched COMSTAR in 1975. The 1980s saw the launching of several hundred transponders for domestic uses by ASC (American Satellite Company), FORDSAT (Ford Aerospace), GALAXY (Hughes), GSTAR (GTE), SPACENET (GTE Spacenet), SBS (Satellite Business Systems), and Telstar (AT&T).

Whalen's contribution also considers Comsat's selection of the ultimate satellite orbit for the single global satellite communications system that the firm was entrusted with creating. The choice was neither simple nor easy. AT&T and DoD favored the tried-and-true medium-altitude satellite, while others, including NASA, favored geosynchronous satellites. Tentatively, Comsat selected the geosynchronous Early Bird satellite, launched in March 1965, and only decided on the geosynchronous orbit after additional study contracts and satellite design studies.

For the most part, however, Whalen's paper sketches the role of private enterprise in U.S. satellite communications development. It is a thread of development, as with that of the military or NASA, that sometimes intertwines with the other threads of development, but often develops on its own. The development of satellite communications thus has not one unified past, but rather it reflects the sometimes connected and sometimes separate relationships among those three threads.

Europe

These three threads, however, illuminate the development of satellite communications only in the United States. Part II, therefore, contains a second section, which addresses the development of European satellite communications and the relationship between U.S. and European satellite communications. The view of European satellite communications presented by these contributions echoes the multifaceted nature of U.S. satellite communications development.

Western Europe necessarily followed a different evolutionary pattern, shaped by each country's distinct culture, politics, and economy, as well as by the move toward European integration that started after World War II. Western European nations created collective organizations to achieve a variety of ends. The Council of Europe, created in 1949 with headquarters in Strasbourg, France, provided member states a political organization, while the European Coal and Steel Community, created in 1952, stimulated the production of coal and steel by reducing trade barriers. The European Atomic Energy Community (known as Euratom) promoted joint exploitation of the peaceful uses of atomic energy, and the European Economic Community (known as the Common Market), created in 1957, sought to eliminate all economic barriers among member states. Despite the move toward European integration, the diversity and passion of European national politics cannot be overlooked.

Although Europe delayed launching its own satellite communications program, some European countries participated in U.S. efforts. For example, investigators at France's national telecommunications research center, the Centre National d'Études des Télécommunications, received signals transmitted from AT&T's Holmdel facility that bounced off Echo in August 1960.[9] Symphonie, a bilateral Franco-German communications satellite, preceded the 1978 launch of the first European communications satellite, the Orbiting Test Satellite. By then, however, Canada and several other countries had satellite capacity, too.

---

9.   Jean P.F. Voge, "Télécommunications spatiales et transmissions à grande distance par satellites artificiels," *L'Industrie Nationale* 13 (1961): 1–16, esp. 11.

Two key aspects of the development of satellite communications in Europe were the drive toward European integration and cooperation with the United States. The nature and extent of that cooperation was not at all obvious to all European governments, especially the French. Should a united Europe develop its own satellite launchers? Should Europe institute its own system of satellite communications? The answers to those questions were undeniably and necessarily coupled to the character of cooperation with the United States. As the 1960s began, the United States held a monopoly on satellite communications—a monopoly that would not ease possibly until the next decade. What AT&T was to U.S. communications, the United States was to European—and global—telecommunications.

A further complication was the creation of a single global satellite communications system, a goal the United States touted. In any unified global system, the United States would have a dominant place because of its lead position in the field and its monopoly on launch vehicles. In negotiating with the United States over the creation of a global satellite communications system, unity served European interests well. The countries of Western Europe decided to negotiate as a bloc, not individually, as had been the practice in past telecommunications negotiations.

The European side of satellite communications development starts with a paper by Arturo Russo on the beginnings of the European effort to launch a communications satellite. Referring to research performed for the European Space Agency (ESA) history project with colleagues John Krige and Lorenza Sebesta, Russo attempts to answer the question: Why did it take such a long time to develop a European communications satellite program?

He finds two major factors that limited Europe's ability to act. The first was the institutional framework. Getting the two multinational European space organizations—the European Space Research Organization (ESRO) and the European Launcher Development Organization (ELDO)—involved in satellite communications implied a change in their purpose and programs, which was a difficult task. Strong disagreements existed among ELDO member states, and a European satellite communications effort required a comprehensive international policy framework in which national economic interests and political goals could be satisfied. The second factor was the question of users—namely, the state telecommunications administrations. While interested in supporting research and development studies of satellite communications, these potential users worried about the economic prospects of a European satellite system. The solution, Russo points out, came in the form of two so-called package deals—a pair of sweeping agreements that aimed to satisfy national interests and transformed ESRO's mission.

In none of the studies conducted did the economics of the European satellite communications program demonstrate its potential to yield revenues in excess of costs. Russo argues that the ultimate justification for the program's approval was a handful of noneconomic factors, namely:

> *(1) the assertion of Europe's political and technological independence from the two superpowers, which was a key element particularly of the French government's space policy; (2) the recognition of the aerospace sector as strategically important for the development of advanced industrial technology; (3) the need to qualify European industry for competitive participation in Intelsat procurement contracts; (4) the understanding that the Orbiting Test Satellite and European Communication Satellite programs were the heart of an evolutionary program leading to other application fields, such as aeronautical and maritime*

telecommunications, direct television broadcasting, and Earth observation; (5) the search for autonomy of political and cultural expression (as the influential French newspaper Le Monde declared in 1967: "The transmission of radio and television programs is one of the most supple and diversified means to assure a presence and influence abroad"); and (6) the general drive toward European economic and technical integration.[10]

Moreover, a key part of the European story was the negative role played by the various national postal and telecommunications administrations. Working through the European Telecommunications Satellite Committee (known by its French acronym CETS), these administrations opposed the placement of a communications satellite over Europe, although not above the Atlantic Ocean, on economic grounds: the system would not nearly pay for itself. ESRO then turned to the European Broadcasting Union (EBU), a frustrated customer of the national postal and telecommunications administrations; as operator of the Eurovision television network, the EBU depended on its network of terrestrial cables. When the EBU proposed to replace that terrestrial network with a European satellite, the deadlock created by the national postal and telecommunications administrations finally ended.[11]

Lorenza Sebesta, also with the European Space Agency history project, examines the impact that the availability of U.S. launchers had on the development of European satellite communications. Tensions between the United States and the Soviet Union, as well as between the United States and France, played a role, as well as the commonly held belief that the technological research necessary for the development of a European launcher would stimulate economic growth. She pays close attention to the influence of U.S. policy on the European nations' decision to design and build their own launchers—and particularly to the shifting U.S. position on the offering of launcher technology and facilities over time.

A number of factors shaping the development of satellite communications in Europe emerge from Sebesta's perceptive study. She demonstrates the interweaving of U.S. and European satellite communications policy. She also discusses how the United States could use its commanding position to dominate the negotiations that led to the creation of Intelsat (the umbrella organization for the unified global satellite communications system) and to garner for U.S. industry, as a consequence of that commanding position, the largest share of Intelsat contracts. The United States sought to impose its will on Europe through these negotiations, especially the clause relating to the creation of regional satellite systems and the geographical definition of the European broadcast space. However, the U.S. position was not based solely on its technological and industrial lead; it was closely associated with security concerns, specifically the transfer of sensitive technology and generally the NATO alliance. Sebesta sheds light, too, on some of the motives behind NASA's attempts to foster cooperation with its European partners, such as the desire to channel funding away from military ends and toward peaceful uses of space. In turn, for example, West Germany hoped to acquire military-related technologies denied it by treaties dating from the end of World War II.

---

10. See Arturo Russo's conclusion in Chapter 10 of this book, titled "Launching the European Telecommunications Satellite Program."

11. René Collette, "Space Communications in Europe: How did we Make it Happen?," *History and Technology* 9 (1992): 83–93.

Sebesta points out that not only was there discordance over a collective satellite communications policy among Western European states, but that within those states discord reigned. At the European level, France and Britain were at odds. France favored an independent European launcher and had taken a number of steps to establish a *force de frappe* separate from the nuclear umbrella held out by the United States through the NATO alliance. Nonetheless, not everybody in France was in favor of a European launcher.

In contrast to the French position, the British generally favored collaborating with the United States and relying on U.S. launch services. As with France, Britain was not unified on satellite communications policy. British disunity is the subject of Nigel Wright's contribution. Scholars, he argues, have tended to portray Britain as an obstructionist player, holding up satellite communications development to protect its substantial investment in underwater telegraph and telephone cables, and they have concentrated on the role of the British Post Office to the exclusion of other government offices. Here, Wright focuses on the part played by the Foreign Office to illuminate the spectrum of views that shaped the British position on satellite communications vis-à-vis Europe and the United States.

While British officials were aware of the potential threat that a U.S. satellite system posed for British cable interests, they did not act "automatically" to frustrate early satellite development. Rather, they viewed satellites and cables as playing complementary roles, as Daniel Headrick notes in the case of cables and radio. Moreover, according to Wright, many people both in and out of British government favored construction of an independent satellite system in collaboration with Europe and the British Commonwealth. The Foreign Office, however, supported cooperation with the United States for diplomatic reasons, mainly to preserve good relations with that country. The Ministry of Aviation, on the other hand, backed an independent satellite system—a position arising from that ministry's role as Britain's voice in ELDO. The position of the British Post Office derived chiefly from its interaction with the membership of the Commonwealth Telecommunications Partnership. The Post Office at first favored an independent Commonwealth-European satellite system, believing that a satellite system controlled by the United States (and serving mainly the highly profitable transatlantic telecommunications routes) would not meet Commonwealth needs. Thus, Wright manages a convincing argument that opposes the established historiography—namely, that with the exception of the Foreign Office, British governmental agencies early on preferred a satellite system independent of the United States.

The lion's share of deliberations between Western Europe and the United States over satellite communications took place within the framework of negotiations over the creation of a unified global satellite communications system—what eventually came to be called Intelsat. In 1972, Jonathan F. Galloway examined those negotiations from a policy studies perspective, shortly after their resolution in the Intelsat Definitive Agreements signed in May 1971.[12] He returns a quarter century later, in this volume, to reconsider those events and his conclusions.

Galloway finds that his original themes "remain very relevant, despite the more colorful and even dramatic new vocabulary introduced by the likes of Newt Gingrich, the Tofflers, and Kenichi Ohmae." These themes are as follows:

---

12. Jonathan F. Galloway, *The Politics and Technology of Satellite Communications* (Lexington, MA: Lexington Books, 1972).

- Revolutionary or evolutionary technological change
- The breakdown of barriers between making and understanding foreign and domestic policy
- Models of rationality in policy-making appropriate to changing contexts

Seeking to understand the decision-making process, Galloway considers the mixture of cooperation, competition, and conflict manifested by the Intelsat negotiations as eluding understanding at the global or comprehensive level, because they operated in a way that was "pragmatic, incremental, and muddled." Indeed, he asserts, "The world is not a tidy place. There is chaos, and at the edge of chaos one new world order is not emerging. So it was in the formative period of satellite communications."[13]

As for the development of satellite communications, Galloway takes the reader through the maze of agencies responsible for establishing U.S. satellite communications policy and for determining the country's bargaining position vis-à-vis the Intelsat negotiations. While Russo, Sebesta, and Wright present the European rationale, Galloway presents that of the United States. A major obstacle in those negotiations was the U.S. penchant for privately owned telecommunications monopolies. If a private company, such as Comsat, were to enter into negotiations with foreign governments over satellite communications policy, what would be the role of government? How would the State Department, the Pentagon, or the FCC participate in those negotiations? Galloway explores these questions and shows how they influenced U.S. satellite communications discussions with Europe. Traditionally, European state telecommunications administrations had contracted on a bilateral basis with AT&T, not the U.S. government. With the creation of Comsat, AT&T was out of the negotiating picture, but Europeans still expected to negotiate with a private business.

Although the U.S. military had been developing passive space communications techniques since the 1950s, DoD use of the civilian satellite system was resolved in 1964. However, according to Galloway, this issue arose again on more than one occasion. In the end, DoD bought satellite access from Comsat. As with the Intelsat negotiations, the Pentagon's use of civilian satellites involved both domestic and foreign considerations. In a final section, Galloway discusses Soviet satellite communications and the diplomatic motives behind the creation of an Intelsat clone known as Intersputnik. In this context, the theme of state versus business interests echoes, as the Soviet Union proposed a world satellite system that only states, not private entities such as Comsat, could join. This position was equally an expression of Cold War diplomacy.

The Unfolding of the World System

Geography, Politics, and Culture

Part III contains the themes of the third stage of satellite communications development and consists of two sections. The first addresses the geographical growth of satellite communications, while the second considers satellite applications. By examining the growth of satellite communications in North America (Canada and Cuba) and Asia (China, India, and Indonesia), the first section draws attention to the influence of geography, politics,

---

13. See the introductory paragraphs in Jonathan F. Galloway, "Originating Communications Satellite Systems: The Interactions of Technological Change, Domestic Politics, and Foreign Policy," which is Chapter 13 of this book.

and culture on satellite communications development in those regions. Above all else, though, the papers implicitly highlight the creation of the global satellite system country by country.

Among the first countries to establish domestic satellite communications systems were those that span vast distances, such as the United States, Canada, the Soviet Union, China, and India. Such vast nation-states require a unified national telecommunications network, and satellites serve that need efficiently and effectively. Bert C. Blevis, drawing on his personal experience in the field, outlines the case of Canada. Appropriately, he begins with the ionospheric research that preceded satellite communications efforts; early Canadian satellites continued that research tradition. As in the United States, Canadian investigators considered lunar and meteor communications techniques. That work merits additional exploration by scholars to provide a greater understanding of the origins of satellite communications.

Political decisions were important in the shaping of Canada's satellite communications program. As Blevis notes, Canada did not follow the U.S. or European lead in developing its own launcher, relying instead on those of NASA and ESA. The government was not unaware of the economic benefits of supporting a space program, however. In 1963, the Canadian government decided to transfer space technology from the Defence Research Telecommunications Establishment to industry. Also, after the 1967 Chapman report recommended encouraging the Canadian space industry, the Communications Technology Satellite program sought to improve Canada's satellite and spacecraft design and manufacturing capability. While Blevis addresses geographical and political questions in his overview of Canadian satellite communications development, the impact of satellites on Canadian culture remains for a future study. The apparent strength of the resurgent French separatist movement in Quebec argues against the notion that satellites tend to homogenize national cultures.

From Canada we turn to China—another country that occupies a large land mass, not to mention a fourth of the world's population. Zhu Yilin, a member of the Chinese Academy of Space Technology, provides an overview of satellite communications development in that country. Geography was a key factor, but, unlike the United States and Canada, China lacked an extensive national telecommunications system until the inauguration of its satellite program. The cultural and political impact of satellites in China, therefore, has been all the more dramatic. Satellite communications served to establish a national educational television network, provided long-distance telecommunications, and assisted in the modernization of the country's banking system. The use of nationwide television broadcasts to foster political unity and to insert the Beijing political line into each and every village cannot be overlooked.

Yet another geographically large country to establish a domestic satellite communications system is India. Participation in the Satellite Instructional Television Experiment (SITE), carried out in cooperation with NASA, and India's first communications satellite, INSAT, form the heart of the contribution by Raman Srinivasan. Drawing from research carried out for his doctoral dissertation, Srinivasan shows how India's earliest experiences with satellite communications served as a fulcrum for social and cultural change and economic development during a critical period in that country's evolution.

SITE was a massive experiment in social engineering. It involved placing televisions in 5,000 remote villages, some of which lacked electricity. One of the major challenges to making the experiment work was to create six hours of programming every day. SITE also proved vital in inculcating technical and managerial expertise necessary for India to create its own domestic satellite system, known as INSAT. This system was geared to serve rural India; a weather system was added specifically to aid farmers.

Brian Shoesmith draws on his research as a media scholar to discuss India and China, as well as Indonesia, and highlights the influence of culture and politics in the Asian "mediascape." The significance of satellites is their ability to undermine state control of media and to force governments to rethink their broadcasting and communications policies. Because Asian satellites have broad footprints, sometimes covering the entire Eurasian land mass, political boundaries and geographical hindrances are surmounted.

The development of satellite communications in Asia, according to Shoesmith, took place in three stages, all of which are linked to the dominant trends in the growth of television on the continent. In the first stage, from 1962 to the 1980s, satellites were perceived as a tool of social engineering in developing economies. Stage two spans the late 1980s and was characterized by the response of Asian governments to the end of the Cold War. The third stage, 1991 to the present, is distinguished by the dominance of commercial considerations.

Shoesmith makes some cogent observations. Rather than killing off old media technologies, such as newspapers, satellites have actually reinforced them, especially the vernacular press. Satellites, moreover, have not promoted the homogenization or standardization supposedly inherent in the technology. In addition, Shoesmith notes that the growth of satellite-related applications in Asia has resulted from a linking of Western technology and local capital.

The introduction of satellite communications into Canada, China, and India served an important internal need by providing telecommunications services efficiently and effectively over a vast expanse of territory. Satellite communications also linked countries to each other. Maintaining communications links with the rest of the world is most critical for island nations. The case of Cuba illustrates this geopolitical requirement, as well as some of the cultural and political aspects of satellite communications in a country that has been a member of both the Intersputnik and Intelsat organizations.

José Altshuler, drawing on his extended research into the history of electricity and his participation in the Cuban space program, begins with some useful background on long-distance shortwave radio communications and an ingenuous attempt to relay the television broadcast of the World Series baseball games to Cuba via an airplane. Radio also linked Cuba to Eastern Europe and other countries before satellites became available. The creation of the Soviet satellite system, separate from that controlled by Intelsat, provided Cuba its first experience with satellite communications, beginning in 1973. However, Intersputnik provided only limited coverage until the replacement of the Molniya series with the geostationary Horizont and Statsionar satellites.

The Molniya satellites had been placed in a highly elliptical orbit, the so-called Molniya A orbit, which made them visible over a wide area of Soviet territories for about eight hours a day. The Molniya A orbit was better suited for coverage of the Soviet Union than a geostationary orbit. It covered areas geostationary satellites could not serve, and launches from high latitudes more easily entered this orbit than a geostationary orbit over the equator. The Molniya satellites were used for telephone and fax services and to distribute television programs from a central point near Moscow. During the 10 years following the launch of Molniya I, twenty-nine more were placed in orbit.

Historian Roberto Díaz-Martín continues Altshuler's discussion of Cuban satellite communications. A technological difficulty that arose as a consequence of Cuba joining the socialist bloc was the switch from NTSC to SECAM. Cuban television, which dated from the 1950s, used the NTSC broadcast standard and operated on 110 volts at 60 hertz, while SECAM receivers required 220 volts at 50 hertz, the European standard. In the end, geographical and cultural factors decided in favor of NTSC, the standard used by most Caribbean countries. Using SECAM would have isolated Cuba from its neighbors. Díaz-Martín also relates how Cuba joined Intelsat, as well as some of the cultural uses (program exchanges) of satellite communications.

Applications

Departing from the preceding discussions of the development of satellite communications systems, the second section of Part III examines satellite applications—that is, the ends to which people use the communications capability made available by satellites. This topic is sufficiently broad in scope as to merit its own volume. However, we present three case studies of satellite applications in the fields of education, medicine, and mobile communications, as well as their integration into corporate strategy.

Joseph N. Pelton discusses Project SHARE (Satellites for Health and Rural Education), which he conceived as Intelsat's director of strategic policy to commemorate the organization's twentieth birthday. The program makes free satellite capacity available to provide health and educational services to rural and remote areas. Project SHARE provided a fresh framework for international cooperation and engendered dozens of projects that affected nearly 100 countries and millions of people. Currently, fifty countries participate in Project SHARE.

Pelton concludes that the most successful programs were those designed and developed by the participating country. The technology that made the program succeed, especially in the rural areas where it was most needed, opposes the dictum that "big is beautiful." Small, low-cost ground pieces of equipment—not the large ground stations and heavy streams of traffic that typify satellite communications—were best suited to delivering educational and medical services to rural and remote districts. This technology was part of a larger satellite communications "revolution" that saw a number of new services and ground technologies emerge.

Among the satellite applications perceived as new is mobile communications. Edward J. Martin draws on his personal experience in this field to discuss the development of mobile satellite communications from the 1950s to the present. He begins with lunar relay communications and Project West Ford. Just as an airplane served to relay the World Series between Miami and Havana, one of the earliest tests of mobile communications involved the use of an airplane.

Martin relates the arduous struggle to create an aircraft-to-satellite navigation and communications system, called Aerosat, between 1963 and 1975. Several factors frustrated and eventually doomed attempts to bring Aerosat to life:

- The lack of a unified policy within the Federal Aviation Administration (FAA)
- European opposition to the prospect of U.S. domination in another satellite application
- Disagreement on which frequency band to use
- The failure to include the airlines in the process

If the airline industry could not realize a satellite system, the maritime trade was more successful, perhaps because the effort began in 1966 under the aegis of the United Nations' International Maritime Consultative Organization, which included delegates from the world's major seafaring countries. The result was Inmarsat (International Maritime Satellite Organization—later the word "Mobile" would be substituted for "Maritime"). After relating the initial problems faced by Inmarsat, Martin considers the organization's extension into aeronautical services.

In the last paper, MCI's historian, Adam L. Gruen, examines the use of communication satellites by that U.S. telecommunications firm. Although not a profitable venture, the company's use of satellites was not a mistake; it was a necessary part of a greater and deliberate business scheme. Indeed, MCI knew that it was acquiring a money-losing enterprise (Satellite Business Systems). Scholars who consider only the immediate profitability of corporate strategies should take heed.

Given life by a court decision, MCI needed to build a telecommunications network quickly. Here, we might see a parallel with China, which acquired a nationwide telecommunications system almost instantaneously by launching a communications satellite. MCI management considered two technologies, satel-

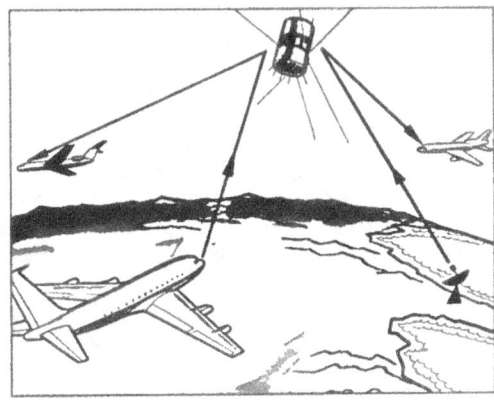

Figure 5
Artist's sketch of how the NASA ATS-E satellite, scheduled for launch in 1969, was to provide communications for transatlantic commercial air traffic. (Courtesy of NASA, photo no. 69-H-1398)

lites and fiber optics, and opted for both. Gruen argues that MCI's purchase of Satellite Business Systems has to be understood within the context of 1980s business practices, when takeovers, corporate raiders, junk bonds, and pension plundering were common. To defend itself against a hostile takeover, MCI acquired "shark repellent" in the form of a stock sale to IBM. MCI's purchase of Satellite Business Systems, partly owned by IBM, gave it access to that firm's business customers. As Gruen concludes, "One has to be willing to expand one's definition of profitability from the narrow frame of return-on-investment to the bigger picture."[14]

14. See the conclusion of Chapter 22 of this book by Adam L. Gruen, "Net Gain: The Use of Satellites at MCI."

Gruen's paper also echoes the cable-versus-radio theme addressed by Headrick, although with MCI, as with Nigel Wright's discussion of British satellite communications, it was a matter of cables versus satellites. The two technologies today have found their niche based on the advantages of each over the other. Whereas cables are best suited to point-to-point circuits, satellites provide point-to-multiple-point service. Also, the cost of cable circuits increases with distance, while satellite circuit costs are independent of distance between ground terminals. Satellites, too, can ford physical and political barriers that impede cable placement, and satellites are uniquely suited to mobile communications.

Conclusion

The contributions to this volume do not begin to exhaust the subject of satellite communications development; that would be impossible given the present state of research. Nonetheless, the papers that follow present a broad, but systematic survey of the development of satellite communications. It is hoped that this book will stimulate readers to undertake fresh research, advancing knowledge in this field and leading eventually to the writing of a synthetic work. In addition, one can hope that a beneficial interaction between scholars and practitioners will be part of that synthetic work.

Appended to this volume are two aids for understanding the development of satellite communications. One is a chronology of key events synthesized from timelines submitted by this book's contributors, as well as from selected works on the subject. The other is a list of suggested additional readings on the development of satellite communications. Those wishing complete technical details on satellites, or a list of satellites currently in service, should consult Jayne's or NASA's *Satellite Situation Report*, published by the Goddard Space Flight Center.

PART I:

Passive Origins

Chapter 1

Radio Versus Cable: International Telecommunications Before Satellites

by Daniel R. Headrick

The current rivalry between satellites and fiber optic cables in long-distance and intercontinental telecommunication is not the first case of two technologies coexisting to meet a similar demand. Consider trucks versus railroads, cable versus broadcast television, and movies versus videos. Sometimes the two rival technologies coexist because they complement each other, such as trucks and railroads. In other cases, they compete until one displaces the other, such as steamships for sailing ships. By studying an earlier rivalry in intercontinental telecommunications—namely, that between submarine telegraph cables and radiotelegraphy in the first half of this century—we may find some parallels and fruitful insights into the current rivalry between satellites and fiber optic cables.[1]

Intercontinental Telecommunications Before 1907

Before 1907, all intercontinental telecommunications used submarine telegraph cables with a copper core insulated with gutta-percha and protected by an armor of iron wires. Submarine cables, in turn, were synonymous with an overwhelming British preponderance in the international telecommunications business. Thus in 1908, British firms owned 56.2 percent of the world's submarine cable network, compared with a U.S. share of 19.5 percent and a French share of 9.4 percent.[2] Taking into account the commercial use of cables, the British share shrinks, because the most heavily used cables across the Atlantic belonged to two American firms, Western Union and Commercial Cable; in 1911, Western Union leased the last two cables owned by British firms.[3] If, however, we focus on the military and political value of cables, Britain had an even greater advantage than its share indicates, for many foreign cables passed through Britain or its colonies; the dominant British firms, Eastern and Associated Telegraph Companies, controlled both the Danish telegraph firm that operated the land lines from Europe across Russia to China and the American company that operated the American transpacific cable.[4]

1.   This paper is based on Daniel R. Headrick, *The Invisible Weapon: Telecommunications and International Politics, 1851–1945* (New York: Oxford University Press, 1991).

2.   Maxime de Margerie, *Le réseau anglais de câbles sous-marins* (Paris: A. Pedone, 1909), pp. 34–35.

3.   Charles Bright, "Extension of Submarine Telegraphy in a Quarter Century," *Engineering Magazine* 16 (December 1898): 420–25; Vary T. Coates and Bernard Finn, *A Retrospective Technology Assessment: Submarine Telegraphy. The Transatlantic Cable of 1866* (San Francisco: San Francisco Press, 1979), ch. 5; Alvin F. Harlow, *Old Wires and New Waves: The History of the Telegraph, Telephone and Wireless* (New York: Appleton-Century, 1936), 425–28; Gerald R. M. Garratt, *One Hundred Years of Submarine Cables* (London: H.M.S.O., 1950), 30; Kenneth R. Haight, *Cableships and Submarine Cables* (Washington, DC: U.S. Underseas Cable Corporation, 1968), pp. 316–21.

4.   On the British cable network, see Hugh Barty-King, *Girdle Round the Earth: The Story of Cable and Wireless and its Predecessors to Mark the Group's Jubilee, 1929–1979* (London: Heinemann, 1979). On their strategic value, see Paul M. Kennedy, "Imperial Cable Communications and Strategy, 1879–1914," *English Historical Review* 86 (1971): 728–52.

It is important to understand this preponderance because it gave Britain the power to scrutinize, censor, or ban foreign telegrams around the world—a power that it first exercised during the Anglo-Boer War of 1899–1902. This power naturally irritated Britain's rivals, France and Germany. Around the turn of the century, both countries tried to build up rival submarine cable networks—a hopeless endeavor, for cables were too costly and remained vulnerable to cutting by British cable ships in the event of war. What the world learned during the turn of the century, and even more clearly during World War I, was that international communication was not just a technology or a business, but an instrument of power in the rivalry between nations.

The Era of Complementarity: Cables and Radio, 1907–1927

Guglielmo Marconi did not invent radio all by himself, but he certainly was the first to think of commercializing it as a communications system. His first customers, not surprisingly, were the British and Italian navies, the major shipping companies, and Lloyd's of London insurance. Thus, he created—and quickly occupied—a new niche in telecommunications. Marconi had greater ambitions, however, for he wanted to create a global telecommunications network that would rival the Eastern Telegraph Company's submarine cable network. In 1907, he opened the first transatlantic stations at Clifden, Ireland, and Glace Bay, Nova Scotia, challenging the cables on their home ground.[6]

Some countries adopted radiotelegraphy without Marconi's help. The German government encouraged its two electrical equipment manufacturers, Siemens and AEG, to found a new company, Telefunken, to provide equipment and service to German ships. By 1906, Telefunken began building a gigantic station near Berlin to communicate with North America and the German colonies in West Africa. Soon thereafter, the U.S. Navy built a station in Arlington, Virginia, which was the first of a chain able to reach any ship in the North Atlantic and Pacific Oceans.[7]

During this period, radio not only reached ships at sea, but it also competed with cables between land locations. The competition, however, was restrained for both economic and technological reasons. It must be remembered that long-distance radio during that period used long waves—that is, wavelengths measured in kilometers with frequencies of 100 kilohertz or less. Propelling such waves across an ocean required enormous and

---

5.   Charles Lesage, *La rivalité franco-britannique: Les câbles sous-marins allemands* (Paris: Plon, 1915); Artur Kunert, *Geschichte der deutschen Fernmeldekabel. II. Telegraphen-Seekabel* (Cologne-Mülheim: Karl Glitscher, 1962). On cables and the great-power rivalries at the turn of the century, see Headrick, *The Invisible Weapon*, chs. 5 and 6.

6.   Hugh G.J. Aitken, *Syntony and Spark: The Origins of Radio* (Princeton: Princeton University Press, 1985), pp. 218–32, 286–90; Rowland F. Pocock, *The Early British Radio Industry* (Manchester: Manchester University Press, 1988); W.P. Jolly, *Marconi* (New York: Stein & Day, 1972), pp. 32–91; W.J. Baker, A History of the Marconi Company (London: Methuen, 1970), pp. 25–51; Rowland F. Pocock and Gerald R.M. Garratt, The Origins of Maritime Radio (London: H.M.S.O., 1972), pp. 34–44.

7.   On the beginnings of German radiotelegraphy, see Hermann Thurn, *Die Funkentelegraphie*, 5th ed. (Leipzig and Berlin: B.G. Teubner, 1918); "Telefunken-Chronik" in *Festschrift zum 50 jährigen Jubiläum der Telefunken Gesellschaft für drahtlose Telegraphie m. b. H.*, special issue of *Telefunken-Zeitung* 26 (May 1953): 133–47. On the early years of American radio and the Navy, see Susan J. Douglas, *Inventing American Broadcasting, 1899–1922* (Baltimore: Johns Hopkins University Press, 1987); Captain Linwood S. Howeth, *History of Communications-Electronics in the United States Navy* (Washington, DC: Bureau of Ships and Office of Naval History, 1963); Hugh G.J. Aitken, *The Continuous Wave: Technology and American Radio, 1900–1932* (Princeton: Princeton University Press, 1985). On the impact of long-wave radio on international politics up to 1919, see Headrick, *The Invisible Weapon*, chs. 7–9.

costly stations. The French station built at Sainte-Assise after World War I, for instance, had an antenna supported by 16 towers, each one of which was almost as high as the Eiffel tower. Moreover, the station consumed 1,000 kilowatts of electricity, as much as a small town. The cost was stupendous. A pair of transatlantic stations cost up to $4 million dollars—at a time when a Model-T automobile cost $300. A transatlantic cable in that same period cost $7 million.

Across the Atlantic, the two technologies were fairly evenly balanced, for if radio was slightly cheaper, cables were more reliable, worked 24 hours a day, and were less vulnerable to storms or equipment failures. As radio technology improved—moving from spark transmitters to arcs and alternators and then to vacuum tubes—so did cable technology—first with regenerative repeaters and automatic printers and then with loaded cables and time-division multiplexing that could carry a higher volume than their prewar predecessors.

During World War I, every cable and radio station was fully occupied with military and war-related traffic. Although the two technologies were evenly matched technologically, after the war radiotelegraphy gradually expanded its market share. By 1923, the Radio Corporation of America (RCA), founded by General Electric and the U.S. Navy, had captured 30 percent of the North Atlantic traffic and 50 percent of the Pacific traffic.[8] The reason was not technical superiority; it was because the cables were all operating to capacity in those years, and it took much longer to lay cables than to build radio transmitters.

Radiotelegraphy politicized telecommunications even more than had been the case before during peacetime, for it changed the playing field between nations. Germany had lost its cables in the war. Britain's radio company, Marconi, was pushed out of the American market by General Electric and the U.S. Navy. The British government, in financial straits and already well connected to its "empire" by cables, hesitated to subsidize radio. In contrast, France, which had few cables, invested heavily in radio to bypass the British cable network. And in the United States, where demand was greatest and capital was abundant, investors were eager to put their money into radio enterprises.[9] The result was that by 1923, the United States had 3,400 kilowatts of high-powered long-distance stations, France had 3,150 kilowatts, and the British Empire had only 700 kilowatts, hardly more than defeated Germany's 600 kilowatts.[10] Britain's erstwhile predominance in global telecommunications had vanished.

The Shortwave Revolution

In 1924, a bombshell overturned the cozy modus vivendi between cable and radio: shortwave radio. Ironically, it was the result of efforts by the same man who had introduced long-wave radio thirty years earlier: Guglielmo Marconi. The technology, which is well known, is not discussed here. Its economic impact, however, is not well understood, and the politics behind it, even less.[11]

8. On the origins of RCA, see Aitken, *The Continuous Wave*, chs. 6–8; Robert Sobel, RCA (New York: Stein & Day, 1986); Kenneth Bilby, *The General: David Sarnoff and the Rise of the Communications Industry* (New York: Harper & Row, 1986).

9. On telecommunications and international conflicts after World War I, see Headrick, *The Invisible Weapon*, ch. 10.

10. Sir Charles Bright, "The Empire's Telegraph and Trade," *Fortnightly Review* 113 (1923): 457–74.

11. See Daniel R. Headrick, "Shortwave Radio and its Impact on International Telecommunications between the Wars," *History and Technology* 11 (1994): 21–32; Headrick, *The Invisible Weapon*, ch. 11.

In the fall of 1924, Marconi dispatched experimental shortwave sets to several places around the world. Transmissions from England were received in Canada (Montreal), Argentina (Buenos Aires), and Australia (Sydney), twenty-three and a half hours a day. Not only was shortwave radio a technical success, it was also incredibly cheap. From the very beginning, a shortwave transmitter cost one-twentieth as much as a long-wave station of equivalent reach, and it used one-fiftieth of the electricity. Furthermore, it could transmit up to 200 words per minute, as fast as the newest cables and much faster than long-wave radio. In a speech to the Institute of Radio Engineers in October 1926, Marconi confessed: "I admit that I am responsible for the adoption of long waves for long-distance communication. Everyone followed me in building stations hundreds of times more powerful than would have been necessary had short waves been used. Now I have realized my mistake."[12]

From 1926 on, manufacturers began building shortwave equipment, and radio communications companies adopted them as quickly as possible. For the first time, radio communication was within the reach of small towns and previously isolated outposts around the world. Because the equipment was cheap to buy and operate, shortwave services could offer rates that cable companies could not begin to match. For instance, radiograms between England and Australia via the British Post Office shortwave service cost four pence per word, one-twelfth as much as a cablegram. Within six months of opening for business, the British Post Office shortwave service captured 65 percent of the cable traffic to India and Australia, as well as 50 percent of the transpacific traffic. Between France and Indochina, shortwave service captured 70 percent of the traffic within the first year.[13]

This was a completely different kind of competition compared with the genteel rivalry of the early 1920s. This time, the cable companies faced ruin. Eastern and Associated announced it would shut down, sell off its cables, and distribute reserves to its shareholders. The British government, in a rare display of energy, merged all British overseas telecommunications systems—public as well as private—into one company called Imperial and International Communications, later renamed Cable and Wireless.[14] The purpose of this merger was to make radio subsidize the preservation of the now-obsolete cable network. Needless to say, this was a heavy burden on the new company. The Depression, which began soon thereafter, weakened it even further. In the United States, Western Union kept only its most profitable cables in operation, while France simply abandoned its cables. Companies that had no cables, such as RCA, profited at the expense of their rivals.

Strategy Versus Business: Cables After 1928

Why did Britain preserve its cables—including barely used ones between such out-of-the-way places as Sierra Leone and Ascension Island or Lagos and Saint-Vincent—at a cost estimated at $2.5 million dollars a year? Because the British government knew, through long experience, that radio was vulnerable to eavesdropping and espionage. Indeed, during World War II, as in World War I, Britain's cable communications with the United States and the rest of the British Empire remained secret, while German and Japanese

---

12.   Quoted from Douglas Coe, *Marconi: Pioneer of Radio* (New York: J. Messner, 1943), p. 237.
13.   Baker, *A History of the Marconi Company*, p. 229; Barty-King, *Girdle Round the Earth*, p. 203; L. Gallin, "Renseignements statistiques sur le développement des communications radiotélégrahiques en Indochine," *Bulletin économique de l'Indochine* 32 (1929): 370–74.
14.   Barty-King, *Girdle Round the Earth*, pp. 203–26; Baker, *A History of the Marconi Company*, pp. 223–31; Headrick, "Shortwave Radio," *passim*.

radio communications, as we now know, were regularly breached. Furthermore, during the war and for several years thereafter, every channel of communication was filled to capacity. Not until the 1950s were the last of the old copper cables finally laid to rest on the ocean floor.

Conclusion

What can the story of cables and radio before World War II tell us about the current rivalry between satellites and fiber optic cables? First, the two modern technologies seem evenly matched, but not identical. Satellites are best at mobile communication with ships and planes and the like. Fiber optic cables, meanwhile, are best at handling high-volume traffic between important urban centers in the developed countries. Both technologies improve daily, demand is strong, and there is plenty of room for both to grow. The situation is uncannily reminiscent of the situation in the early 1920s: a bit of competition, much complementarity, and good times for all. Second, if international telecommunications were considered strategic resources before World War I, and even more so between the wars, there is no reason to think that they are less strategic today, although such matters are never discussed publicly.

But does history help predict the future? Based on the history of telecommunications before World War II, one can predict not one, but two futures.

The first of these two alternative futures can be called "the scenario of continuity." It assumes refinements in technology, but no revolutionary changes. There will be more and more satellites until every American, European, and Japanese can go anywhere with a pocket or wristwatch phone and call up anyone anywhere in the world by his or her personal identification code. Cables, meanwhile, will be fully occupied with computer data, video images, and the "chatter" of deskbound people, talking to friends overseas for next to nothing. Already, dozens of fiber optic cables cross the Atlantic and Pacific Oceans. According to the British journal *Public Network Europe*, the "Fiberoptic Link Around the Globe," slated to open in 1997, stretching from Britain to Japan, will contain 120,000 digital circuits, each capable of carrying images and video as well as voice.[15] Other cables are planned around Africa, between Europe and Southeast Asia, and on almost every other major route. Although satellites always will be needed, cables are likely to increase their market share into the early twenty-first century. Fiber optic cables have the same advantage as their copper-cored ancestors: they are practically impossible to eavesdrop on, and they are much less vulnerable than radio or satellites to electromagnetic interference, including nuclear explosions.

The other future, "the scenario of discontinuity," involves an unprecedented and revolutionary technology that might make cables and/or satellites obsolete, just as shortwave radio ruined the submarine telegraph cables. Discontinuities, by definition, cannot be predicted.

Which of the two scenarios is likely to take place? It is anyone's guess.

---

15. *Public Network Europe* (July–August 1995): 31, quoted in the monthly bulletin of the International Telecommunications Union, *Teleclippings* 929 (August 1995): 33.

Chapter 2

Moon in Their Eyes: Moon Communication Relay at the Naval Research Laboratory, 1951–1962

by David K. van Keuren

On 24 July 1954, James H. Trexler, an engineer in the Radio Countermeasures Branch at the Naval Research Laboratory (NRL), spoke carefully into a microphone at the laboratory's Stump Neck radio antenna facility in Maryland. Two and a half seconds later, his words speeded back to him at Stump Neck, after traveling 500,000 miles via an Earth-Moon circuit.[1] For the first time ever, the sound of a human voice had been transmitted beyond the ionosphere and returned to Earth.

Trexler's achievement marked an early watershed in the Department of the Navy's Communication Moon Relay project (also known as "Operation Moon Bounce"). The ultimate goal was to create the longest communications circuit in human history, with the Earth's satellite acting as a passive relay. Military strategists had long considered secure and reliable communications lines to be a tactical necessity. During the 1950s, the heyday of the Cold War, with the U.S. Navy's fleets encircling the globe, secure and reliable communications links were considered critical to national security. Ionospheric storms had recently cut off radio transmissions to the U.S. fleet in the Indian Ocean, thereby demonstrating dramatically the vulnerability of communication lines.[2] The objective of the Communication Moon Relay project was to add another high-tech option within the Navy's inventory of secure global communications technologies.

The origins of Communication Moon Relay, however, lie not in postwar communications research and development, but rather within the secure world of electrical intelligence gathering. The project was the spinoff of a deeply classified program centered on the surveillance of Soviet radar technologies, known as PAMOR (Passive Moon Relay).[3] The early history of the two programs demonstrates just how close the linkages between classified and unclassified research and development programs often were within the American military laboratories during the Cold War.

---

1. A reference to the 1954 voice transmission is found in the James H. Trexler biographical file, NRL Historian's Reference Collection, Naval Research Laboratory, Washington, DC. The first transmission on 24 July, because of security concerns, consisted only of the repetition of vowel sounds. The broadcast of actual words followed on 22 August. James H. Trexler, private communication, 15 October 1995.

2. Discussed in James H. Trexler, interview with David K. Allison, 30 October 1980, Historian's Office, Naval Research Laboratory, Washington, DC.

3. This discussion of PAMOR is based on previously classified documents within the NRL record collection now in storage at the Federal Records Center, Suitland, MD. Important documents are James H. Trexler, comp., "A Chronological History of U.S. Naval Radio Research Station (NRRS), from 1946 to 14 April 1962, prepared 22 July 1962"; anonymous, "Unique Aspects of the Elint [electrical intelligence] Collection Potential of the U.S. Naval Radio Research Station, Sugar Grove, West Virginia, January 1962." This reference is to edited versions in the files of the Historian's Office, NRL. Additional key documents include "Communications by Satellite Relay," March 1959, NRL Historian's Files; James H. Trexler, "Proposed URSI Paper for May 1955: Lunar Radio Circuits," collection of Countermeasures Branch memoranda dating to 1954, NRL records, Federal Records Center, Suitland, MD.

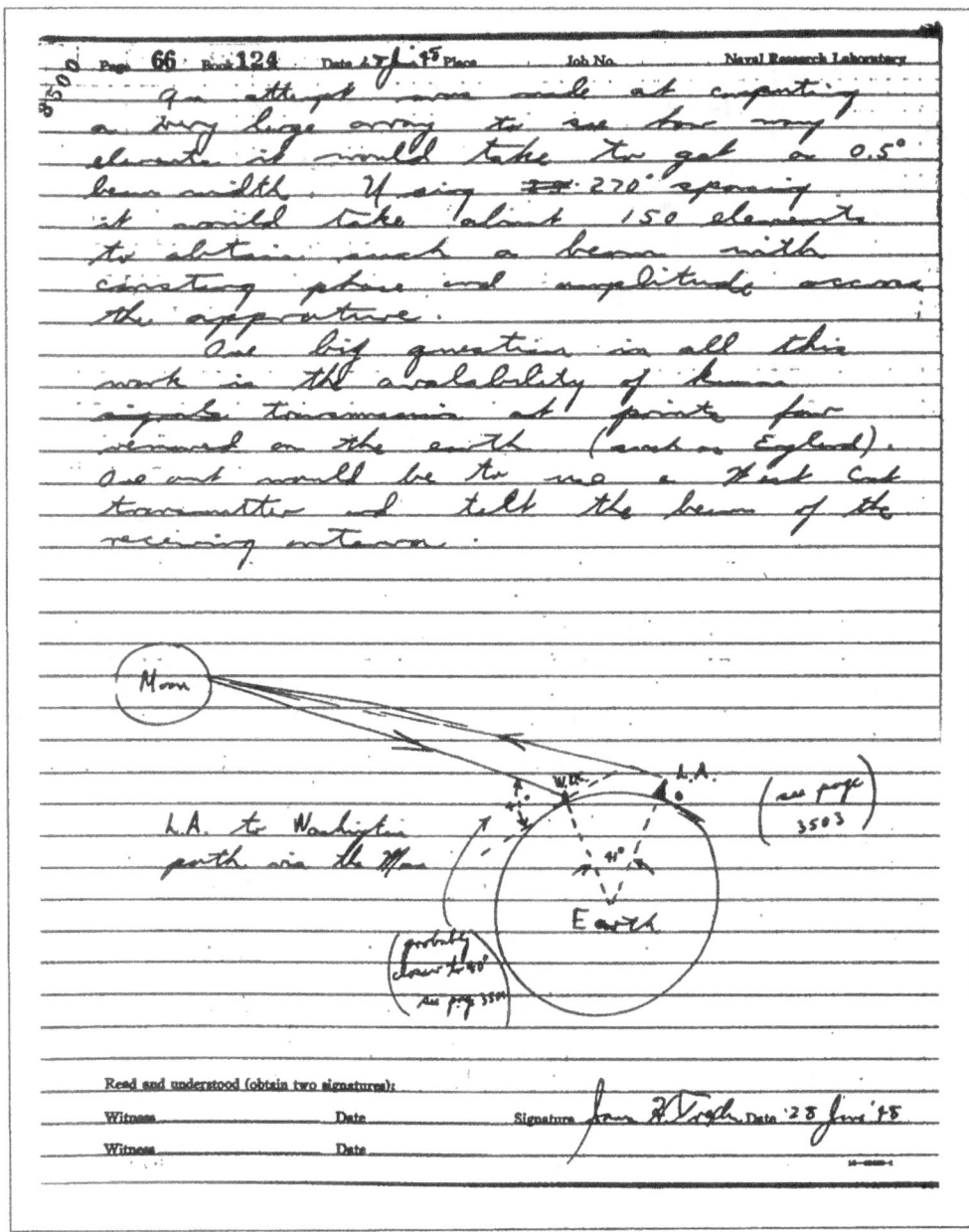

Figure 6
Notebook entry of James H. Trexler, dated 28 January 1945, showing calculations for a long-distance communications link between Los Angeles, California, and Washington, D.C., via the Moon. (Courtesy of the Naval Research Laboratory)

Naval Radio and Radar Research

Radio communications, particularly high-frequency communications, had long been a topic of interest at NRL. One of the original two laboratory divisions, when it opened in 1923, had been Radio, under the direction of A. Hoyt Taylor. Under Taylor's direction, NRL personnel throughout the 1920s and 1930s explored the application of high-frequency radio to Navy communications. One side effect of these studies was the accidental discovery in 1929, by Leo Young and Lawrence Hyland, of the underlying principle of what later came to be known as radar. Subsequent work by NRL personnel in the 1930s led to the development of operational radar detection sets by the end of the decade.[4]

Similarly, collaborative work in the 1920s by Taylor and physicist E.O. Hulburt on the propagation of high-frequency radio waves in the upper atmosphere led to the discovery of radio skip distance.[5] Consequently, by the beginning of World War II, NRL had evolved into a leading center for research on the application of high-frequency radio to long-distance communications and detection.

World War II brought most laboratory research programs to a temporary halt, as personnel turned their attention to incremental improvements in existing technologies, as well as the testing and evaluation of contractor-produced war materials.[6] However, the war *did* spark research and development in applied programs, including radar countermeasures. A dual-pronged agenda of technical development aimed at simultaneously countering German and Japanese radar, while improving the effectiveness of American equipment, was put into place in early 1942. In 1945, these programs merged into a newly established Countermeasures Branch within the Ship-Shore Radio Division. Branch research and development continued after the war, with a shift in focus toward the electronic capabilities of an increasingly belligerent Soviet Union.

One program of particular interest to the Countermeasures Branch was the interception of what was referred to as "anomalous propagation."[7] The study of random, anomalous signals from around the world had been of interest to Radio Division researchers as far back as the mid-1920s.[8] Interest in this phenomenon heightened during World War II, as increasingly powerful and sensitive Navy radar receivers picked up stray radio signals from Europe and Japan. After the war, under Trexler's direction, several German Wuerzburg antennas were shipped to Washington and scavenged for parts. Wuerzburg antenna arrays were subsequently erected at NRL's Blue Plains field station in 1947 for a program aimed at developing intercept direction-finding equipment for anomalous signals originating in Europe and the Soviet Union.[9]

4.   David K. Allison, *New Eye for The Navy: The Origin of Radar at the Naval Research Laboratory* (Washington, DC: Naval Research Laboratory, 1981); David K. van Keuren, "The Military Context of Early American Radar, 1930–1940," in Oskar Blumtritt, Hartmut Petzold, and William Aspray, eds., *Tracking The History of Radar* (Piscataway, NJ: Institute of Electrical and Electronics Engineers, 1994).

5.   Allison, *New Eye for the Navy*, pp. 56–57; Bruce William Hevly, "Basic Research Within a Military Context: The Naval Research Laboratory and the Foundations of Extreme Ultraviolet and X-Ray Astronomy, 1923–1960," Ph.D. diss., Johns Hopkins University, 1987, pp. 11–53.

6.   Alfred T. Drury, "War History of The Naval Research Laboratory," 1946, unpublished history in the series, "U.S. Naval Administrative Histories of World War II," deposited in the Department of the Navy's library.

7.   Trexler interview, 30 October 1980.

8.   See Allison, *New Eye for the Navy, passim;* Hevly, "Basic Research Within a Military Context," *passim.*

9.   Trexler interview, 30 October 1980.

Consequently, with this history of research in the long-range propagation of radio waves, the Army Signal Corps's detection of radar waves bounced off the Moon in March 1946 did not come as a surprise to NRL researchers.[10] Indeed, there is evidence that laboratory researchers had unsuccessfully attempted to retrieve echoes from the Moon as early as 1928.[11] Efforts ceased during World War II and did not resume until 1948. Nevertheless, it appears that the Signal Corps program did attract the attention of Dr. Donald Menzel of the Harvard College Observatory. In 1946, Menzel, a commander with the U.S. Naval Reserve during World War II, as well as a member and the chair of the radio propagation committee of the Joint and Combined Chiefs of Staff, proposed to the Department of the Navy that it use Moon-reflected radio signals for secure communications.[12]

The response to Menzel's suggestion within the Department of the Navy is unclear. There does not seem to be a copy of the Menzel proposal or a Navy Department response within departmental records. A Navy program for using the Moon for active communications, as suggested by Menzel and pioneered by the Army, was not immediately forthcoming. However, the idea of using the Moon for communications and radar intercept purposes came under active consideration within two years. The moving force in this case was NRL engineer James Trexler.

James Trexler and Electronic Surveillance

Trexler had studied electrical engineering at Southern Methodist University (SMU), where his father taught in the Political Science Department. He had demonstrated little interest in traditional academic subjects, to his family's disappointment. However, he did prove himself to be an excellent, hands-on electrical engineer and amateur radio technician, thereby able to support himself during his undergraduate days.

In 1942, he came to NRL as a junior radio engineer, following in the footsteps of a former SMU engineering professor, Dr. Samuel Lutz, and simultaneously avoiding imminent induction into the U.S. Army. At NRL, Trexler was assigned to the Measurement and Direction-Finding Unit, where he spent much of the war working with various forms of high-frequency direction-finding units, before being assigned to the new electronic countermeasures group in 1945.[13]

While at SMU, Trexler had experimented with the reflection of high-frequency radio waves off meteor ionization trails as part of a study on the impact of atmospheric ionization on radio propagation. He continued in the mid-1940s to follow the work of investigators probing the upper atmosphere and near space with high-frequency radio transmissions. A paper by D.D. Grieg, S. Metzger, and R. Waer of ITT's Federal Telecommunication Laboratories in New York City, published in May 1948, proved to be particularly intriguing.[14] Trexler noted in his scientific notebook on 24 June 1948 that in Grieg, Metzger, and Waer's paper:

---

10.   John H. DeWitt, Jr., and E. King Stodola, "Detection of Radio Signals Reflected from the Moon," *Proceedings of the IRE* 37 (1949): 229–42; Jack Mofenson, "Radar Echoes from the Moon," *Electronics* 19 (1946): 92–98; Herbert Kauffman, "A DX Record: To the Moon and Back," QST 30 (1946): 65–68; James Trexler, "Lunar Radio Echoes," *Proceedings of the IRE* (January 1958): 286–92.

11.   See correspondence between Kenneth B. Warner and A. Hoyt Taylor, 31 January–7 February 1946, NRL Historian's Files, NRL, Washington, DC; Trexler interview, 30 October 1980.

12.   See Trexler, "A Chronological History"; Trexler interview, 30 October 1980.

13.   Trexler interview, 30 October 1980.

14.   D.D. Grieg, S. Metzger, and R. Waer, "Considerations of Moon-Relay Communication," *Proceedings of the IRE* 36 (May 1948): 652–63.

[I]t was pointed out that there is a possibility that the Moon has an ionosphere. If this is true then there is a possibility that certain radio frequencies will be reflected from the Moon's ionosphere with considerably higher efficiency than from the rugged surface of the Moon. The possibility of this being true was suggested by the recent experiments in Australia report-ed by [Grote] Reber at the Radio Astronomy meeting recently. In these tests standard trans-mitting and receiving equipments were used in a pulse system whereby the identity of the reflected signal was determined by coding. Delays of the correct magnitude were noticed. If there is any possibility of this mode of transmitting being useful to RCM [Radio Counter Measures] it should be given some very careful thought.[15]

Trexler pondered how a test might be set up to explore this possibility and what sort of equipment might be required. "The method of test could consist of using a beamed antenna having a sharp East-West pattern and a broad North-South shape," he wrote in his scientific notebook. "The intensity of the signal would be noted continuously and an attempt would be made to correlate it with the position of the Moon."[16] He computed antenna size to determine how many antenna elements were needed to achieve a 0.5-degree beam width. Using 270-degree spacing, Trexler determined it would take "about 150 elements" to obtain such a beam width with constant phase and amplitude across the aperture. Such an instrument would be expensive to build, but worth the expense, he noted.

His laboratory notebook indicates that Trexler continued his calculations throughout the week of 24 June 1948. By 29 June, Trexler had come to the following conclusion:

From the RCM [Radio Counter Measures] point of view this system hold[s] promise as a communication and radar intercept device for signals that cannot be studied at close range where normal propagation is possible. It might be well to point out that many radars are very close to the theoretical possibility of contacting the Moon (the MEW [Ballistic Missile Early Warning System, BMEWS] for example) and hence the practicability of building a system capable of intercepting these systems by reflections from the Moon is not beyond the realm of possibility. . . . The strictest security should be maintained as to the existence of such intercept devices since the enemy could with little difficulty restrict the operations of these sets so as to avoid Moon contact. One immediate application of the system would be the detec-tion and analysis of the Russian Radar signals that have been monitored at 500 MC [megahertz] near Alaska.[17]

Trexler's efforts over the following two years were directed toward demonstrating that such a Moon-intercept program was technically viable.

In late 1948 and early 1949, NRL constructed two large long-wire antennas designed for the high-angle observation of the Moon at its Blue Plains field facility.[18] The antennas were designed to carry out observations in bands where Soviet radar signals were known to exist. By August 1949, regularly scheduled observations of the Moon were under way, operating under the code name "Joe," with the intent of intercepting Soviet radar signals. In January 1950, the Navy consulted Group Captain Dunsford, the Royal Air Force's elec-tronic warfare coordinator with the U.S. Navy; he helped NRL researchers hone their search with improved Soviet target parameters.

---

15.   James H. Trexler, Laboratory Notebook No. N-124, 14 June–15 July 1948, p. 61, NRL records, Federal Records Center.

16.   *Ibid.,* p. 65.

17.   *Ibid.,* pp. 70–71.

18.   Trexler, "A Chronological History."

Although information on this early period of the program is sketchy, it seems that the results were sufficiently promising to allow Trexler and his immediate supervisor, Howard Lorenzen, to request approval from NRL's commanding officer, Captain Frederick Furth, for extended tests with more powerful radars.[19] By June 1950, the Chief of Naval Operations coordinator for electronic countermeasures had been briefed and had issued an official military requirement for Moon-intercept intelligence. The Chief of Naval Research, Rear Admiral T.A. Solberg, subsequently provided $100,000 for an experimental program, which included money for a new radar.[20] The effort, now an official Navy intelligence program, was renamed project PAMOR (Passive Moon Relay).

Communication Moon Relay

A site was chosen for the new antenna at Stump Neck, Maryland, on the annex grounds of the Navy's Indian Head Propellant Plant. Construction began in late December 1950 and was completed by the following September.[21] The design chosen for the Stump Neck antenna was a parabola having an elliptical opening 220 by 263 feet (sixty-seven by eighty meters). Earth-moving equipment scooped a hole out of the ground and paved it with asphalt; then a galvanized iron grid with three-inch-by-three-inch (7.5-centimeter-by-7.5-centimeter) openings was attached to provide a reflecting surface good for wavelengths of one meter or more. A cable-supported boom housed the focal point feed structure, which could then be steered in celestial coordinates of right ascension and declination by adjusting the cable length. The antenna was oriented so as to maximize observations of the "Sino-Soviet Block," although only a few hours per month of observation time were available.[22]

The first short-pulse radar contact with the Moon was made on 21 October 1951. The 750-watt transmitter sent ten-microsecond, 198-megahertz pulses.[23] The results surprised even Trexler. The fidelity of the received echo proved to be unexpectedly high. It had been assumed that the echo from a short pulse would have a fast rise, but a short fall. Theories predicted that energy would be returned from the entire illuminated sphere; however, the majority of the energy from the reflected pulse was received during the first 100 microseconds, meaning that half the power in the echo had to come from a circle on the Moon only 210 miles (338 kilometers) in diameter, or almost exactly one-tenth the diameter of the Moon.[24] The consequence was a much more coherent signal than originally expected. An immediate aftermath of the early test results was that Navy officials placed a higher priority and security status on the project.[25] The intelligence potential of passive Moon reflection was greater than originally surmised. A second consequence was the inauguration of the Communication Moon Relay project.

19.   *Ibid.*
20.   James H. Trexler, Laboratory Notebook No. N-411, 12 June–13 October 1950, p. 47, NRL records, Federal Records Center.
21.   Trexler, "A Chronological History"; anonymous, "Unique Aspects"; Trexler, "Lunar Radio Echoes"; B.E. Trotter and A.B. Youmans, "Communication Moon Relay (CMR)," 21 June 1957, declassified NRL Secret Report; Trexler, "Proposed URSI Paper for May 1955."
22.   Anonymous, "Unique Aspects."
23.   Trexler, "A Chronological History"; Trexler, "Lunar Radio Echoes."
24.   Trotter and Youmans, "Communication Moon Relay (CMR)," p. 1; Trexler, "Lunar Radio Echoes," p. 290.
25.   Trexler, "Chronological History."

The high fidelity of the reflected transmissions presented Trexler and his coworkers with an unexpected spinoff of their project. The quality of the received signal was potentially good enough to be manipulated for communications purposes. As B.E. Trotter and A.B. Youmans, Trexler's coworkers in the Countermeasures Branch, reported later, the results of the 1951 trials demonstrated that "the fidelity of this circuit was higher than suggested" and "implied that the circuit would be usable in modern communication systems."[26] Experiments using continuous wave, modulated continuous wave, and audio-frequency-modulated signals followed.

Lunar communications signal and equipment testing continued over the next three years. On 15 June 1954, Trexler summarized the status of the work done with the Stump Neck and earlier NRL radio telescopes in a memorandum to Louis Gebhard, the NRL Radio Division's superintendent. "From the experience gained over several years of work," he noted, "it appears that the fidelity of the Moon circuit is much better than predicted resulting in the possible use of many types of circuits such as high-speed teletype, facsimile and voice." The potential uses for such a telescope included functions associated with radar intercept, jamming, communications, navigation, satellite search, and ionospheric and atmospheric research. Under "Communications," Trexler proposed using the Moon as a passive reflector to "broadcast to half the world at any one time" at very high frequencies (VHF), as well as for two-way communications between the United States and ships, submarines, or large aircraft.[27] Early work had already advanced in this direction.

By the end of 1954, Stump Neck was nearing the end of its usefulness to the PAMOR project. A much larger antenna was needed to actually collect the weak Soviet radar signals. As noted in his 15 June 1954 memorandum, Trexler felt that "to utilize many of the possibilities of Moon relay an antenna having a 600-foot [183-meter] diameter would be required at VHF. For many applications, this diameter does not change fast with frequency since it is the absolute area that is important in receiving."[28] This marked the beginning of an extensive lobbying and development effort for a 600-foot radio antenna at Sugar Grove, West Virginia. It also marked the effective separation of PAMOR from what became the Communication Moon Relay project.

An advantage of the communications project was that simple antennas could be used at the receiving end. This advantage was particularly enhanced by the use of a 10-kilowatt klystron amplifier covering the ultrahigh frequency (UHF) band.[29] With the Stump Neck parabola serving as the transmitter, the receiver used an array of antennas, the basic element of which was a standard Model SK-2 radar antenna.[30] Early testing included both teletype and voice transmission, with Trexler's voice, as mentioned previously, being the first to make the round-trip lunar circuit.

After preliminary tests between the Stump Neck site and Washington, D.C., the first transcontinental test was set for the week of 20 November 1955. The receivers were established at the Navy Electronics Laboratory in San Diego, California, and after orientation of the field equipment, the Communication Moon Relay circuit began operating at

26. Trotter and Youmans, "Communication Moon Relay (CMR)," p. 1.

27. James Trexler to NRL Code 5400, Memorandum, 15 June 1954, in James H. Trexler, Laboratory Notebook No. N-756, 14 April–20 September 1954, p. 91, NRL records, Federal Records Center. Quoted text is from facsimile in NRL Historian's files.

28. *Ibid.*

29. Louis A. Gebhard, *Evolution of Naval Radio-Electronics and Contributions of the Naval Research Laboratory* (Washington, DC: NRL, 1979), p. 115.

30. Trotter and Youmans, "Communication Moon Relay (CMR)," p. 3.

31. *Ibid.*, pp. 7–10.

301 megahertz on 27 November. Because of the low declination of the Moon, initial performance was weak. Further adjustment of the equipment allowed a successful teletype message to be sent and received at 11:51 p.m., Pacific Standard Time, on 29 November. Dr. Robert Morris Page, the associate director of research at NRL, as well as one of the American inventors of radar, signaled Dr. Franz Kurie, the technical director of the Navy Electronics Laboratory, to "lift up your eyes and behold a new horizon." NRL conducted further experiments during December 1955 and early January 1956 to understand and to counter the signal fading that had been observed during the November tests.[32]

Beginning on 21 January 1956, the experimental baseline was extended to Hawaii, where an array of eight SK-2 receivers was set up at Wahiawa, Oahu. Teletype signals were sent at 300 megahertz from a ten-kilowatt transmitter. On 23 January, the system received its U.S. Navy christening when Admiral Arleigh Burke, Chief of Naval Operations, signaled a message of congratulations to Admiral Felix B. Stump, Commander-in-Chief of the Pacific Fleet.[33]

The reaction of the Department of the Navy to the experimental system was quick. Assistant Secretary of Defense for Research and Development Donald A. Quarles, who witnessed the tests, became a strong supporter and provided special Department of Defense funds to cover development costs. Also, Admiral Burke directed the Bureau of Ships to develop a demonstration model of a reliable, long-range communications system using the new technique. By May 1956, a Department of the Navy contract had been issued to the Developmental Engineering Corporation of Washington, D.C., for system development.[34] The costs for total development (including construction) were approximately $5.5 million.[35] An indication of the popularity of the Communication Moon Relay system may be found in the National Academy of Science's Advisory Committee on Undersea Warfare, which recommended in December 1956 that future American submarines use Moon-reflection path signaling for ship-to-shore communications.[36]

As the Communication Moon Relay system went into its production phase, the Communications Section of the NRL's Radar Division, which had inherited the project from the Radio Countermeasures Branch, began emphasizing improvements in receiver and transmitter design, including more powerful transmitters and transmission at higher frequencies. By mid-1957, lunar echo experiments were being conducted in the UHF band at 290 megahertz.

The experimental system produced by the Developmental Engineering Corporation for the Bureau of Ships quickly led to the development of a fully operational satellite communications system between Washington, D.C., and Hawaii. The system, functional by 1959, was inaugurated publicly in January 1960.[37] As part of the inaugural ceremonies, pictures of the aircraft carrier U.S.S. Hancock were beamed from Honolulu to Washington via the Communication Moon Relay system. The transmitted facsimile featured thousands of Hancock officers and seamen spelling out "Moon Relay" to a worldwide audience.

---

32.   Eventually, frequency diversity operations and the use of circular polarization were recommended. *Ibid.*, p. 10.

33.   *Ibid.*, p. 16.

34.   "U.S. Navy Communications Moon Relay (CMR) System," *Naval Research Reviews* (March 1960): 17–20.

35.   Department of the Navy press release, 26 January 1960, in NRL document file labeled "CMR—thru 1960," NRL records, Federal Records Center.

36.   National Academy of Sciences-National Research Council, Advisory Committee on Undersea Warfare, *Project Nobska, The Implications of Advanced Design on Undersea Warfare, Final Report, Volume 1: Assumptions, Conclusions, and Recommendations* (Washington, DC: National Academy of Sciences, 1 December 1956), pp. 16–19.

37.   Gebhard, *Evolution of Naval Radio-Electronics*, pp. 115–16.

Figure 7
Facsimile picture of the U.S.S. Hancock *with ship officers and crew spelling out "Moon Relay." This picture was transmitted via the Moon from Honolulu, Hawaii, to Washington, D.C., on 28 January 1960.* (Courtesy of the Naval Research Laboratory)

The completed system used eighty-four-foot-diameter (twenty-eight-meter-diameter) steerable parabolic antennas and 100-kilowatt transmitters installed at Annapolis, Maryland, and Opana, Oahu, with receivers at Cheltenham, Maryland, and Wahiawa, Oahu. The system operated at frequencies around 400 megahertz, it could accommodate up to sixteen teleprinter channels operating at the rate of sixty words per minute, and it was capable of processing teletype and photographic facsimiles.[38]

During the next two years, the Communication Moon Relay system expanded to include ship-to-shore communications. A sixteen-foot (five-meter) steerable parabolic antenna and receiving equipment installed on the *U.S.S. Oxford* in 1961 permitted one-way shore-to-ship lunar satellite communications for the first time.[39] The addition of a one-kilowatt transmitter to the *Oxford* in 1962 permitted two-way communications, as the ship sailed in South American waters. These successful trial experiments with the *U.S.S. Oxford* led to the establishment of the Navy's worldwide artificial satellite communications system later in the decade.

---

38.   *Ibid.*, pp. 117–18.
39.   *Ibid.*, pp. 121–22; L.H. Feher, V.W. Graham, W.E. Leavitt, and M.L. Musselman, "Satellite Communication Research—Communications by Moon Relay (CMR)," *Report of NRL Progress* (Washington, DC: NRL, May 1962), pp. 36–37; "Moon Used to Transmit Shore-to-Ship Radio Messages," *Naval Research Reviews*, February 1962, pp. 21–22.

Conclusion

The Communication Moon Relay system was the unexpected outgrowth of research and development in electronics intelligence—an allied but distinct field. The perceived need by the U.S. Navy and the American military as a whole to constantly assess Soviet technical capabilities rationalized and facilitated the diversion of funds and talent to fields that otherwise might not have been developed at such a comparatively early date. The construction of large-scale antenna facilities at Stump Neck, as well as the provision of technical support functions, such as the then-cutting-edge computational capabilities of the NAREC computer, provided the technical and scientific background that made the Communication Moon Relay system possible. Indeed, it was not only the Communication Moon Relay project that benefited, but also Navy radio astronomers who had access to the facilities during those substantial time periods when the Moon's position did not permit the use of the facilities for intelligence gathering.

If anything, the history of the Communication Moon Relay project demonstrates the complex and often hidden history of early space communications. It clearly illustrates, moreover, that during the Cold War, even the most basic research, such as radar studies of the lunar surface, often had a national security component. As the declassification of documents from this era progresses, the intricate and interwoven history of national security needs, science, and technological development should become clearer.

Chapter 3

Moon Relay Experiments at Jodrell Bank

by Jon Agar

The following describes research carried out at Jodrell Bank on the reflection of radio waves from the Moon. These experiments formed part of wider research programs pursued at the radio astronomy establishment and had implications for military and civil communications. Certain aspects of the experiments can be related to issues of spectacle and the public presentation of science that placed Jodrell Bank as one of the most significant scientific projects of postwar British science. After the descriptions of the radar research programs, there is discussion of the initial reactions at Jodrell Bank to the Echo balloon satellite project, as well as some of the later lunar bounce experiments at another important British laboratory, the Royal Aircraft Establishment at Farnborough.

Radar Astronomy at Jodrell Bank

After Patrick Blackett succeeded William L. Bragg as the Langworthy Professor of Physics at Manchester University in the fall of 1937, he immediately switched the research direction of the physics department from crystallography to his interest, cosmic rays.[1] With this redirection, many staff changes occurred, among them the appointment of a new assistant lecturer, the young Bernard Lovell. Barely had this research gathered momentum when another event, the outbreak of hostilities with Germany in 1939, scattered the physicists into altogether new environments and responsibilities. World War II mixed and transformed the three traditional locales of scientific research: the military, academia, and industry. In the radar, aeronautical, and code-breaking projects, academic scientists became aware of the extent of available resources, as well as the possibilities of goal-oriented, large-scale research. Links were forged between academic scientists and the government, in both the military and civil service. Many physicists, including Lovell from Manchester and Martin Ryle from Cambridge, entered radar research at the Telecommunications Research Establishment, as it was called by the end of the war. Located at Malvern, this establishment was in many ways the British analogue of the Radiation Laboratory at the Massachusetts Institute of Technology (MIT).[2]

1. A.C. Bernard Lovell, *P.M.S. Blackett: A Biographical Memoir* (Bristol: John Wright & Sons, 1976), p. 29.
2. A. Calder, *The People's War: Britain 1939–45* (London: Cape, 1969), pp. 457–77; A.P. Rowe, *One Story of Radar* (Cambridge: Cambridge University Press, 1948); Edward G. Bowen, *Radar Days* (Bristol: Adam Hilger, 1987); A.C. Bernard Lovell, *Echoes of War: The Story of H2S Radar* (Bristol: Adam Hilger, 1991).

Figure 8

Sir Bernard Lovell, director and founder of the Jodrell Bank radio astronomy observatory, seated at his desk. Partially visible in the background is the observatory's 250-foot (seventy-six-meter) telescope, which, with equipment from Pye Telecommunications Ltd., served to establish a long-distance communications link via the Moon. (Courtesy of NASA)

The physicists demobilized at the end of World War II were skilled in radar techniques; they were intent on reentering academia, but they were also well connected with the military and government. Committees considering postwar reconstruction and policy gave high priority to science in the universities. With the Barlow Committee in 1946 insisting on a doubling of "scientific manpower," there was pressure on the universities to expand their scientific departments.[5] Blackett, who was one of a number of scientists to be appointed to important government advisory positions during the war, as well as a member of the Barlow Committee, thus returned to his Manchester physics department, and he was prepared for growth.

Blackett and Lovell recruited radar researchers and secured war surplus equipment to continue the department's prewar work on cosmic rays, this time hoping to use radar

---

3.   Philip Gummett, *Scientists in Whitehall* (Manchester: Manchester University Press, 1980), pp. 218–20.

techniques to detect and investigate cosmic ray showers.[4] The radar group established itself at a university-owned location, a botanical research site named Jodrell Bank and located twenty miles (thirty-two kilometers) south of Manchester. Blackett's policy of departmental expansion was to encourage rapid growth in a diversity of projects under team leaders, rather than channeling all of the department's resources into cosmic rays. The Jodrell Bank Research Station grew. When the echoes displayed by the radar sets were identified with meteors in 1946, cosmic ray research branched into an expanding meteor astronomy program.

One can discern two phases of meteor astronomy at Jodrell Bank. First, from 1946 to about 1955, meteor research expanded, with the development of new techniques and the acquisition of new staff. A radar technique devised to deduce orbital characteristics—in particular, whether an orbit was closed (elliptical) or open (hyperbolic)—allowed radar astronomy to comment on a recognized problem already raised, but not resolved, by optical astronomy: the origins of sporadic meteors. Jodrell Bank astronomers argued that they had demonstrated closed orbits for meteors that did not form part of showers, and they concluded that all meteors form part of the solar system. During this first phase, research support came mostly through Manchester University and student fellowships awarded by the Department of Scientific and Industrial Research (DSIR).

During the second phase, from about 1955 to the late 1960s, meteor astronomy was no longer a program of central importance at Jodrell Bank. As the 250-foot (seventy-six-meter) radio telescope neared completion, new subjects and techniques took center stage: research on the scintillation of radio sources; the use of long baseline interferometry techniques; and radar studies of the Moon, planets, and, after 1957, satellites. Meteor research in this second phase was directed less toward astronomy and more toward ionospheric phenomena that might affect missiles. Research support came from the U.S. Air Force.

Studies of radar echoes from the Moon emerged from Jodrell Bank's meteor astronomy program. Observations of lunar echoes had been achieved in 1946 by Lt. Col. John H. DeWitt at the U.S. Army Signal Corps's Evans Signal Laboratory, using a continuous-wave transmitter, and by Zoltán Bay in Hungary, using an unusual chemical electrolysis receiver.[5] During the late 1940s, Frank Kerr, Alex Shain, and Charles Higgins, Australian physicists at the Division of Radiophysics of the Commonwealth Scientific and Industrial Research Organization, also measured the strength of lunar echoes. They confirmed DeWitt's observation of occasional signal fading, but they differed in terms of the explanation. While DeWitt ascribed rises in signal strength to smooth bounce points, the Australians distinguished between rapid fading (caused by Moon libration) and slow fading (from some other, possibly ionospheric, cause).[6]

At Jodrell Bank, William Murray, supported by a DSIR student fellowship, and J.K. Hargreaves began their study of lunar echoes in connection with a meteor research team led by Tom Kaiser. By 1953, they reported that "50,000 echoes were photographed and

4.   P.M.S. Blackett and A.C. Bernard Lovell, "Radio Echoes and Cosmic Ray Showers," *Proceedings of the Royal Society of London*, ser. A, vol. 177 (1941): 183–86. See also A.C. Bernard Lovell, "The Blackett-Eckersley-Lovell Correspondence of World War Two and the Origins of Jodrell Bank," *Notes and Records of the Royal Society of London* 47 (1993): 119–31.

5.   Andrew J. Butrica, *To See the Unseen: A History of Planetary Radar Astronomy* (Washington, DC: NASA SP-4218, 1996), pp. 6–12; James S. Hey, *The Evolution of Radio Astronomy* (New York: Science History Publications, 1973), pp. 25–26.

6.   Frank Kerr, Alex Shain, and Charles Higgins, "Moon Echoes and Penetration of the Ionosphere," *Nature* 163 (1949): 310–13; Frank Kerr and Alex Shain, "Moon Echoes and Transmission through the Ionosphere," *Proceedings of the IRE* 39 (1951): 230–242; Hey, *The Evolution of Radio Astronomy*, pp. 118–19; Butrica, *To See the Unseen*, pp. 21–22.

analyzed." Murray and Hargreaves noted that the slow fading had a diurnal pattern, strongly suggesting an ionospheric origin. They concluded, after fresh observations of both the horizontally and vertically polarized components of lunar echoes, that "the long period fading" arose from a slow rotation of the plane of polarization of the radio wave as it passed through the ionosphere—a phenomenon known as the "Faraday effect."[7]

Soon after the Faraday effect experiments, Murray and Hargreaves took positions elsewhere, at the Radar Research Station (the renamed Telecommunications Research Establishment) and the DSIR's Radio Research Station at Slough.[8] Lunar echo research at Jodrell Bank now became the work of John V. Evans, who arrived as a postgraduate student in 1954 and whose supervisor was Ian C. Browne, a colleague of Tom Kaiser working on radar echoes from meteors.[9] Finding the apparatus of Murray and Hargreaves a "poor instrument," Evans rebuilt it, increasing the transmitter peak power from one to ten kilowatts and improving the receiver sensitivity.[10] The aerial, which was used for both transmission and reception, was "an echelon of ten elements arranged one behind the other along a North-South baseline"; each element had a "reflecting screen tilted back at 45°."[11]

Evans's postgraduate stipend was for three years. The lunar echo research needed further financial support to continue, and that support was to come from the U.S. Air Force. The route whereby it was secured reveals the crucial assistance of links with American astronomers, as well as the implications that the new source of funding had for Manchester University and the British military. Lovell's link in the United States was Fred Whipple, a Harvard astronomer, a personal friend, and a colleague in meteor astronomy. Whipple reassured Lovell that an Air Force grant did not mean burdensome publishing "restrictions," and he offered to act as initial "intermediary" in negotiations.[12]

Lovell responded that "important aspects of our program were in the process of being shelved because of financial stringency and consequent lack of research workers." This situation described meteor height determination work, as well as "the lunar echo apparatus," Lovell explained, which was "now in a state where it can be used to measure the total electron content of the ionosphere. . . . Here again, the financial situation is such that my man-power on this valuable program is reduced to one Second Year research student." If extra funds were available from the U.S. Air Force, Lovell envisaged substituting "our present fixed aerial with a steerable one in order that we could do the measurements at all times of the day and night—not merely at lunar transit."[13]

Further negotiations occurred face to face when Gerald Hawkins, a Jodrell Bank astronomer who had moved to the Harvard College Observatory, visited Lovell in October 1955. In a nice example of a gift exchange cementing scientific networks, Harvard got

---

7.   On Kaiser as a team leader, see David O. Edge and Michael J. Mulkay, *Astronomy Transformed: The Emergence of Radio Astronomy in Britain* (New York: John Wiley & Sons, 1976), p. 311; Also see William Murray and J.K. Hargreaves, "Lunar Radio Echoes and the Faraday Effect in the Ionosphere," *Nature* 173 (1954): 944–45.

8.   I.C. Browne, John V. Evans, J.K. Hargreaves, and William Murray, "Radio Echoes from the Moon," *Proceedings of the Physical Society* B69 (1956): 901–20. Hargreaves later moved to the High Altitude Observatory of the University of Colorado at Boulder, before returning to Lancaster University in Britain. Edge and Mulkay, *Astronomy Transformed*, p. 414.

9.   Ian C. Browne and Tom Kaiser, "The Radio Echo from the Head of Meteor Trails," *Journal of Atmospheric and Terrestrial Physics* 4 (1953): 1–4.

10.   Butrica, *To See the Unseen*, p. 23.

11.   Browne, Evans, Hargreaves, and Murray, "Radio Echoes from the Moon," p. 909.

12.   Fred Whipple to Bernard Lovell, letter, 18 May 1955, Jodrell Bank Archives ACC/56/5.

13.   Bernard Lovell to Fred Whipple, letter, 9 September 1955, Jodrell Bank Archives ACC/56/5. Lovell also wrote that he was "impressed with the possibilities opened up by the availability of the sixty-foot paraboloid in America."

Hawkins, and Jodrell Bank received Harvard's knowledge of Air Force contacts. Lovell was told to contact Jules Aarons of the Electronics Research Directorate of the Air Force Cambridge Research Center (AFCRC) in Bedford, Massachusetts.[14] Although Aarons turned down the proposal to supply Jodrell Bank with a sixty-foot radio telescope for lunar echo work, he did indicate Air Force interest in funding staff and operating costs for certain meteor and lunar echo research. Lovell requested $10,000 to boost the lunar transmitter from ten to ninety kilowatts and $5,000 to rebuild a steerable radio telescope to replace the transit array.[15]

Whipple reported that he had heard at Christmas in 1955, "via the grapevine," that an AFCRC grant had been awarded to Jodrell Bank. He reassured Lovell that this was good news. The Air Force was "really quite easy to work with," so long as reports were in on time, and once Jodrell Bank had "gotten started with them, you can count on continuation."[16] However, Manchester University's bursar, R.A. Rainford, expressed reservations. "I am very worried about possible repercussions," particularly about "security," he wrote to Lovell. "Permission from the Ministry of Defence," he thought, would be necessary.[17]

The Ministry of Supply, in fact, already had considered the matter. This ministry was responsible for British research on long-range ballistic rockets and their countermeasures and already supported meteor research at Jodrell Bank. In a manner that seemed typical of his lobbying for support for Jodrell Bank projects, Lovell then tried to use American interest to gain further help from the Ministry of Supply. Lovell wrote to Sir Owen Wansbrough-Jones, the Ministry of Supply's chief scientific advisor, that the meteor and lunar echo programs had both "potential military value" and "very great" fundamental interest. Moreover:

> *If the experiments were not so severely handicapped by lack of money I would have no hesitation in refusing approaches from America, since I am sure that there will be many minor difficulties even if the University felt able to give their permission. One easy way out might be for the Ministry of Supply to increase its support to these programs.*[18]

The Ministry of Supply indeed consented to the AFCRC grant, because a formal consultative procedure between the two military research bodies already existed. Contracts to expand the lunar echo research and to complement the Ministry of Supply–supported work on upper atmosphere winds were agreed on.[19] By February 1956, the Engineering Laboratories of the Army Signal Corps joined the AFCRC in developing the Jodrell Bank lunar echo apparatus, particularly with respect to "the problem of transatlantic communication via the Moon."[20]

Meanwhile, lunar echo research continued. In December 1956, Evans used his rebuilt equipment to carry out what he called the "Double Pulse Experiment," the results of which, he suggested, showed "that the Moon is very limb 'dark,'" and that the effective scattering region is at the center of the visible dis[k] having a radius of about one-third that

---

14. Bernard Lovell to Jules Aarons, letter, 27 October 1955, Jodrell Bank Archives ACC/56/5.

15. Bernard Lovell to Jules Aarons, letter, 21 December 1955, Jodrell Bank Archives ACC/56/5.

16. Fred Whipple to Bernard Lovell, letter, 29 December 1955, Jodrell Bank Archives ACC/56/5.

17. R.A. Rainford to Bernard Lovell, letter, 30 December 1955, Jodrell Bank Archives ACC/56/5.

18. Bernard Lovell to Owen Wansbrough-Jones, letter, 12 January 1956, Jodrell Bank Archives ACC/56/5.

19. Copy of contract AF61(514)-947 for "research of Moon echo phenomena," Jodrell Bank Archives ACC/56/5.

20. Memorandum, "Meteor and lunar echo programme," 7 February 1956, Jodrell Bank Archives ACC/56/5.

of the lunar radius."[21] Further experiments the following March agreed with this reflection hypothesis. The conclusion that only part and not all of the Moon's disk was a strong reflector of radio waves had implications for the use of the Moon as a passive communication device.

Already in 1951, teams at the Naval Research Laboratory and the Central Radio Propagation Laboratory of the National Bureau of Standards had used the Moon as an experimental communications relay. The Naval Research Laboratory transmitted and received Morse later that year; it then operated a voice relay in 1954 and teleprinter connections between Washington, D.C., and San Diego in 1955 and between Washington and Oahu in 1956.[22] Evans thought that his results raised the possibility of a higher quality Moon relay system than expected, because a message would be less troubled by echoes reflected from the edges of the Moon's disk. Suggesting that he was curiously unaware of the American achievements, Evans wrote: "[S]ince the effective depth of the Moon is 1 msec or less it becomes possible to use the Moon in a communication circuit with modulation frequencies up to ~1000 c/s. This is probably just sufficient for intelligible speech and could be used for teletype."[23]

Lovell immediately used Evans's success to increase U.S. Air Force financial support. Because of the "important repercussions on the consideration of the use of the Moon as a relay station," he argued, the work was "so important that a doubling of [the original support] could easily be justified."[24] After resolving whether outside authorities, such as the U.S. Air Force, had the right to inspect university accounts[25]—a central debate in British higher education in the 1950s[26]—an extension was granted, along with renewals of the meteor programs.[27]

In the fall of 1957, the Jodrell Bank's 250-foot-diameter, fully steerable radio telescope finally became operational, just in time for the October 1957 launch of the Soviet satellite Sputnik. After pressure from the consultant engineer for the instrument, H. Charles Husband, Lovell agreed to demonstrate the efficacy of the troubled radio telescope with a publicized tracking of both the satellite and its carrier rocket.[28] The radio telescope was hurriedly fitted with a small seventy-five-watt transmitter, borrowed from the Air Ministry,[29] and radar echoes were recorded, first in a test run on the Moon and then from Sputnik and its rocket a week after their launch.

The following autumn, John Evans assisted Lovell in making recordings of voices reflected off the Moon using the radio telescope. Lovell incorporated these recordings as part of his 1958 Reith Lectures. After the radio broadcast of these talks, J.R. Brinkley, a

---

21.   J.V. Evans, "The Scattering of Radio Waves by the Moon," *Proceedings of the Physical Society* B70 (1957): 1105–12.

22.   Butrica, *To See the Unseen*, pp. 24–25.

23.   Evans, "The Scattering of Radio Waves by the Moon," p. 1112.

24.   Bernard Lovell to Jules Aarons, letter, 28 May 1957, Jodrell Bank Archives CS7/35/5.

25.   Jules Aarons to Bernard Lovell, letter, 10 June 1957, Jodrell Bank Archives CS7/35/5; R.A. Rainford to Bernard Lovell, letter, 18 June 1957. Rainford wrote: "[T]he University has no objection to taking dollars from the United States for research programmes which are approved by the Ministry of Supply and the DSIR so long as they do not include terms which allow the books of the University to be open for inspection by the United States Government."

26.   Jon Agar, "The New Price and Place of University Research: Jodrell Bank in the Economies of Postwar British Science," forthcoming.

27.   "Memo on meeting with Colonel Trakowski, Captains Berge and Derrick (USAAF Brussels) on Wednesday, 19 March 1958," 19 March 1958, Jodrell Bank Archives CS7/35/5.

28.   H. Charles Husband to Bernard Lovell, letter, 8 October 1957, Jodrell Bank Archives CS7/31/3. Husband wrote: "I do beg of you for the sake of both our reputations that some immediate joint action be taken."

29.   Bernard Lovell to J.R. Brinkley, letter, 5 December 1958, Jodrell Bank Archives CS7/33/4.

director of Pye Telecommunications Ltd., a British electronics firm based in Cambridge, contacted Lovell and expressed interest in developing a lunar relay communications system.[30] Pye offered Jodrell Bank the free loan of equipment. In Pye's view, transatlantic short-wave radio links were too "few, unpredictable and distorted." Moreover, although the service had improved with the opening of the first transatlantic telephone cable (TAT-1) in 1956, submarine cable was no answer to the exponential rise in demand. "[S]ubmarine cable will not solve all the problems," Brinkley declared. "The production and laying of tens of thousands of miles of such cable is a major undertaking, costing great sums of money and taking up much time . . . many of the territories to be reached are inland, with no modern trunk routes connecting them to the sea. The only medium which offers itself is communication via space."[31]

Pye installed lunar relay transmitters and receivers at Jodrell Bank[32] and supplied receivers for the eighty-four-foot (twenty-eight-meter) AFCRC radio telescope at Sagamore Hill, Massachusetts.[33] Transmissions from the Jodrell Bank radio telescope using a one-kilowatt transmitter succeeded in May 1959. As Lovell wrote later: "Within a short time clear voice circuits via the Moon had been established between Jodrell Bank and an American telescope, and it was even found possible to transmit recognizable music."[34] However, Jules Aarons, Herbert Whitney, and Hugh Peters of the AFCRC considered that "[a]n increase of approximately 15 db gain must be realized in the overall system before good intelligible voice or music can be received."[35] The experiment was reversed in November 1959, when Millstone Hill radar at MIT's Lincoln Laboratory transmitted and Jodrell Bank received.[36] Plans followed to transmit between Jodrell Bank and Australia and between Jodrell Bank and New Zealand.[37]

In June 1960, Brinkley told the Commonwealth Press Union that after the Jodrell Bank lunar relay experiments, he was "left in no doubt that practical circuits via the Moon are feasible now and will, in the space of a few years, become economic."[38] Although lunar relays could only be used at certain times each day, the "vast number of teleprinter channels that Moon transmission can make available," argued Brinkley, would "relieve existing overloaded short-wave circuits" as well as have uses in "new fields." Pye set out plans for a commercial Moon relay communication involving the defense firm Vickers, using small thirty-foot (ten-meter) dishes made by Vickers and operating around 7,000 megahertz.[39]

30.   J.R. Brinkley to Bernard Lovell, letter, 2 December 1958, Jodrell Bank Archives CS7/33/4. Brinkley had "discussions with Dr. Ryle and Dr. Smith of the Radio Astronomy Laboratory [in Pye's town, Cambridge] on the exploitation of moon reflection." See also A.C. Bernard Lovell, *Out of the Zenith: Jodrell Bank, 1957–1970* (New York: Harper & Row, 1973), p. 212.

31.   Talk by J.R. Brinkley to Commonwealth Press Union, "Telecommunications and the Press," 13 June 1960, Jodrell Bank Archives CS7/31/4.

32.   J.R. Brinkley to Bernard Lovell, letter, 8 December 1958, Jodrell Bank Archives CS7/33/4. The transmitter immediately available had one kilowatt of power, was highly stable, and worked within the band 100–200 megahertz. The receivers were "mass production and high grade."

33.   Jules Aarons, Herbert Whitney, and Hugh Peters, "Bistatic Transatlantic Moon Reflections," undated report, Jodrell Bank Archives CS7/67/1.

34.   Lovell, *Out of the Zenith*, p. 212.

35.   Aarons, Whitney, and Peters, "Bistatic Transatlantic Moon Reflections."

36.   John Thomson, "Moon and Venus Radar: Passive Satellite Observations," Technical (Final) Report under contract AF61(052)-172, October 1958–December 1960, Jodrell Bank Archives CS7/33/4.

37.   J.R. Brinkley to Bernard Lovell, letter, 7 February 1961, Jodrell Bank Archives CS7/31/4; J.R. Brinkley to Clifton D. Ellyett, 21 May 1959, Jodrell Bank Archives CS7/33/4. J.L. Pawsey of the Division of Radiophysics group in Sydney coordinated the Australian end of the experiment.

38.   Brinkley, "Telecommunications and the Press."

39.   J.R. Brinkley to Bernard Lovell, letter, 16 September 1959, Jodrell Bank Archives CS7/33/4.

A year later, part of the commercial opportunity Pye saw was the manufacture of large steerable aerials, in conjunction with aircraft manufacturer Hawker Siddeley.[40] However, Pye abandoned the plans, according to Lovell, when "the successful use of the initial low orbit satellites, and later of the stationary orbit satellites removed the commercial incentive for the Moon communication link."[41] The comments of Pye's managing director, Brinkley, in October 1961 corroborate this argument. Pye would "hesitate to commit ourselves to heavy expenditure on further research and steerable aerials until [Pye was] reasonably sure the expenditure would fit into space communication too." And space communication to Brinkley meant "space satellites."[42]

After Evans's departure from Jodrell Bank in 1960 to join MIT's Lincoln Laboratory,[43] John Thomson took over the leadership of the lunar and other planetary radar experiments. John E.B. Ponsonby, an Imperial College graduate, soon joined him. Under Thomson, lunar radar studies continued through the 1960s and involved, in particular, the development of lunar aperture synthesis techniques to produce high-definition radar maps of the Moon's surface. This research program declined after the Arecibo radio telescope began operating in November 1963.[44]

The Great Public Spectacle of British Science

By far, the biggest public attraction of 1951 was an exhibition at a specially constructed site on the south bank of the River Thames. This extravaganza was the Festival of Britain, a government-planned display of a revival of British culture after the postwar years of austerity. The festival's aim was to make "visible a brave New World"—a vision continued in the proclamations after the death of George VI in 1952 of a New Elizabethan Age, harking back to the days of glory, undisputed British sea power, and the beginnings of empire under Elizabeth I.[45] At the festival's center was the Dome of Discovery, designed around a "narrative." Viewers were told a story of "creditable British exploration, invention and industrial capacity."[46] As the highlight to the story, there stood a "radio telescope . . . operated from the Dome of Discovery, with its 'dish' aerial mounted on the top of the Shot Tower. This was beamed on the Moon and visitors could see on a cathode ray tube signals being transmitted there and their reflection back about two and a half seconds later."[47] The visiting public, of which there were millions, carried away associations between radio telescopes and public prestige as early as 1951.

A second instance of the use of the Moon to demonstrate radio telescopes can be found in the planned opening ceremony of the 250-foot Jodrell Bank radio telescope. As it neared completion, Sir Charles Renold, chair of the telescope's Site Committee, wrote that once the instrument was in suitable "condition"—that is, "capable of being rotated by power" and receiving radio signals—then a "dramatic and impressive" public display of

40.   J.R. Brinkley to Bernard Lovell, letter, 9 December 1960, Jodrell Bank Archives CS7/31/4.
41.   Lovell, *Out of the Zenith*, p. 212.
42.   J.R. Brinkley, "The Economics of Space Communications," 22 September 1961, Jodrell Bank Archives CS7/55/2.
43.   John Evans, interview with Andrew Butrica, NASA Headquarters, Washington, DC, 9 September 1993, JPL Archives. See also Lovell, *Out of the Zenith*, pp. 194–96.
44.   See Lovell, *Out of the Zenith*, p. 207, for Arecibo's effect on the Venus radar program.
45.   Roy Strong, "Prologue," in Mary Banham and Bevis Hillier, eds., *A Tonic to the Nation: The Festival of Britain 1951* (London: Thames & Hudson, 1976), p. 8.
46.   Misha Black, "Architecture, Art and Design in Unison," in Banham and Hillier, eds., *A Tonic to the Nation*, p. 84.
47.   Ian Cox, "Three Years a-Growing," in Banham and Hillier, eds., *A Tonic to the Nation*, p. 69.

the efficacy of the telescope should be performed.[48] Sir Ben Lockspeiser of the DSIR and Sir John Stopford, who was vice-chancellor of Manchester University, judged that the Duke of Edinburgh would be a suitably symbolic person to perform the opening ceremony.[49] Lovell suggested a possible display: "[O]n pressing the button . . . the telescope [would] sweep over one or more of the remote radio sources in the depths of the universe. The resulting signal could be displayed on a number of pen-recording instruments, and these could be used to initiate a local series of events such as the unfurling of flags."[50] "Even better" than a remote, invisible radio source, Lovell suggested, was a "radar demonstration" using the visible Moon—a "target likely to create an impression."[51]

In any event, Sputnik provided a demonstration that the telescope worked—and without an opening ceremony. However, radar echoes from the Moon were shown again in public at the Reith Lectures. The BBC's annual Reith Lectures embodied the principles of its celebrant, the corporation's stern patriarch and defender of elite culture, Lord Reith. From the first lecture, given by Bertrand Russell, an invitation was offered each year to an authoritative public figure "to undertake some study or original research on a given subject and to give listeners the results in a series of broadcasts."[52] Lovell was invited, only months after Sputnik, to be the Reith lecturer of 1958. He made good use of his series of talks, called *The Individual and the Universe*, to defend the cause of big telescopes.[53] To Lovell it was "a mortifying thought that the largest [optical] telescope in Great Britain today is considerably smaller than the telescope which Herschel built." However, the national shame of "the steady decay of British influence in astronomy" had "been arrested by remarkable developments in . . . radio astronomy." He located the current position of the "great powers" (the United States and the Soviet Union) as stemming from their support of pure science. According to Lovell, "the technical devices which form the basis of the present economic and cultural strength of the Great Powers can be traced back within a few generations to fundamental scientific investigations which were carried out in the abstract, supported without thought of direct practical benefit."[54]

Radio telescopes were portrayed as a root of national resurgence, feeding on the cultural association of instrument with nation in the Festival of Britain. To bring the capabilities of the instrument to its listening audience, in the lecture named "Astronomy and the State," Lovell played the recordings of voices relayed via the Moon.

I'm sorry the echo was so weak, but that's not really important. After all, our transmitter had only a thousandth of the power of some of the transmitters which are broadcasting my voice now. The important point is that the voice when received back from the Moon is perfectly intelligible and that the telescope was working on wavelengths which could never be disturbed by atmospheric or ionospheric conditions. Well the result of that investigation is a free gift of the radio astronomer, to all the commercial and military organizations who will no doubt use it in future.[55]

---

48. Renold, "Radio Telescope: Opening Date," memorandum, 6 December 1956, Jodrell Bank Archives CS7/31/2.

49. Bernard Lovell to John Stopford, letter, 8 December 1954, Jodrell Bank Archives CS7/39/5.

50. Bernard Lovell to Charles Husband, letter, 13 December 1956, Jodrell Bank Archives CS7/31/2.

51. Bernard Lovell to Charles Husband, letter, 4 January 1957, Jodrell Bank Archives CS7/31/3.

52. Preface to Bertrand Russell, *Authority and the Individual*, BBC Reith Lectures 1948 (London: George Allen & Unwin, 1949).

53. Bernard Lovell, *The Individual and the Universe*, BBC Reith Lectures 1958 (London: Oxford University Press, 1959).

54. *Ibid.*, pp. 66–67.

55. *Ibid.*, p. 69.

This was no free gift. Instead, it was part of the ongoing campaign by Lovell to hold together the projects at Jodrell Bank—in particular, the colossal radio telescope.

In the same summer of 1951, during the Festival of Britain, the DSIR considered a request from Manchester University for £279,140 for a giant, steerable radio telescope. The DSIR agreed to fund the project for at least four interlocking reasons. First, it was a project that satisfied the civil service's twin desiderata of scientific "timeliness and promise." Second, the project was proposed by scientists who had built up extensive credit within Whitehall during the war by contributing to research programs, such as radar (for example, Lovell), or by acting as scientific advisors (Blackett). Third, as John Krige has argued in the case of Britain's entry into CERN, the DSIR felt that it faced competition and future challenges to its function, as other departments expanded their own research programs. Consequently, the DSIR was sympathetic to a prestige project.[56] Finally, radio astronomy was identified with British capability and leadership, and the giant radio telescope, through its scale and visibility, was promoted as an icon of British progress for consumption at home and abroad. Writing about a quality documentary film to be funded by the British Foreign Office, a DSIR civil servant enthusiastically stated that it had "already proved of greater public interest than any other project" he had handled; "it would bring credit to Britain . . . it should prove more effective propaganda than the films on our social system, housing [or] justice." The radio telescope was, he wrote, the "great public spectacle" of British science.[57]

However, the spectacle soon slid into debt, and the radio telescope had to be reinterpreted for different audiences in an effort to secure funds.[58] For example, it was "financial stringency" that made the U.S. Air Force grants for lunar echo work attractive and that overrode the small doubts about freedom to publish.[59] The radio telescope came to dominate work at Jodrell Bank. For example, meteors were translated as "nature's missiles," when Lovell wrote to defense firms seeking contributions to clear the telescope's debt.[60] While the promotion of the telescope could prove troublesome in attracting "thousands of visitors from all over the country,"[61] identification with national progress could be mobilized in the efforts to pay off the debt. The audio spectacle of voices reflected from the Moon in the Reith Lecture "Astronomy and the State" must be understood within this context.

Working With America

In parallel with the work supported by the U.S. Air Force discussed above, Jodrell Bank also was involved in the satellite program of NASA. Besides the scientific and prestige interests, collaboration with NASA, mostly involving the use of the big radio telescope

---

56.   John Krige, "Britain and the European Laboratory Project: 1951–mid-1952," in Armin Hermann, John Krige, Ulrike Mersits, and Dominique Pestre, eds., *History of CERN* (Amsterdam and New York: North-Holland Physics Publishing, 1987–1990), vol. 1, pp. 431–74.

57.   Hingston, "Film of the Radio Telescope," memorandum, December 1953, Jodrell Bank Archives CS7/15/1. See also Jon Agar, "Screening Science: Spatiality and Authority at Jodrell Bank," in Jon Agar and C.W. Smith, eds., *Making Space for Science* (London: Macmillan, forthcoming).

58. See Jon Agar, "Making a Meal of the Big Dish: The Construction of the Jodrell Bank Mark I Radio Telescope as a Stable Edifice, 1946–57," *British Journal for the History of Science* 27 (1994): 3–21.

59.   As the Manchester University's bursar wrote in 1958: "Money is now very tight and the only way to keep the Station going fully is to make sure that outside 'users' pay their due share of the total costs." J.R. Rainford to Bernard Lovell, letter, 7 August 1958, Jodrell Bank Archives CS7/33/4.

60.   Bernard Lovell to Espley (GEC), letter, 25 November 1955, Jodrell Bank Archives CS7/3/3. Likewise, as Lovell wrote to de Ferranti, the Radio Telescope became "really a gigantic radar scanner." Bernard Lovell to de Ferranti, letter, 22 November 1955, Jodrell Bank Archives CS7/3/3.

61.   Letter, Lovell to Mansfield Cooper, 6 June 1957, Jodrell Bank Archives CS7/41/3.

to track satellites and probes, brought much needed money to Jodrell Bank. One contract alone with NASA paid $179,200 to the radio astronomy observatory.[62] The techniques involved in tracking the two Echo mylar balloons were similar to those used in, and were continuous with, work in the Manchester radar program, such as the lunar relay experiments.

The initial response to Echo at Jodrell Bank is revealing. Echo was an attempt to "establish communications between the Bell Telephone Laboratory facility at Holmdel, New Jersey, the Jet Propulsion Laboratory [JPL] facility at Goldstone, California, and the 250-foot dish facility at Jodrell Bank, England."[63] Messages would be transmitted and received from the Bell Telephone Laboratory and JPL, but Jodrell Bank would only receive. Evans visited the Bell Telephone Laboratory in October 1959 and discussed the project with Bell scientists. Summarizing his visit for Lovell on his return, Evans wrote: "This experiment seems to be essentially a 'stunt.' There are no good reasons for our participation unless a voice message were to be transmitted, which would give the 'stunt' its maximum publicity value."[64]

These "stunts," or public displays of scientific spectacle, were highly significant in holding together the Jodrell Bank project, so it is not surprising that the proposed Echo involvement was viewed in this light. Lovell's response was to "suggest we do nothing about the communication part of the business unless we are pressed to do so by NASA. We have already promised NASA to help with the preliminary radar tracking."[65]

NASA was indeed interested, and Lovell agreed to Jodrell Bank's participation in the Echo communication experiment, so long as the observatory received assistance from Space Technology Laboratories personnel.[66] Collaboration with NASA garnered for Jodrell Bank public and national prestige, as well as the possibility of technology transfer. The same pattern emerged from involvement in Echo 2, when Lovell commented to Pye's Brinkley, "I cannot imagine that much other than the establishment of working relations with the Russians will come out of this experiment, but even for that reason we must obviously do our best to join in."[67]

Government laboratories in Britain were also following the American satellite programs closely. Staff at sites such as the Royal Aircraft Establishment in Farnborough, the General Post Office's growing Goonhilly ground station, the Royal Radar Establishment in Malvern, and the Signals Research and Development Establishment all suggested their own satellite programs[68] or sought to collaborate with the Americans.[69] The use of the Moon for military communications was carefully examined within the Royal Aircraft Establishment,[70] along with other passive systems, such as Echo and Project West Ford, that promised resistance to jamming and interception.[71] Although one initial reaction to the proposal to use Moon relay circuits was that the band width would be too restrictive unless

62.   NASA Contract No. NASw-68, effective date 14 April 1959, Jodrell Bank Archives ACC/53/2.

63.   "NASA SPACECONN Project Echo S-42 OPLAN 4-60," report, Jodrell Bank Archives CS3/9/4.

64.   John Evans, "Notes upon my Visit to Bell Telephone Lbs.," memorandum, October or November 1959, Jodrell Bank Archives CS1/3/1.

65.   Bernard Lovell to John Evans, internal note, 5 November 1959, Jodrell Bank Archives CS1/3/1.

66.   Bernard Lovell, *Out of the Zenith*, p. 213. Lovell quotes from his letter to Leonard Jaffe. Space Technology Laboratories was a wholly owned subsidiary of Ramo-Wooldridge (later TRW).

67.   Bernard Lovell to J.R. Brinkley, letter, 4 March 1963, Jodrell Bank Archives CS3/13/1.

68.   For example, T.W.G. Dawson of the Royal Aircraft Establishment suggested "hair satellite" and "flashing satellites" systems.

69.   The General Post Office was NASA's British partner in such proposed experiments as Relay, TSX, and Rebound. The Royal Aircraft Establishment was anxious that it should also gain experience with satellites.

suitable frequency multiplexing systems were developed,[72] a "Moon relay service" became a Foreign Office "requirement" by 1962[73] and was probably used elsewhere as a secure link.

Conclusion

This paper considers both private and public lunar echo experiments at Jodrell Bank. The private research was supported initially by British civil and military bodies and later by the U.S. Air Force. The military's interest in the experiments was twofold. First, the Faraday effect measurements imparted knowledge about the content of the ionosphere through which rockets and guided missiles might pass. Second, the Moon could be used as a relay in a passive communications system. The military preferred passive systems, because they were understood to be less susceptible to jamming, interference, and interception.

Companies—in this case, Pye—also were interested in the Jodrell Bank lunar echo experiments because of their significance for communications. Pye briefly planned a commercial Moon relay; it seemed to offer competition with transatlantic telephone cables, as well as a new service where demand was high, but communications unreliable (Brinkley stressed Europe-to-Africa links). It is also possible, given that Pye and its potential collaborators (Vickers and Hawker Siddeley) were defense contractors, that Pye hoped to find a market with the military.

To the scientists at Jodrell Bank, the experiments offered an interesting new field and the possibility of gaining support useful in other projects (for example, new equipment in the form of transmitters and telescopes, as well as new staff). The public lunar echo experiments in these ongoing research programs and in the promotion of radio astronomy in Britain—and the radio telescope in particular—were certainly national spectacle. This spectacle was useful to its sponsors in government in presentations of British progress, as well as for the DSIR internally within Whitehall. To scientists such as Lovell, demonstrations of lunar echoes formed part of the wider campaign to keep the Jodrell Bank project together.

---

70.   Lunar echo experiments were carried out at the Royal Aircraft Establishment in the early 1960s to prepare techniques for passive satellite communications systems.

71.   John E. Clegg, "A Note on the Use of the Moon and Passive Satellites for Long Distance Communication," 22 June 1961, Public Records Office, AVIA, 13 1292. See also the comment in "Military Communications—Satellites/Moon," meeting minutes, 20 October 1960: "[I]t was becoming clear that the GPO [General Post Office] will confine themselves to active satellites in co-operation with NASA (to ensure adequate bandwidth for a civil system) but that NASA would continue with both passive and active satellites. Passive satellites have some properties of particular importance to possible military communications systems and it was important for Ministry of Aviation to study such satellites."

72.   C. Williams, Royal Aircraft Establishment, Radio Department, note, Public Records Office, AVIA, 13 1292.

73.   "Satellite Communications Research," meeting minutes, 12 February 1962, Public Records Office, AVIA, 13 1292.

PART II:

Creating the Global, Regional, and National Systems

————⇒►○◄———

The United States and Canada

Chapter 4

Something of Value: Echo and the Beginnings of Satellite Communications

by Donald C. Elder

The world has changed in many ways since 1960, but few areas have undergone as radical a transformation as the field of telecommunications. From the virtually instantaneous transmission of television images to the relaying of telephone messages across vast distances, people have access to capabilities only dreamed of 35 years ago. If the world is indeed becoming a "global village," the revolution in the field of telecommunications is in large measure responsible for that development.

Although a number of technological innovations help explain this progress, the significant breakthrough involved the advent of communications satellites. They display a remarkable degree of technological sophistication today, yet it is instructive to note that they also have undergone a process of evolution. Indeed, the first such device seems almost simplistic in comparison to the communications satellites of today. That satellite, christened Echo, was in fact a mylar sphere coated with vaporized aluminum, and it could reflect only signals directed at it. Still, the story of Echo I does have great significance in the history of today's telecommunications revolution. It proved the viability of the concept of the communications satellite and allowed interested parties to conduct experiments that presaged the uses to which others would apply the ensuing generations of satellites.

The Theoretical Basis

The story of Echo actually begins during the days just after the end of the World War II. In October 1945, Arthur C. Clarke, already on his way to becoming one of the preeminent figures of science fiction, wrote an article suggesting that a device placed in orbit around the Earth could relay messages transmitted to it from one point on the planet to another.[1] His idea found resonance with others, who in various forms kept the concept of a communications satellite alive during the following few years. Indeed, individuals in a number of U.S. government agencies noted in reports the value of such a venture to both the public and private sectors of the country.[2] However, the postwar Truman administration never gave any official backing for the development of communications satellites, thus keeping the concept in the theoretical realm.

---

1. Arthur C. Clarke, "Extra-Terrestrial Relays: Can Rocket Stations Give World-Wide Radio Coverage?," *Wireless World* 51 (October 1945): 305–08.
2. U.S. Senate, "Policy Planning for Space Communications," *Staff Report Prepared for the Committee on Aeronautical and Space Sciences*, 86th Cong., 2d sess. (Washington, DC: U.S. Government Printing Office, 1960), p. 3.

Figure 9
Echo balloon fully inflated for testing purposes. (Courtesy of NASA, photo no. 60-E-6)

This situation began to change in 1952. During that year, John Robinson Pierce, who was the director of research at the American Telephone and Telegraph (AT&T) Bell Telephone Laboratories, wrote a story for *Astounding Science Fiction,* in which he discussed the potential benefits of communications satellites.[3] In 1954, he further refined his thinking on the subject in a speech he delivered to the annual meeting of the Institute of Radio Engineers. In this address, Pierce examined possible communications satellite configurations and suggested that such a device could either actively repeat or passively reflect signals transmitted to it. An active repeater would require an internal power source to allow the retransmission of signals broadcast to it. While noting that this type of satellite had many advantages, Pierce concluded that it possessed one highly significant drawback: the limited lifetime of power sources available at the time would give such a device only a relatively brief period of usefulness. For that reason, active repeater satellites were impractical for private industry as an alternative to existing methods of long-distance communications relays.

Having rejected the viability of active repeater satellites because of technological limitations, Pierce turned to the concept of passive reflectors. Such a satellite would have no need for an internal power source, making it of more immediate utility than an active repeater. After establishing the superiority of the concept, Pierce then examined the potential configurations for a passive reflector. He rejected the practicality of both a plane mirror and a corner reflector for technical reasons. Instead, Pierce concentrated on the

---

3.   J.J. Coupling, "Don't Write: Telegraph," *Astounding Science Fiction* 49 (March 1952): 82–96. John Robinson Pierce often used the pseudonym "J.J. Coupling," an electrical engineering term, for his writings in the field of science fiction.

concept of a uniformly reflective sphere. First, if such a satellite 100 feet (about thirty-one meters) in diameter were placed in an orbit 1,000 miles (1,600 kilometers) above the surface of the Earth, it would afford the best possibility for successfully relaying messages. A sphere of that size would not require a special alignment in orbit to reflect signals. Second, its characteristics would allow the satellite to relay the widest range of signal frequencies from one point on the Earth to another. Pierce concluded his speech by suggesting aluminum foil as a possible material for the construction of the satellite, as long as "one could inflate [the sphere] gently."

Pierce's speech received a favorable response. Encouraged by individuals in the audience to publish his presentation, he submitted a modified version to the journal *Jet Propulsion*, which published his paper in its issue of April 1955—an issue whose focus was on possible outer space ventures. In this article, Pierce correctly predicted all of the components that would make up the successful Project Echo venture five years later, but he also noted the one factor that prevented the immediate implementation of his proposal: such a venture would need information "from rocket men about constructing and placing satellites" in orbit.[4] Indeed, at the time, no government on the Earth had committed itself to launching such a craft. Until that situation changed, the prospects for communications satellites remained dim.

The Crucial Breakthrough

Nonetheless, in April 1955, official backing for a satellite was closer than Pierce could have imagined. For some time, President Dwight D. Eisenhower had known that the intelligence-gathering agencies of the United States would soon have the ability to photograph the Earth with remarkable resolution from high altitude; his advisors also informed him that a satellite had excellent potential as a platform for basing such an observation system. Eisenhower understood, however, that under existing practice, any nation could consider a satellite passing over its territory as an invasion of its airspace and therefore could legally shoot the surveillance satellite down—if it possessed that capability. Unless the United States could somehow convince the nations of the world to regard satellites as having a different legal status than airplanes, the value of surveillance satellites would be severely limited.

At this critical juncture, the world's scientific community had offered Eisenhower a possible way to establish the principle of legal satellite overflight. To enhance the gathering of useful data during the International Geophysical Year (IGY) 1957–58, these individuals in 1954 called on the governments of the world to launch satellites for conducting scientific experiments. American scientists did not hesitate to propose such a venture to the Eisenhower administration. Eager to provide a precedent for the overflight of other nations by surveillance satellites, Eisenhower approved the launching of an American scientific satellite in conjunction with the IGY. The official announcement came in July 1955.[5]

Soon after, the American coordinators of the IGY formed a committee to select experiments to include in the payload of a venture named Vanguard. One proposal presented to this group came from William J. O'Sullivan, Jr., of the National Advisory Committee for Aeronautics (NACA). He envisioned ejecting a small balloon out of the final stage of a launch vehicle, inflating it, and then observing the effects of atmospheric drag on the

---

4.   John Robinson Pierce, "Orbital Radio Relays," *Jet Propulsion* 25 (April 1955): 77–78, quotation from p. 78.

5.   For a full examination of Eisenhower's decision to initiate an American satellite program, see R. Cargill Hall, "The Origins of U.S. Space Policy: Eisenhower, Open Skies, and Freedom of Space," *Colloquy* 14 (December 1993): 5–6, 19–24.

sphere. O'Sullivan suggested that this balloon could consist of either metallic foil or plastic with a metallic coating. The committee considered his proposal and shortly gave its approval.[6]

O'Sullivan and a small staff began work on the balloon project. While developing a prototype of the 30-inch (75-centimeter) sphere, O'Sullivan began to realize that larger balloons placed in orbit would offer greater opportunities for experimentation. He therefore began work on a balloon 100 feet (about 31 meters) in diameter, and in 1958 a picture of this larger sphere began to appear in journals and magazines. As fate would have it, John Pierce saw a copy of this picture and immediately realized that a sphere of that size met precisely the requirements of the passive reflector he had proposed back in 1954. After securing the backing of AT&T, he contacted Hugh Dryden, the director of NACA, and suggested the possibility of using the larger version of the balloon for a communications experiment. Dryden gave his enthusiastic approval; he then asked O'Sullivan for his opinion. After deliberating for two days, he, too, responded affirmatively. Thus, in April 1958, two separate and distinct ideas for utilizing a sphere placed into the Earth's orbit had become one.

The Technological Component

Certain developments soon aided the likelihood of a successful communications satellite venture. In July 1958, the director of the Jet Propulsion Laboratory (JPL), William H. Pickering, offered the services of his institution to Pierce for his venture. JPL already had begun work at the Goldstone Dry Lake in southern California on a parabolic antenna 85 feet (26 meters) in diameter to track and receive telemetry from the military's Pioneer probes. This so-called HA-DEC antenna, so named because its axes were arranged to measure angles in terms of local hour angle (HA) and declination (DEC), could receive messages using the proposed passive reflector sphere.[7] JPL soon constructed a second 85-foot (26-meter) antenna at the Goldstone location to broadcast signals as well.[8]

Meanwhile, Rudolph Kompfner, an associate of Pierce at AT&T who had aided him in developing the communications experiment proposal, found money in the budget of that corporation in late 1958 to build a large, steerable horn antenna at Holmdel, New Jersey, for receiving messages relayed from outer space.[9] AT&T later arranged for the construction of a transmitting antenna, giving the corporation the same two-way capability that JPL enjoyed. Finally, Pierce and Kompfner recognized that the perfection of the maser (an acronym for **microwave amplification by stimulated emission of radiation**) would reduce the power levels needed for the successful transmission of audible radio waves. The maser was a new type of solid-state microwave amplifying device vaunted by one author as "the greatest single technological step in radio physics for many years" and had become available outside the laboratory only earlier in 1958.[10] From a technical standpoint, then, as 1958 drew to a close, Pierce's balloon venture seemed very promising.[11]

---

6.   Don Murray, "O'Sullivan's Wonderful Lead Balloon," *Popular Science* 178 (February 1961): 74–77.

7.   Calvin Tomkins, "Woomera Has It!," *New Yorker* 39 (21 September 1963): 85; William R. Corliss, *A History of the Deep Space Network,* CR-151915 (Washington, DC: NASA, 1976), pp. 16–17, 20–25.

8.   Corliss, *Deep Space Network,* pp. 25–27.

9.   Tomkins, "Woomera Has It!," p. 87.

10.   Quotation is from J.V. Jelley, "The Potentialities and Present Status of Masers and Parametric Amplifiers in Radio Astronomy," *Proceedings of the IEEE* 51 (1963): 31, 36, esp. 30. For the invention of the maser, see Paul Forman, "Inventing the Maser in Postwar America," *Osiris* ser. 2, vol. 7 (1992): 105–34.

11.   Donald C. Elder, *Out From Behind the Eight-Ball: A History of Project Echo,* AAS History Series, vol. 16 (San Diego: American Astronautical Society, 1995), pp. 25–26.

Figure 10

The Bell Telephone Laboratories satellite communications center at Crawford Hill in Holmdel, New Jersey, circa 1960. The steerable horn antenna used for receiving is to the right, while the sixty-foot (twenty-meter) transmitting dish is in the upper left corner. (Courtesy of NASA, photo no. 227031)

Those associated with the communications satellite project soon found reason to rejoice in the political realm as well. NACA had approved the venture in April 1958, but that organization ceased to exist on 1 October 1958 because of a process of events set in motion months earlier. Spurred by the launch of Sputnik, the world's first satellite, by the Soviet Union in October 1957, President Eisenhower sought to demonstrate his administration's commitment to the development of a vibrant American space program. Accordingly, the president advocated the creation of a civilian space agency, the National Aeronautics and Space Administration (NASA). As part of the legislation authorizing NASA, Congress authorized the new agency to absorb NACA. The demise of the original governmental champion of the communications satellite project momentarily left the fate of the project in doubt. However, T. Keith Glennan, the first NASA administrator, saw merit in the concept and convinced the Eisenhower administration to give it official approval. Glennan made this news public on 19 February 1959.[12]

---

12. "Highlights of the Inflatable Satellite Program," undated, Folder XII, Satellites, Echo—Project Echo, NASA History Office, Washington, DC.

Aware of the work under way on the signal transmitting and receiving equipment, and secure in the knowledge that the government stood solidly behind the project, the individuals interested in the communications satellite project turned their attention in 1959 to the progress being made on the construction of the sphere. O'Sullivan had given the contract for the sphere to the General Mills Company. At the time, General Mills was a leader in the field of research balloon manufacturing, and that firm had built a prototype of the passive reflector sphere by the summer of 1959. General Mills, however, had never fabricated a balloon of that size before, and O'Sullivan had no facility large enough to observe the results of inflation procedures under the conditions that would exist at an altitude of 1,000 miles (1,600 kilometers). Until NASA could conduct suborbital tests with the General Mills balloon, he would not know whether the sphere could maintain its integrity in orbit.[13]

When NASA finally conducted the first test in October 1959, the results confirmed O'Sullivan's worst fears: the balloon disintegrated upon inflation, creating a dazzling spectacle of sparkling light in the sky over the eastern coast of the United States. General Mills in fact had warned O'Sullivan beforehand of a potential problem. The Minnesota company informed him that it did not believe the substance its technicians had used to bind the 82 separate panels of the balloon together would allow the sphere to withstand the tremendous pressure generated by rapid inflation in a near vacuum. The first test dramatically validated the concern of General Mills and put the future of the project in doubt.[14]

As it turned out, General Mills already had put into motion a plan to rectify the problem. Before the first suborbital test, the firm had asked a balloon-making rival, the G.T. Schjeldahl Company of Northfield, Minnesota, for help in creating an effective sealing procedure. After six weeks of intensive study, G.T. Schjeldahl himself, the company's founder, developed a satisfactory adhesive, and one of his employees devised a technique for applying it to the panels of the disassembled sphere.[15] After inconclusive results during two subsequent suborbital missions, a sphere built through the combined efforts of General Mills and the G.T. Schjeldahl Company performed flawlessly in a test flight on 1 April 1960. Now confident about the quality of the product, NASA officials began to plan for a full-scale mission.

The Dawning of a New Age

NASA scheduled the launch of the giant balloon for May 1960. The rocket left Cape Canaveral successfully, but control jets in the second stage did not function properly. NASA officials surmised that the Thor-Delta rocket had plunged into the Atlantic Ocean. Nonetheless, they did not let the failure prevent them from initiating plans to schedule another launch immediately.[16] After one additional successful suborbital launch, NASA selected August 1960 for the next attempt.

This mission, after a number of postponements, did succeed. The Thor-Delta performed flawlessly, lifting the payload on the morning of 12 August 1960 to the desired

---

13.   *DuPont Magazine* 53 (May–June 1961): 14.

14.   William J. O'Sullivan, Jr., interview with Edward Morse, 28 August 1964, "Historical Origins of Echo I," William J. O'Sullivan Jr., file, NASA History Office.

15.   Sheldahl Company, *The Fine Line*, special ed., 1985, pp. 4–7, G. T. Schjeldahl Papers, Minnetonka, Minnesota. After G.T. Schjeldahl severed his ties with the G.T. Schjeldahl Company, its directors changed both the spelling and the name of the firm.

16.   NASA Statement to the Press, 13 May 1960, Documentation Echo Folder, NASA History Office.

altitude. The ejection mechanism in the final stage sent the tightly folded collapsed sphere into the near vacuum of the Earth's orbit; then a combination of chemicals inside the balloon underwent a process of sublimation and released gas that gently inflated the sphere. The satellite—at that point officially named Echo by NASA—had achieved orbit.[17]

The crews at Holmdel and Goldstone learned of the successful deployment of the sphere in orbit and immediately prepared for the first communications experiment. When both stations located the satellite in the sky, the JPL facility at Goldstone sent the first message. With remarkable clarity, Bell Telephone Laboratories personnel heard the voice of President Eisenhower, who previously had recorded a short speech for this occasion.[18] In this dramatic fashion, a new era in the history of communications began.

In the days following the launch, many different groups conducted experiments using Echo. JPL and Bell Telephone Laboratories successfully conducted the first two-way transmission using a satellite; after the California team again broadcast the message of President Eisenhower, the Holmdel personnel transmitted a recording supplied by Senate Majority Leader Lyndon B. Johnson. The Collins Radio Company completed a live, two-way radio conversation between its home location in Cedar Rapids, Iowa, and a subsidiary in Richardson, Texas. This firm also turned a teletype machine on by means of a signal bounced off the sphere. The Naval Research Laboratory joined with the original participants to "double bounce" a message from Maryland to New Jersey, then to California. Clearly, Echo had allowed interested parties to demonstrate the potential benefits of a communications satellite.[19]

While the results to that point would have been enough to satisfy all concerned with Project Echo, two later experiments gave them an even greater appreciation of what a communications satellite could offer in the future. First, on 15 August 1960, JPL and Bell Telephone Laboratories project managers used the satellite, and their transmitting and receiving equipment, in the words of a Bell publicity release, to engage in a "historic [telephone] conversation, exchanging pleasantries and carrying on small talk."[20] Second, in April 1962, the Massachusetts Institute of Technology and the U.S. Air Force joined in a successful effort to relay a live television transmission via Echo. With good reason, then, but perhaps with excessive modesty, too, Pierce later would refer to the satellite as "something of value."

Echo in Retrospect

Even at the time of the successful relaying of television signals, NASA officials realized that technological developments had made the passive repeater satellite obsolete. Indeed, in July 1962, NASA launched Telstar, an active repeater satellite. People soon remembered the original communications satellite more for its visibility to the naked eye at night than for the experiments individuals used it to conduct. The national news media noted the demise of Echo in May 1968, as it returned through the Earth's atmosphere, but recognized in their coverage that events had long since stripped it of its relevance.[21]

---

17.   Elder, *Out From Behind the Eight-Ball*, pp. 99–110.
18.   *Washington Star*, 14 August 1960, p. 9; *New York Herald Tribune*, 13 August 1960, p. 7.
19.   Elder, pp. 113–116.
20.   Bell News Release, 15 August 1960, Folder XXII, Satellites Echo 1, NASA History Office.
21.   NASA News Release, No. 75-217, 10 August 1975, Folder XXII, Satellites Echo 1, NASA History Office.

Still, it is important to remember that a successful communications satellite effort, however simple in design and execution, was necessary for individuals to plan a more ambitious generation of devices. The telecommunications industry today may be "the world's largest economic sector," as the Los Angeles Times has proclaimed, but few companies would have allocated resources for a field that had yielded no apparent hope of success until August 1960.[22] Viewing the results yielded by Echo, individuals could envision more ambitious telecommunications projects for the future. Echo, then, represents the proverbial single step in a journey in which the world is still participating today.

---

22.   *Los Angeles Times*, 26 July 1994, p. H-2.

Chapter 5

Project Echo, Goldstone, and Holmdel: Satellite Communications as Viewed From the Ground Station[1]

by Craig B. Waff

As the geostationary orbit reaches saturation, and as carriers begin to look again at satellites in medium and low orbits, such as the proposed Iridium system, the pioneering work of NASA's Jet Propulsion Laboratory (JPL) and AT&T's Bell Telephone Laboratories in acquiring and tracking the first Echo balloon satellite has taken on new meaning and importance. Donald Elder has provided an overall description of the Echo project, which includes the design, construction, testing, and launching of the inflatable balloon, as well as the political impact of the experiment.[2] The focus here is on the acquisition and positioning of the ground equipment necessary to undertake the experiment; the conduct of preliminary acquisition, tracking, and communication tests prior to launch; the performance of operations immediately after launch; and the conclusions reached by engineers of the two main participant organizations, Bell Telephone Laboratories in New Jersey and JPL in southern California.[3]

In an October 1958 internal technical memorandum outlining "A Program of Research Directed Toward Transoceanic Communication by Means of Satellites," Bell Telephone Laboratories engineers John Robinson Pierce and Rudolf Kompfner suggested that a worthwhile preliminary step would be the establishment of "an experimental narrow-band communication link . . . between two points on the American mainland, far enough apart to preclude the possibility of any other signal path." The main objectives of the experiment were to observe atmospheric refractive effects, to study the influence of satellite shape, and to make signal-to-noise and bandwidth measurements from 100-foot-diameter spheres launched into orbit. Pierce and Kompfner believed that the required

---

1.   Most of the research for this paper was conducted as part of the author's work from 1989 to 1992 as Deep Space Network contract historian at the Jet Propulsion Laboratory (JPL) in Pasadena, CA.

2.   See Donald C. Elder, "Something of Value: Echo I and the Beginnings of Satellite Communications," chapter 4 in this publication; Donald C. Elder, *Out From Behind the Eight-Ball: A History of Project Echo*, AAS History Series, Vol. 16 (San Diego: American Astronautical Society, 1995).

3.   The overall roles played by JPL and Bell Telephone Laboratories in Project Echo are summarized in Walter K. Victor and Robertson Stevens, "The Role of the Jet Propulsion Laboratory in Project Echo," *IRE Transactions on Space Electronics and Telemetry* SET-7 (March 1961): 20–28; William C. Jakes, Jr., "Participation of Bell Telephone Laboratories in Project Echo and Experimental Results," *The Bell System Technical Journal* 40 (July 1961): 975–1028; William C. Jakes, Jr., and Walter K. Victor, "Tracking Echo I at Bell Telephone Laboratories and the Jet Propulsion Laboratory," in H.C. van de Hulst, C. de Jager, and A.F. Moore, eds., *Space Research II, Proceedings of the Second International Space Science Symposium, Florence, April 10–14 1961* (Amsterdam: North-Holland Publishing Company, 1963), pp. 206–14. JPL reported its involvement in Project Echo more fully in a series of progress reports appearing in issues (nos. 6 and 37-1 to 37-6) of its bimonthly *Space Programs Summary* series covering the period from 15 September 1959 to 15 November 1960, as well as in its final report on the project, Robertson Stevens and Walter K. Victor, eds., *The Goldstone Station Communications and Tracking System for Project Echo*, JPL Technical Report No. 32-59 (Pasadena, CA: JPL, 1 December 1960).

ground equipment for the experiment would be two sixty-foot steerable antennas connected to high-power modulators and amplifiers, low-noise receivers, band compressors, servo-tracking apparatus, and computer facilities.[4] "[T]o interest some agency that could launch such a [communications] satellite" and "to convince ourselves and others of the value of such a passive satellite experiment," Pierce and Kompfner presented a paper based on their proposal to the National Symposium on Extended Range and Space Communication, held in Washington, D.C., on 6 and 7 October 1958. That paper subsequently was published in March 1959.[5] Their proposal, of course, was the genesis of NASA's Project Echo.

Recruitment of a Partner and Site Selection

Performing the passive satellite communications experiment called for in the Pierce-Kompfner proposal required two antenna stations, one on each coast of the North American continent. Their proposal did not suggest any specific sites for the two stations. Given their belief that the Bell system would play a leading role in the future development of satellite communications, however, they undoubtedly desired that the east coast station be located at one of the Bell Telephone Laboratories facilities, all of which were in New Jersey.

When they wrote their proposal, Pierce and Kompfner already had in mind both a desirable site and an interested partner organization for the required west coast station. At an Air Force–sponsored meeting on communications satellites held at Woods Hole, Massachusetts, on 13–14 July 1958, Pierce had discussed the feasibility of using 100-foot-diameter (about thirty-one-meter-diameter) balloons as part of a "passive satellite relay system." Among the conference attendees was William H. Pickering, the director of JPL, at that time an Army contract facility operated by the California Institute of Technology (Caltech). Following Pierce's talk, Pickering "raised the question of using the currently proposed balloon experiments for some initial work on the possibility of satellite relay systems."[6] Pickering was referring to the twelve-foot-diameter (about four-meter-diameter) orbiting balloons conceived by William J. O'Sullivan, Jr., an engineer at the Langley Memorial Aeronautical Laboratory (under NASA, it became the Langley Research Center) in Hampton, Virginia—a facility of the National Advisory Committee for Aeronautics (NACA)—as a means of measuring air resistance in the Earth's upper atmosphere. The U.S. Department of Defense's Advanced Research Projects Agency (ARPA) had approved that balloon project in late March 1958 for launches in late 1958 and in 1959, and JPL was supplying the upper stages for the Juno II launch vehicles.[7]

4.   John Robinson Pierce and Rudolf Kompfner, "A Program of Research Directed Toward Transoceanic Communication by Means of Satellites," manuscript, Technical Memorandum MM-58-135-24, 22 September 1958, p. 20, Vol. KK, Filecase No. 20564, AT&T Archives, Warren, NJ. The author is grateful to Sheldon Hochheiser, AT&T archivist, for assistance in locating this paper and granting permission to quote from it.

5.   John Robinson Pierce and Rudolf Kompfner, "Transoceanic Communication by Means of Satellites," *Proceedings of the IRE* 47 (March 1959): 372–380.

6.   William H. Pickering, "Notes on Satellite Conference, July 13–14, 1958," 22 July 1958, doc. no. 15, microfilm roll 33-4, JPL Archives, Pasadena, CA.

7.   The launch attempts of the twelve-foot-diameter balloons were made on 22 October 1958 and 14 August 1959, but both ended in failure. See Elder, *Out From Behind the Eight-Ball*, pp. 45, 65.

Pierce, who had known Pickering since the mid-1930s when they were fellow graduate students at Caltech, quickly informed his former classmate of the proposal that he had already made to NACA director Hugh Dryden to use a 100-foot-diameter version of O'Sullivan's balloon for a passive communications satellite experiment. Pierce recalled many years later that Pickering agreed at the Air Force conference that such an experiment "would be a profitable one" and "offered his encouragement and support."[8] The support that Pickering undoubtedly offered was the use of an eighty-five-foot-diameter (twenty-six-meter-diameter) polar-mounted antenna that JPL had ordered from Blaw Knox in April 1958. JPL installed the so-called Pioneer antenna in July near Goldstone Dry Lake in California's Mojave Desert in support of two ARPA-approved Army lunar probe launches (subsequently named Pioneers 3 and 4) in late 1958 and early 1959.

Pickering and his colleagues at JPL, unlike their counterparts at Bell Telephone Laboratories, had little interest in the future development of communications satellites, but they foresaw that much of the additional ground-support equipment that JPL would need to conduct the passive satellite communications experiments at Goldstone could be applied subsequently to the tracking of, and communication with, space probes. By early 1958, JPL already was hoping to devote itself to those activities in the post-Sputnik era. In January 1959, JPL proposed to NASA a list of space probes that called for launches of a circumlunar flight in early July 1960 and two flybys of Mars in mid-October 1960. JPL participation in a NASA-sponsored passive satellite communications experiment in particular might enable its engineers to acquire more quickly the transmitter that eventually would be needed for issuing commands to probes to perform mid-course corrections and for determining probe positions more accurately.

The use of the Goldstone antenna in the proposed satellite communications experiment became more feasible during the latter half of 1958 as a result of the following three developments:

- The formation on 1 October 1958 of NASA, the new civilian space agency, as well as the appointment of Dryden, an enthusiastic supporter of Pierce's proposal who had played an active role in securing ARPA approval of the project, as NASA's first deputy administrator
- The transfer, probably with Dryden's encouragement, of the satellite communications experiment (soon to receive the name "Echo") in early October from ARPA to NASA
- The transfer in early December 1958 of JPL from the Army to NASA

At a meeting on 22 January 1959, representatives from NASA headquarters, Bell Telephone Laboratories, and JPL negotiated an agreement that outlined the equipment that the latter two organizations would be responsible for acquiring, installing, and testing prior to the initial launch attempt of the Echo balloon satellite, then tentatively scheduled for September 1959. At that meeting, Bell Telephone Laboratories engineers announced that they planned to erect, at a site known as Crawford Hill in Holmdel, New Jersey, a twenty-foot-by-twenty-foot-aperture (a six-meter-by-six-meter-aperture) horn-reflector antenna. Horn antennas were known for their demonstrated low-noise properties.[9]

---

8.   John Robinson Pierce, *The Beginnings of Satellite Communication* (San Francisco: San Francisco Press, 1968), pp. 11–12.
9.   A description of this antenna is given by A.B. Crawford, D.C. Hogg, and L.E. Hunt, "A Horn-Reflector Antenna for Space Communication," *The Bell System Technical Journal* 40 (1961): 1095–1116. The authors noted that this type of antenna had originated at Bell Telephone Laboratories in the early 1940s. The antenna employed in the Echo experiment later gained fame as the instrument used by Arno Penzias and Robert Wilson in 1965 to discover the cosmic microwave background radiation.

This particular horn antenna was to receive signals from Goldstone, while a sixty-foot-diameter (eighteen-meter-diameter) parabolic antenna purchased from the D.S. Kennedy Company would transmit signals to Goldstone via the Echo balloon. Possibly because of funding uncertainties in early January 1959, JPL indicated that it would both transmit and receive signals through duplexing, using the existing eighty-five-foot-diameter (twenty-six-meter) polar-mounted antenna at Goldstone.[10] JPL project funding was uncertain because at that time NASA was operating on funds transferred from the Department of Defense and other agencies and was in the midst of formulating its first budget request for new funds from Congress.

After NASA's budgetary situation began to stabilize in the early months of 1959, however, JPL engineers opted, as did their Bell Telephone Laboratories counterparts, to use separate antennas for receiving and transmitting. For the second (transmitting) antenna, they chose an azimuth-elevation (Az-El), rather than polar, mounting. This choice was not dictated by Echo requirements. Rather it resulted from a desire by JPL engineers Robertson Stevens (chief of the Communication Elements Research Section), William Merrick (in charge of antenna construction), and Walter Victor (chief of the Communications Systems Research Section) to make a comparative study between the two types of mounts, as well as a scale study for larger antennas that the JPL engineers hoped to design and construct in the future for supporting more sophisticated and more distant space probes. Early design studies had indicated that an Az-El-mounted large antenna would cost less and weigh less than a polar-mounted antenna built to the same specifications of maximum frequency, wind-loading design, and tracking rates. JPL engineers also were aware that counterweight problems inherent in the polar-mount configuration probably would make it easier to scale up an Az-El-mounted antenna to a size comparable to the 210-foot-diameter (sixty-four-meter-diameter) Commonwealth Scientific and Industrial Research Organisation (CSIRO) radio telescope then under construction at Parkes, Australia.

Unlike the situation at Holmdel, where the two Bell Telephone Laboratories antennas would be in sight of each other, JPL engineers in 1958 had selected a site at Goldstone for an anticipated transmitter antenna that would be approximately seven miles from the Pioneer antenna and separated from the latter by a range of intervening mountains. The site had been chosen to preclude the possibility of the transmitter generating radio interference that might prevent the reception of weak radio signals from space probes at the receiver antenna.

JPL's prior and continuing involvement in space probe communications and tracking strongly influenced the selection of frequencies for the Echo experiment. JPL and Bell engineers chose 960.05 megahertz for east-to-west transmissions, because the JPL receiver at Goldstone was operating at this frequency in support of the Pioneer lunar program. For the west-to-east transmissions, they selected 2,390 megahertz because "it was the correct frequency band for future satellite and space-probe experiments."[11] In fact, the radio frequency used by NASA's Deep Space Network (managed by JPL and comprised initially of antennas at Goldstone; Woomera, Australia; and Hartebeesthoek near Johannesburg, South Africa) changed from 960 to 2,390 megahertz in the mid-1960s.

---

10.   The agreement reached at the 22 January 1959 meeting is outlined in Rudolf Kompfner, "A Proposed Plan for a Joint JPL-BTL Experiment in Communication by Means of a Passive Satellite," 9 February 1959, attached to Rudolf Kompfner to Leonard Jaffe (chief of the Communications Satellite Program in NASA's Office of Advanced Technology), 10 February 1959 (copy to Pickering), "Satellite Tracking 1959" section, microfilm roll 614-93, JPL Archives.

11.   Victor and Stevens, "The Role of the Jet Propulsion Laboratory," p. 22.

Figure 11

The twenty-six-meter Az-El dish antenna built for Project Echo by the Jet Propulsion Laboratory (JPL) at Goldstone Dry Lake in the Mojave Desert, about 160 kilometers from the laboratory in Pasadena, California. JPL also used an extant twenty-six-meter HA-DEC antenna, erected in late 1958 to track and receive telemetry from the military's Pioneer probes, to communicate via the Echo balloon. (Courtesy of NASA)

Designing the Acquisition, Tracking, and Communications Systems

As noted previously, the principal objective of Project Echo was to demonstrate the feasibility of long-range communications using a reflecting sphere as a passive satellite. To accomplish this objective, JPL and Bell engineers had to develop systems that could not only perform the communications experiment itself, but also rapidly acquire the fast-moving satellite soon after it rose above the western horizon and track it during subsequent periods of visibility. Indeed, in the view of JPL engineers, the acquisition and tracking of Echo would be more challenging than the communications aspect of the project. Once the acquisition and tracking problems were solved, according to JPL's Walter Victor and Robertson Stevens, "the communications problem would be no more difficult than a laboratory experiment with comparable signal and noise powers." Not surprisingly, therefore, most of the new equipment procured by JPL for the project was used for acquisition and tracking.[12] By comparison, the acquisition and tracking of later geostationary satellites were in certain ways less daunting, because these essentially remained stationary in the sky.

At its Goldstone facility, JPL engineers developed for the Echo experiment two methods of acquisition: "slaving" the antennas to a precalculated pointing program using digital techniques and local optical sighting. They also developed three methods of tracking:

---

12.   *Ibid.*

digital slave, optical, and automatic radar. When the digital-slave method (for both acquisition and tracking) was required, the NASA Minitrack network of stations generated and transmitted primary tracking data to a general-purpose digital computer at NASA's Goddard Space Flight Center in Beltsville, Maryland. This computer then issued antenna-pointing commands via teletype to Goldstone. Because of the two different types of antennas at Goldstone, JPL engineers developed a digital encoding and computing system that converted antenna-pointing data from one antenna site and coordinate system to the other.

The principal optical antenna-positioning method used a television camera and lens subsystem located on the structures of both antennas, along with the regular boresight telescope and boresight camera. The images provided by this system were displayed on a control console directly in front of the servo operator's position at each antenna site. This subsystem was not suitable for the initial acquisition of the satellite, however, because of its very narrow field of view (about 0.5 degree). Therefore, JPL engineers provided broad field-of-view acquisition telescopes at each site.

Because of a blind spot in the coverage of the polar-mounted receiver antenna, Goldstone personnel used a third optical means to acquire the satellite during its first orbit. On its first orbital pass, the satellite was expected to approach Goldstone from the northwest, but the mounting of this antenna (which had been set to track and communicate with space probes having declinations varying from about thirty degrees north to about thirty degrees south) prevented it from being pointed to the horizon in this direction. As a result, the antenna would be unable to monitor the satellite for the first eight minutes of the first pass. However, positional data on the first orbital pass were considered critical for initial orbit determination, so Goldstone personnel used a stand-alone Contraves phototheodolite to provide time-tagged Az-El data during the initial part of this pass. The main limitation of all three optical methods was, of course, that the satellite had to be visible. Thus, satellite observation had to occur at night, in clear weather, and when the Sun illuminated the satellite.

The third method of tracking the satellite developed by JPL engineers was a continuous-wave radar subsystem. Once either the digital-slave or one of the optical methods acquired the satellite, this subsystem tracked the satellite automatically. Control signals generated from a simultaneous lobing antenna feed and receiver positioned the polar-mounted receiver antenna, to which the Az-El-mounted transmitter antenna was slaved via the coordinate-converter computer and related equipment.

Bell Telephone Laboratories engineers employed similar tracking methods at their Holmdel station. The antennas were usually slaved to the tracking information provided by the drive tape from Goddard. Differences between the predicted and true positions were compensated by manual corrections obtained from either optics, radar, or (when a west-to-east transmission was being made) the strength of the received signal. Alternatively, the antennas could be slaved to the positional readouts of an optical tracker operated manually to track the satellite.

The optical telescope (borrowed by Bell engineers from a surplus M-33 fire-control radar system) consisted of a large trailer carrying a periscope-type optical train leading down to convenient operator positions inside the trailer. A ten-kilowatt transmitter (used for both communication and the radar system and purchased from ITT) was installed on the sixty-foot-diameter (eighteen-meter-diameter) antenna. The radar signals reflected off the satellite were received by a separate eighteen-foot-diameter (about five-meter-diameter) antenna located about a mile and a half away from Crawford Hill so as to increase the separation between the transmitted signals and the radar receiver.

Prelaunch Tests

After the installation of the new equipment, JPL and Bell Telephone Laboratories engineers began a testing program of the equipment and personnel at Goldstone and Holmdel prior to the launch of the first Echo balloon. The testing program consisted of a series of exercises that simulated as nearly as possible an actual Echo mission.[13] The most striking of these exercises was a series of seventeen so-called "Moon Bounce" communications experiments that began 23 November 1959 and continued until 7 August 1960, just five days before the launch of Echo. These experiments involved using the Moon as a passive reflector (in the manner of the Echo balloon) between the Holmdel and Goldstone stations. Such tests were useful because the combination of the Moon's ability to reflect energy to the receiving antenna (known as its radar cross section) and its distance from the Earth provided a transmitter-to-receiver path loss (that is, the amount of radio signal power lost between the transmitter and receiver) nearly equal to that expected for the Echo satellite.[14]

As for acquisition and tracking problems, however, the Moon was too easy a target compared to Echo because of its highly predictable orbit, good visibility, and relatively slow motion across the sky. Therefore, Goldstone and Holmdel personnel participated in two types of tests with a "dark" satellite in a relatively high-altitude, stable orbit—namely, the TIROS I polar-orbiting meteorological satellite launched 1 April 1960. In one type of test, JPL engineers used basic TIROS tracking data generated by the Minitrack network to position the Goldstone antennas; then they tested the accuracy of the ephemeris by attempting to obtain a radar echo from the satellite. Another type of test involved bouncing continuous-wave radio signals off the satellite. The satellite's polar orbit made these tests more difficult, because it was visible over the Goldstone and Holmdel antennas for a period of only a few minutes. As a result, station operators had to rely on Minitrack-generated orbital data to position their antennas.[15]

---

13.   Regarding the overall philosophy behind these tests, see the section "Systems Tests for Project Echo" in Stevens and Victor, eds., *The Goldstone System*, pp. 48–52; Jakes, "Participation of Bell Telephone Laboratories," pp. 980–82.

14.   For detailed descriptions of the "Moon Bounce" tests, see "Moon Bounce Experiments," in JPL, *Space Programs Summary No. 37-1 for the period 15 November 1959 to 15 January 1960*, 1 February 1960, pp. 40–41; "Project Echo," in JPL, *Space Programs Summary No. 37-2 for the period 15 January 1960 to 15 March 1960*, 1 April 1960, pp. 44–45; "Project Echo," in JPL, *Research Summary No. 36-2 for the period 1 February 1960 to 1 April 1960*, vol. I, pt. 1, 15 April 1960, pp. 1–3; "Moon-Bounce Experiments," in JPL, *Space Programs Summary No. 37-3 for the period 15 March 1960 to 15 May 1960*, 1 June 1960, p. 39; Jakes, "Participation of Bell Telephone Laboratories," p. 981. Press coverage of the first publicized "Moon Bounce" communication on 3 August 1960 is in Marvin Miles, "Attempt to Bounce Voice Off Moon Scheduled Today; Two-Way Phone Talk Scheduled," *Los Angeles Times*, 3 August 1960; "JPL Sets First Long Distance 2-Way Phone Call Via Moon," *Pasadena Independent*, 3 August 1960; "Moon Used to Relay Phone Calls," *Los Angeles Mirror*, 3 August 1960; Marvin Miles, "East and West Coasts Converse in Phone Call Bounced Off Moon," *Los Angeles Times*, 4 August 1960; "JPL Call Via Moon Success," *Pasadena Independent*, 4 August 1960; Bill Sumner, "No Hamming in Historical Phone Call" and "Something Wroughten in Space" ("Daily Report" column), *Pasadena Independent*, 5 and 8 August 1960, all in *JPL News Clips*, 3 August 1960, JPL Archives.

15.   For detailed descriptions of these tests, see "Dark Satellite Tracking Experiment," in JPL, *Space Programs Summary No. 37-3 for the period 15 March 1960 to 15 May 1960*, 1 June 1960, pp. 38–39; "Project Echo," in JPL, *Space Programs Summary No. 37-4 for the period 15 May 1960 to 15 July 1960*, 1 August 1960, pp. 43–45; Richard van Osten, "Goldstone Uses Now-silent Tiros I for Bouncing Signals," *Missiles and Rockets*, 15 August 1960, cited in *JPL News Clips*, 15 August 1960, JPL Archives.

Other tests conducted at Goldstone addressed the difficulty of tracking targets optically. Some tests involved tracking stars optically to check the absolute and relative alignment of the two antennas and to measure the sensitivity of the television cameras mounted on them. Other tests used a helicopter with an optical target light and a 2,388-megahertz radio beacon to exercise the optical equipment and to train antenna operators in the acquisition and tracking of visible satellite targets.

Postlaunch Operations

After conducting a series of five suborbital ballistic ("Shotput") tests of the balloon payload in late 1959 and early 1960, NASA ignored superstition and made its first attempt to launch the Echo satellite into orbit on Friday, 13 May 1960. The day indeed turned out to be unlucky for the Echo project. The attitude control jets on the second stage of the Thor-Delta rocket failed to fire during a coast period after the main engine of that stage burned out. Incapable of maintaining the proper angle for orbital insertion, the final stage and payload ultimately plunged into the Atlantic Ocean.

A second launch attempt on 12 August 1960 succeeded in placing Echo 1 into the desired 1,000-mile-attitude orbit. On the first orbital pass, the phototheodolite at Goldstone optically spotted the satellite at 4:31 a.m. (Pacific Daylight Time) as it came over the northwest horizon. Three minutes later, the receiver antenna began collecting signals from the satellite's radio beacons. The radar acquired the balloon at 4:37 a.m. and tracked it automatically over the next 15 minutes until it disappeared below the southeast horizon. In the meantime, beginning at 4:36 a.m., the transmitter antenna began sending a 2,390-megahertz radio signal toward the satellite.

At Holmdel, the acquisition of the Echo target was more difficult. Station operators lacked optical visibility, and their efforts were hampered by the use of a radar that was "still unproven." Bell engineers therefore used a drive tape supplied by the Goddard Space Flight Center prior to the launch. It was based on one of the nominal trajectories and adjusted approximately to the actual launch time. This method allowed the horn antenna to begin receiving Goldstone's reflected signal at 4:41 a.m. (Pacific Daylight Time). One minute later, engineers at Goldstone used the transmitter antenna to begin sending a prerecorded message from President Eisenhower, and the Holmdel antenna clearly received it after its reflection off the balloon.

Incorrect data points on the drive tape, however, caused the horn antenna to slew away from the actual satellite track during this first pass. Although angular offsets were quickly implemented to compensate for these errors, the reception of the Goldstone signal was split into three separate periods lasting from one to three minutes. Bell Project Echo engineer William Jakes candidly acknowledged that "had the launching not been virtually perfect, there would have been no reception at all on the first pass because of the severe acquisition problem" at Holmdel.[16]

In summarizing JPL's participation in project Echo, Stevens and Victor concluded that acquisition and tracking of the balloon with sufficient accuracy (0.1 degree) to make communication possible was "the most difficult task." By far the most reliable method, they pointed out, was optical acquisition and tracking. The desired accuracy was easily obtained by this method "as long as [the servo operators] were not fatigued." Slaving the

---

16.   For the most detailed description of the Shotput tests and the launches of 13 May and 12 August, see Elder, *Out From Behind the Eight-Ball,* pp. 68–70, 79–80, 84–85, 88–97, 103–09.

antennas digitally by means of a perforated teletype tape resulted in an average difference between the slave command and the measured Goldstone angles of between 0.1 and 0.15 degree "as long as the orbital parameters were updated daily." Atmospheric drag on the balloon, which neither remained constant over time nor varied in a predictable manner, caused the largest uncertainty in predicting the orbit.[17]

Victor and Jakes noted that the Echo project, through a large amount of organizational coordination, had accomplished the "unprecedented feat of simultaneously slaving narrow beam antennas located a continent apart to a computer so that a relatively fast-moving satellite could be accurately tracked." They concluded, however, that the procedure "probably does not represent the most efficient use of a large scale, general purpose computer or the personnel involved."[18] JPL's Stevens and Victor also felt that the reliability of this method needed to be improved before it could be considered operationally feasible. They reported that radar tracking of the balloon had been accurate to 0.03 degree for the receiving antenna and about 0.1 degree for the transmitting antenna at Goldstone.[19]

Looking to the future, Stevens and Victor suggested that "standard television might be relayed on a reasonably practical basis via passive satellites" if 1,000-foot-diameter balloons could be launched instead and if transmitter power and antenna gain could be boosted. While these improvements were "technically feasible," the JPL engineers warned that they were not necessarily "economically feasible or desirable"; other satellite techniques might prove more suitable for the transmission of wide-band video signals. On the other hand, Stevens and Victor judged, passive satellites "may be a good solution for narrower bandwidth data-transmission systems."[20]

17. Stevens and Victor, eds., *The Goldstone System*, p. 55.
18. Jakes and Victor, "Tracking Echo I," p. 214.
19. Stevens and Victor, eds., *The Goldstone System*, p. 55.
20. Victor and Stevens, "The Role of the Jet Propulsion Laboratory," p. 28.

Chapter 6

NASA Experimental Communications Satellites, 1958–1995

by Daniel R. Glover

As the civilian agency exercising control over U.S. space activities, NASA has had a program of technology development for satellite communications since the agency was established in 1958. Part of this program has involved flying experimental communications satellites. NASA's first communications satellite project was Echo. Launched on 12 August 1960, Echo 1 was a passive satellite that reflected radio waves back to the ground.

Echo started out in 1956 as a National Advisory Committee for Aeronautics (NACA) experiment to probe the upper reaches of the atmosphere and the effects on large light-weight structures in orbit. John Robinson Pierce and Rudolf Kompfner of AT&T's Bell Telephone Laboratories had been working on ideas for communications satellites, including passive systems, for some time. They realized that the Echo sphere would provide an excellent test mirror and proposed a communications experiment. The National Academy of Sciences sponsored a meeting, held on 28 August 1958, to define the project.[1] In 1958, when NASA was created and NACA dissolved, Echo became a NASA project.

The Echo satellite was a 100-foot-diameter (thirty-one-meter-diameter) aluminized-polyester balloon that inflated after insertion into orbit. The G.T. Schjeldahl Company built the Echo 1 balloon, and Grumman built the dispenser, for NASA's Langley Research Center in Hampton, Virginia. Two-way voice links of "good" quality were set up between Bell Telephone Laboratories in Holmdel, New Jersey, and NASA's Jet Propulsion Laboratory facility at Goldstone, California. Some transmissions from the United States were received in England at Jodrell Bank.

Echo demonstrated satellite tracking and ground station technology that later applied to active satellite systems. Leonard Jaffe, director of the NASA satellite communications program at headquarters, wrote: "Echo [I] not only proved that microwave transmission to and from satellites in space was understood and there would be no surprises but it dramatically demonstrated the promise of communication[s] satellites. The success of Echo [I] had more to do with the motivations of following communications satellite research than any other single event."[2]

Echo 2, managed by NASA's Goddard Space Flight Center in Beltsville, Maryland, left the launch pad on 25 January 1964. It had a better inflation system, which improved the balloon's smoothness and sphericity. Echo 2 investigations were concerned less with communications and more with the dynamics of large spacecraft. After Echo 2, NASA abandoned passive communications systems in favor of active satellites. The superior performance of the U.S. Department of Defense's SCORE (Signal Communication by Orbiting Relay Equipment) satellite, launched almost two years before Echo I, already had demonstrated the viability of the active approach.

---

1. Leonard Jaffe, *Communications in Space* (New York: Holt, Rinehart and Winston, 1966), p. 67.
2. *Ibid.*, p. 80.

Telstar

In the fall of 1960, as Echo 1 was achieving its first successes, AT&T began developing an active communications satellite system called Telstar. Although some observers felt that AT&T's early interest in communications satellites was part of a defensive maneuver to protect its commitment to cable technology, the company was investing large quantities of its own capital to create and launch its own communications satellite program.[3] Initially, the operational system was to consist of fifty to 120 active satellites in orbits approximately 7,000 miles (about 9,310 kilometers) high. Using the large launch vehicles then in development, Pierce envisioned that "a dozen or more of these satellites could be placed in orbit in a single launching." With the satellites in random orbits, Bell Telephone Laboratories figured that a "system of 40 satellites in polar orbits and 15 in equatorial orbits would provide service 99.9 per cent [sic] of the time between any two points on earth." As Pierce explained, "AT&T has proposed that the system contain about 25 ground stations so placed as to provide global coverage."[4]

The cost of such a system would be high. In 1961, Pierce estimated the expense at $500 million, but that high price tag was not a detriment from AT&T's standpoint. As a telecommunications monopoly, AT&T's rates were regulated, and those rates included an amount that allowed AT&T to recover its costs as well as to make a profit. Thus, the cost of the proposed Telstar satellite system would be passed on to consumers, just as the high costs of undersea cables were, so AT&T found the system attractive.

Bell Telephone Laboratories designed and built the Telstar spacecraft with corporate funds. The first Telstars were prototypes intended to prove various concepts behind the large constellation of orbiting satellites. Moreover, of the six Telstar spacecraft built, only two were launched. NASA's contribution to the project was limited to launch services, as well as some tracking and telemetry duties. AT&T reimbursed NASA $6 million for those services. NASA was able to negotiate such an excellent deal with AT&T, even though Telstar was not really a NASA project, because NASA held the monopoly for launch services. Moreover, NASA claimed Telstar as a NASA-supported project and even published the results of the communications experiments, originally issued as articles in the Bell Telephone technical journal, as a NASA publication (NASA Special Publication [SP]-32). In addition, NASA obtained the rights to any patentable inventions arising from the experiments.

On 10 July 1962, a Delta launcher placed the first Telstar spacecraft into orbit. The faceted 171-pound (about seventy-seven-kilogram) sphere had a diameter of a little more than thirty-four inches (about one meter). Telstar was the first satellite to use a traveling-wave-tube amplifier; transistor technology at the time was not capable of the three watts of power output at the required microwave frequencies.[5] Bell Telephone Laboratories also developed much of the technology required for satellite communication, including transistors, solar cells, and traveling-wave-tube amplifiers. To handle Telstar communications, AT&T built ground stations at Andover, Maine; Pleumeur-Bodou, France; and Goonhilly Downs, Britain. These were similar to, but larger than, the ground station used for project Echo. The French station used a duplicate of the AT&T Holmdel horn antenna, while the British antenna was a parabolic dish.

---

3.   Delbert D. Smith, *Communication Via Satellite: A Vision in Retrospect* (Boston: A.W. Sijthoff, 1976), p. 71.

4.   John Robinson Pierce, "Communication Satellites," *Scientific American* 205 (October 1961): 101.

5.   *Telstar I*, 3 vols. (Washington, DC: NASA SP-32, Goddard Space Flight Center, June 1963). Also published as A.C. Dickieson, *et al.*, "Telstar I," *Bell System Technical Journal* 42 (July 1963). In December 1965, Goddard issued volume four of NASA SP-32, which related Goddard Telstar experiments. The four-volume set of NASA SP-32 consequently provides a useful compendium of Telstar information.

Telstar was a tremendous technical success, and the international reaction was spectacular. A U.S. Information Agency (USIA) poll showed that Telstar was better known in Great Britain than Sputnik had been in 1957. Rather than launching a useless bauble, the Americans had put into orbit a satellite that promised to tie together the ears and eyes of the world. Interestingly, the world saw Telstar as an undertaking of the U.S. government (the USIA publicity may have helped). President Kennedy hailed Telstar as "our American communications satellite" and "this outstanding symbol of America's space achievements."[6]

Regarding Telstar, Jaffe, head of communications satellite programs at NASA, wrote in 1966: "Although not the first communications satellite, Telstar is the best known of all and is probably considered by most observers to have ushered in the era of satellite communications."[7] This impression resulted from the tremendous public impact of the first transmission of live television across the Atlantic Ocean from the United States to France by Telstar I on 10 July 1962, the very same day it was launched. In addition to television broadcasts, Telstar relayed telephone calls, data transmissions, and picture facsimiles.

Telstar was AT&T's major move into satellite communications. That move failed to extend AT&T's monopoly of terrestrial communications into space, however; changing telecommunications policy from one presidential administration to the next, and a government desire to avert a monopoly of satellite communications, kept AT&T's monopolistic aspirations in check. At the same time, NASA contracted its communications satellite work to firms other than AT&T.

When AT&T began working on Telstar, the Eisenhower administration seemed willing to allow it to extend its monopoly into space. A statement by President Eisenhower in December 1960, in which he presented his administration's policy on space communications, stressed the traditional U.S. policy of placing telecommunications in the hands of private enterprise subject to governmental licensing and regulation and the achievement of "communications facilities second to none among the nations of the world." The role of NASA was "to take the lead within the executive branch both to advance the needed research and development and to encourage private industry to apply its resources toward the earliest practicable utilization of space technology for commercial civil communication requirements."[8]

The election of President Kennedy ushered in a new policy on satellite communications that was openly antagonistic to monopolies, particularly to the extension of AT&T's monopoly in terrestrial communications to space communications. President Kennedy released a policy statement on 24 July 1961 that favored private ownership of satellite systems, but with regulatory and other features aimed at avoiding a monopoly.[9]

AT&T's preeminent position as the largest U.S. common carrier and sole international telephone carrier, together with its willingness and ability to commit large sums of money to the development of communications satellites, convincingly suggested that commercial satellite utilization would very likely become AT&T utilization. Concern over the possibility of an AT&T monopoly in space was one factor that prompted a later reorientation of the direction that commercialization seemed to be following.[10]

6.   Typed manuscript, Peter Cunniffe, "Misreading History: Government Intervention in the Development of Commercial Communications Satellites," Report no. 24, Program in Science and Technology for International Security, Massachusetts Institute of Technology, May 1991, p. 29, ACTS Project Office, NASA Lewis Research Center, Cleveland, OH.

7.   Jaffe, *Communications in Space*, p. 107.

8.   Cited in Lloyd D. Musolf, ed., *Communications Satellites in Political Orbit* (San Francisco: Chandler, 1968), pp. 17–18.

9.   John F. Kennedy, *Public Papers of the Presidents of the United States* (Washington, DC: Office of the Federal Register, National Archives and Records Service, 1961), p. 530.

10.   Smith, *Communication Via Satellite*, p. 74.

On 31 August 1962, President Kennedy signed the Communications Satellite Act. The government assigned the monopoly of international satellite communications to a new corporation called Comsat. AT&T went ahead with Telstar 2, completing its experimental program. Of the six flightworthy spacecraft built by AT&T with corporate funds, only two were launched, but Telstar's publicity served AT&T very well. Nonetheless, between the success of Telstar 1 and the launch of Telstar 2 on 7 May 1963, AT&T lost its chance to control commercial satellite communications.

NASA's role in communications satellites was changing, too. A 1958 agreement between NASA and the Department of Defense gave responsibility for the development of active communications satellites to the military, leaving NASA with the development of passive satellites. In August 1960, however, NASA decided to pursue active satellite research, but not synchronous satellites. The military already had an active synchronous satellite, Project Advent, in place. NASA began developing medium-altitude satellite systems and issued a request for proposals on 4 January 1961 for an experimental communications satellite to be known as Relay. Both AT&T and Hughes approached NASA with their design concepts, but in May 1961, NASA selected RCA to build the two Relay spacecraft, instead of AT&T or Hughes. The Goddard Space Flight Center oversaw the project.

Project Relay

Although AT&T did not win the contract to build them, the Relay satellites used the same primary ground stations as those used by Bell Telephone Laboratories' Telstar 1 satellite. These were located in the United States (in Maine, New Jersey, and California) and overseas (in West Germany, Italy, Brazil, and Japan). Relay was an experimental satellite program; however, the satellites transmitted television signals between the United States and Europe and Japan. The Tokyo 1964 Olympics, however, were passed from Tokyo to the United States, and then on to Europe via Relay.

NASA launched Relay 1 on 13 December 1962 into an elliptical orbit with an apogee of 4,012 nautical miles (about seven kilometers). The orbit took Relay through the Earth's inner radiation belt, so that the spacecraft could measure the levels of radiation and study its effects on satellite electronics. Relay taught many lessons in communications spacecraft design. The idea of flying experimental communications spacecraft is to try new things and to determine whether they work. Failures are expected and provide the learning experience necessary for technology advancement. Relay was no exception.

While in orbit, the power supply for Relay 1's primary transponder failed, and the spacecraft had to switch to its backup transponder, which performed well. Another problem was spurious commands. The satellite recorded 401 anomalies (errors) during its first year. Ground stations observed anomalies when the satellite was in view, which was during only 15 percent of its orbit. The main culprit was interference from the wideband subsystem. Consequently, as a corrective measure, Relay 2 carried a filter on the command receiver's transmission line and had improved circuitry to better differentiate between noise and command signals. As a result, Relay 2 recorded only sixty-two command anomalies.

Among the other problems faced by the first Relay experimental satellite was the failure of the charge controller for one of three battery packs after about three months. Yet another was the long time required for the traveling-wave tube to warm up. Normally, the tube took three minutes to warm up, but the malfunctioning tube could take as long as sixteen minutes. This delay reduced the time the satellite was usable, as Relay 1's orbit placed it in any particular ground station's view for only about thirty minutes. Relay 2, launched 21 January 1964, had increased radiation resistance plus measures that improved reliability. Finally, Relay 1 had a design life of one year, but when its turnoff switch failed, it continued to operate for a second year.

Syncom

The objective of the Syncom satellite project was to demonstrate synchronous-orbit communications satellite technology. In the early 1960s, achieving a synchronous orbit was a challenge. According to Lawrence Lessing, an observer at the time (1962):

> *Nearly everyone agrees that for a short-range, experimental first venture, the medium alti-tude active repeater satellite, such as Telstar or Relay, is the best bet . . . . Lockheed . . . is confident that before the complex problems of operating 50 or more satellites at lower alti-tudes are solved . . . the U.S. will be able to put up a full-powered, simpler, high-altitude sys-tem. Other experts, however, say that the synchronous high-altitude satellite is still some order of magnitude beyond present technology.*[11]

A synchronous orbit is one in which a satellite makes one orbit per day, the same peri-od as the Earth's rotation around its axis. As a result, the satellite hovers over the same area of the Earth's surface continuously. The altitude of a synchronous orbit is 22,235 miles (19,322 nautical miles or 35,784 kilometers). At lower altitudes, satellites orbit the Earth more than once per day. For example, the Space Shuttle, at a nominal altitude of 180 miles (290 kilometers), orbits the Earth in an hour and a half. The Moon, on the other hand, at a distance of around 240,000 miles (nearly 390,000 kilometers), takes a month to orbit the Earth.

A key advantage of a synchronous satellite is that ground stations have a much easier job of tracking the satellite and pointing the transmitting and receiving antennas at it, because the satellite is always in view. With spacecraft in lower orbits, tracking stations must acquire the satellite as it comes into view above one horizon, then track it across the sky as the antenna slews completely to the opposite horizon, where the satellite disappears until its next pass. For continuous coverage, a ground station might need two antennas to acquire the first satellite, then connect with the next satellite passing overhead. In addi-tion, continuous coverage requires the placement of ground stations distributed around the globe, so that any given satellite is rising over the viewing horizon of one ground sta-tion, while it is setting in relation to another station.

The chief communications advantage of the geosynchronous satellite, however, is its wide coverage of the Earth's surface. About 42 percent of the Earth's surface is visible from a synchronous orbit. Three properly placed satellites can provide coverage for the entire globe. Although Arthur C. Clarke published the first idea of a synchronous com-munications satellite in 1945, the first such synchronous-orbit spacecraft, Syncom 1, was not launched until 14 February 1963. However, when the motor for circularizing the orbit fired, the spacecraft fell silent. To demonstrate attitude control for antenna pointing and station keeping, Syncom had two separate attitude control-jet propellants: nitrogen and hydrogen peroxide. The most likely cause was a failure of the high-pressure nitrogen tank.[12]

---

11.   Lawrence Lessing, "Launching a Communications System in Space," in The Editors of Fortune, eds., *The Space Industry: America's Newest Giant* (Englewood Cliffs, NJ: Prentice-Hall, 1962), pp. 140–41.
12.   Richard M. Bentley and Albert T. Owens, "SYNCOM Satellite Program," *Journal of Spacecraft and Rockets* 1 (July–August 1964): 395.

Syncom 2 addressed these critical problems from the first attempt at making a geosynchronous communications satellite. Launched on 26 July 1963, after improvements in the nitrogen tank design, Syncom 2 successfully achieved synchronous orbit and transmitted data, telephone, facsimile, and video signals. Its successor, Syncom 3, launched 19 August 1964, had the addition of a wideband channel for television and provided coverage of the 1964 Tokyo Olympics. Syncom 3 was different from its predecessor in other ways, notably in its orbital pattern. A particular type of synchronous orbit is the geostationary orbit— namely, a synchronous orbit around the equator. Geostationary satellites seem to be stationary over a point on the surface, as distinguished from an area of the surface. Syncom 3 had a geostationary orbit, while the orbit of Syncom 2 was inclined thirty-three degrees to the equator, so that over a twenty-four-hour period, it appeared to move thirty-three degrees north and thirty-three degrees south in a "figure 8" pattern as observed from the ground.

Figure 12

Television image transmitted from Japan to the United States of the 1964 Olympic Games in Tokyo via the Relay experimental communications satellite launched in 1962. (Courtesy of NASA)

In addition to communications experiments, the Syncom satellites contributed to a determination of the Earth's gravitational field. They were capable of measuring range at synchronous altitude to an accuracy of less than fifty meters. The high altitude of their orbits minimized perturbations arising from local topology changes on the Earth's surface.

The Applications Technology Satellite Program

The Syncom spacecraft, built by Hughes for NASA's Goddard Space Flight Center, marked the end of NASA's experimental satellites of the early 1960s. NASA turned both Syncom satellites over to the Department of Defense in April 1965, and they were turned off in April 1969. As a continuation of its successful program of experimental communications satellites, NASA inaugurated the Applications Technology Satellite (ATS) series. These spacecraft demonstrated communications technologies and conducted weather observations and space research in response to congressional pressure. NASA and Hughes had hoped to continue the success of the Syncom project with an advanced Syncom satellite. Some members of Congress, however, feared that NASA was developing technology for the benefit of a single private company—namely, Comsat. Therefore, the advanced Syncom's objectives were broadened to include meteorology and other experiments, and the program became the ATS series.

The five first-generation ATS satellites, built by Hughes for Goddard, tested a range of new communications electronics in the Earth's orbit, as well as technology for gravity-gradient stabilization (on ATS-2, ATS-4, and ATS-5) and for medium-altitude orbits (ATS-2) on behalf of the Department of Defense. All of these first-generation ATS spacecraft were capable of carrying more signal traffic than any of their predecessors.

The first of these ATS satellites, launched 7 December 1966, carried out an impressive array of communications experiments and collected weather data. ATS-1 was the first satellite to take independently uplinked signals and convert them for downlink on a single carrier. This technique, called "frequency division multiple access," conserves uplink spectrum and also provides efficient power utilization on the downlink. ATS-1 also carried a black-and-white weather camera, which transmitted the first full-disk Earth images from geosynchronous orbit. The communications hardware functioned for another two decades until 1985, when the spacecraft failed to respond to commands.

The second of these ATS satellites, in addition to communications experiments and space environment research, was to conduct technological testing of gravity-gradient stabilization for the Department of Defense. Launched 5 April 1967 atop an Atlas-Agena D rocket, ATS-2 never achieved circular orbit, because the Agena upper stage malfunctioned. Only a few experiments were able to return data. ATS-2 reentered the atmosphere on 2 September 1968.

The following ATS is the oldest active communications satellite by a wide margin. Launched in November 1967, it is still in service more than 28 years later. Among its widest known achievements are the first full-disk, color Earth images transmitted from a satellite. Its imaging capability has served during disaster situations, from the Mexico earthquake to the Mount St. Helens eruption. ATS-3 experiments included VHF and C-band communications, a color spin-scan camera, an image dissector camera, a mechanically despun antenna, resistojet thrusters, hydrazine propulsion, optical surface experiments, and the measurement of the electron content of the ionosphere and magnetosphere. Because of failures in the hydrogen peroxide systems on ATS-1, ATS-3 was equipped with a hydrazine propulsion system. Its success led to its incorporation on ATS-4 and ATS-5 as the sole propulsion system.[13]

The ATS-4 and ATS-5 satellites, because of the unsuccessful ATS-2 mission, again attempted to test technology for gravity-gradient stabilization for the Department of Defense—a key objective of the first generation of the ATS series. Gravity-gradient stabilization was chosen to maintain satellite stability, because it uses low levels of onboard power and propellant. The real goal, however, was to move away from spin-stabilized spacecraft to three-axis stabilization. Spin stabilization has the advantage of simplifying the method of keeping a spacecraft pointed in a given direction. A spinning spacecraft resists perturbing forces, similar to a gyroscope or a top. In space, forces that slow the rate of spin are very small, so that once the spacecraft is set spinning, it keeps going.

Spin stabilization, however, is inherently inefficient. Only some of the satellite's solar cells are illuminated at any one instant. Also, because the radio energy from the nondirectional antennas radiates in all directions, only a fraction of that energy is directed toward the Earth. Three-axis stabilization allows the solar panels to be always pointed at the Sun and enables the use of a directional antenna that not only remains pointed toward the Earth, but concentrates the radio energy into a beam, rather than a scattering pattern.

How, then, does one achieve three-axis stabilization? Gravity-gradient stabilization uses the Earth's gravitational field to keep the spacecraft aligned in the desired orientation. The spacecraft is designed so that one end is closer to the Earth than the other. The spacecraft end farther from the Earth is in a slightly weaker gravity field than the end closer to the Earth. Although this technique had been used in low orbit before the advent of the ATS program, the question to be addressed was whether or not the difference in gravity fields (the gradient) was too weak to be useful at higher altitudes. That was the objective of both ATS-4 and ATS-5.

13. Paul J. McCeney, "Applications Technology Satellite Program," *Acta Astronautica* 5 (1978): 299–325.

Figure 13
*ATS-3, the third in NASA's Applications Technology Satellite (ATS) series of experimental communications satellites.
Launched in November 1967, ATS-3 is still in service today. Among its most acclaimed successes is the first full-disk, color
image of the Earth transmitted from a satellite.* (Courtesy of NASA, photo no. 67-HC-612)

ATS-4 was launched 10 August 1968 atop a powerful Atlas-Centaur rocket, but it reentered the atmosphere on 17 October 1968 because the Centaur upper stage failed to reignite. The ATS-5, then, was the final attempt at a synchronous gravity-gradient spacecraft. Launched 12 August 1969 on an Atlas-Centaur rocket, ATS-5 developed problems in its parking orbit and expended large amounts of propellant to stabilize itself. To try to salvage the mission, NASA injected the satellite into its final orbit ahead of schedule.

Although ATS-5 was to be a gravity-gradient stabilized satellite, spin stabilization was used during orbit insertion (a common practice). The spacecraft carried a device to remove the spin after it reached its final orbit. The device deployed booms to slow the spin, which is very similar to spinning figure skaters who extend their arms to slow down. Thus, ATS-5 successfully achieved a synchronous orbit, but the spacecraft's spin was in the wrong direction for this device to work. As a consequence, the gravity-gradient stabilization experiment was useless. The communications experiments were severely handicapped because the antennas were spinning with the spacecraft and could only work as a lighthouse beacon, rather than as a spotlight. Some communications experiments were later carried out in a pulse mode, and some secondary experiments were conducted as late as 1977.[14] Among those experiments were an L-band aeronautical communications package, an ion engine, a charge neutralizer, solar cell tests, and research on particles, electric and magnetic fields, and solar radio waves.

Applications Technology Satellites: The Next Generation?

The first of the second generation of the ATS program, known as ATS-6, also was the last ATS mission. Congress canceled the program in 1973 as a budget-cutting measure and to allow the commercial communications satellite industry to underwrite its own research and development. In 1974, NASA unsuccessfully attempted to reinstate the ATS program. Thus, the impressive ATS-6 spacecraft, launched 30 May 1974, marked the end of an era and the beginning of a dry spell for NASA experimental communications satellites.

Built by Fairchild Space and Electronics Company for Goddard, the ATS-6 spacecraft was much larger than its predecessors, weighing 1,336 kilograms (compared with 431 kilograms for ATS-5) and standing just over eight and a half meters tall and sixteen meters across its booms (ATS-5 was 1.8 meters tall and 1.4 meters in diameter). In addition to being the largest geosynchronous communications satellite launched to date, it was the first three-axis stabilized communications satellite. ATS-6 incorporated many significant design firsts, such as a 9.14-meter parabolic reflector, a digital computer for attitude control, solid-state high-power radio frequency transmitters, a primary structure made of graphite composite material, heat pipes for primary thermal control, monopulse tracking for attitude control, and a radio frequency interferometer for attitude determination and control.[15]

Equally significant was the demonstration of technology for tracking and data relay satellites that led to the Tracking and Data Relay Satellite System (TDRSS) program. In the TDRSS, a tracking and data relay satellite uses the geosynchronous orbital vantage point to look down on low-altitude satellites. Data are relayed from the low-altitude

---

14.   *Ibid.*, p. 324.
15.   Robert O. Wales, ed., *ATS-6 Final Engineering Performance Report* (Washington, DC: NASA Research Publication [RP]-1080, 1981).

satellite to a ground station through the geosynchronous satellite. Without this space relay capability, NASA needed ground stations all over the globe to collect data from satellites as they passed overhead. Because a low-altitude satellite orbits the Earth in a matter of a few hours, it is only in view of a single ground station for typically 20 minutes at a time. ATS-6 tracked the Nimbus 5 and 6 and the GEOS 3 (Geodynamics Experimental Ocean Satellite) satellites with a roll-and-pitch accuracy of better than 0.2 degree.

The nine-meter antenna enabled small ground receivers to pick up a good quality signal. A demonstration in India, in 1975, relayed television signals from a six-gigahertz uplink through the ATS-6 spacecraft and back to Earth at 860 megahertz, directly to three-meter antennas installed in approximately 2,000 villages. The large deployable antenna required tight pointing by the spacecraft, which is why it used three-axis stabilization. ATS-6 carried out radio-wave propagation studies at frequencies up to thirty gigahertz; it also established L-band (1,550 to 1,650 megahertz) relay links to aircraft and demonstrated multiple aircraft tracking.

ATS-6 experienced a failure of three of its four orbit control jets in May 1979. That failure led to the decision to power down the spacecraft on 3 August 1979. Subsequently, its telemetry system was activated between November 1979 and February 1980 to collect particle data for correlation with similar data being collected by other satellites.[16]

NASA Quits

Although the launch of the ATS-6 spacecraft in 1974 marked the end of NASA's program of experimental communications satellites, the space agency also participated at the same time in a Canadian satellite venture known initially as "Cooperative Applications Satellite C" and renamed Hermes. This joint effort involved NASA and the Canadian Department of Communications. NASA's Lewis Research Center provided the satellite's high-power communications payload. Canada designed and built the spacecraft; NASA tested, launched, and operated it. Also, the European Space Agency provided one of the low-power traveling-wave tubes and other equipment.[17] Hermes was launched 17 January 1976 and operated until October 1979.

Canada also created a telecommunications policy that the United States would emulate, and this would lead to the end of NASA's communications satellite research and development program. In late 1969, Canada announced that any financially qualified organization could apply for, and expect to be granted, authority to operate a domestic satellite system. As a result, in November 1972, that country put into orbit the world's first domestic satellite. In addition, the Canadian government abandoned sponsored research in the hope of motivating competition in the development of satellite technology. This Canadian "Open Skies" policy represented a striking contrast to past U.S. policy,[18] but it was in tune with the Nixon administration's advocacy of competition.

---

16.   In 1985, the ATS program was transferred from Goddard to the Lewis Research Center in Cleveland to consolidate NASA's communications program. ATS-3 is still operational. Michael A. Cauley, "ATS-3: Celebrating 25 Years of Service in Space," unpublished manuscript, no date, p. 1, ACTS Project Office, NASA Lewis Research Center.

17.   Harold R. Raine, "The Communications Technology Satellite Flight Performance," *Acta Astronautica* 5 (1978): 343–368.

18.   A.D. Wheelon, "Von Karman Lecture: The Rocky Road to Communication Satellites," in *AIAA 24th Aerospace Sciences Meeting: January 6–9, 1986, Reno, Nevada,* Paper No. AIAA-86-0293, Vol. 4 (New York: American Institute of Aeronautics and Astronautics, 1986), p. 21.

Subsequently, in January 1973, budget pressures caused NASA essentially to eliminate its communications satellite research and development program, much of which was carried out at the Goddard Space Flight Center, although the Lewis Research Center was working on advances in traveling-wave tube design and was participating in the Canadian Hermes project. Goddard had been responsible for most NASA experimental (as well as operational) communications satellites, including the ATS series.

Meanwhile, the danger of foreign competition, especially from Japan and Europe, loomed large. The Japanese launched the first commercial Ka-band operational satellite, called Sakura 2a, on 4 February 1983. Built by Ford (now Loral) and Mitsubishi, and launched on a Japanese N2 rocket, Sakura 2a also was Japan's first commercial communications satellite. It was replaced by Sakura 3a, which was launched 19 February 1988. In the 1990s, Loral also built the Superbird and N-Star satellites for Japan, and with Japanese contractors led by Toshiba, Japan's National Space Development Agency designed and built the ETS 6 (Kiku 6) spacecraft.[19]

At the same time, the Europeans were catching up. The Olympus satellite began development in 1979 as L-Sat and was built by European aerospace companies, of which British Aerospace was the prime contractor. Launched on 12 July 1989 on an Ariane 3 rocket, Olympus was a large multipurpose satellite demonstrating and promoting new applications in television broadcasting, intercity telephone routing, and the use of the Ka-band for videoconferencing and low-rate data transfer for business communication.[20]

Foreign competition provided NASA a strong argument for reinstating its commercial satellite development program. The question was: what technology ought to be developed? Market studies conducted during the 1970s revealed the crowding of synchronous orbits. The obvious solution to overcrowding was to use higher frequency Ka-band communications satellites.

A compelling synergy exists among the use of Ka-band frequencies, spot beams, and onboard processors. In general, higher frequencies produce smaller beam widths with a given antenna, and so it is easier to make antennas that produce spot beams at Ka-band frequencies. A spot beam covers a smaller area, such as a major metropolitan area, compared to typical beams covering the entire country. These spot beams improve the problem of rain fade at Ka-band by concentrating the signal strength to punch through clouds. Once there are spot beams, it is a natural extension to switch signals between various spot coverage areas (such as routing one signal from New York to Chicago and another from New York to Los Angeles) aboard the spacecraft.

Despite the virtual shutdown of the NASA communications satellite research and development program, NASA systems engineers throughout the mid-1970s sought ways to revive their canceled program. Work continued on satellites that were still operating in space, as well as on projects that were too far along to stop. Both the Lewis Research Center and the Goddard Space Flight Center advocated reviving the space communications programs, but along very different lines. Goddard championed public service satellites directly in competition with industry. The technology development program at Lewis supported U.S. industry, although some companies saw the Lewis approach as subsidizing competitors. The Nixon administration's advocacy of competition thus favored the Lewis approach.

The question of federal funding of communications satellite technology development by NASA came before the National Research Council, whose Committee on Satellite

19.   Andrew Wilson, ed., *Jayne's Space Directory 1993–1994* (Surrey: Jayne's, 1993), pp. 66, 359–60.
20.   *Ibid.*, pp. 330–32.

Communications released a report on the subject in 1977. The committee considered several options and recommended funding a NASA experimental communications satellite technology flight program and an experimental public service communications satellite system. The committee opposed creating an operational public service system on the grounds that it was "inappropriate for NASA."[21]

In 1978, five years after budget pressures forced NASA to eliminate its commercial communications satellite research and development program, the space agency reentered the field.[22] The Lewis Research Center, not the Goddard Space Flight Center, acted as the lead center, in light of the program's emphasis on technology. Lewis worked on the next NASA experimental communications satellite and began a $45 million program of technology development using duplicate contracts, to have the new designs needed for a radically new spacecraft. NASA involved the five major builders of communications satellites—TRW, Hughes, Ford, General Electric, and RCA—by awarding each a study contract. These contracts ranged in cost from $264,000 (RCA) to $1,213,000 (TRW), and all were completed in the summer of 1981.

Joe Sivo, chief of the communications division at Lewis, brought U.S. communications carriers—the users of the technology—into the program to develop a consensus on advanced technology requirements. Between November 1979 and May 1983, the Carrier Working Group, formed by Sivo, met nine times to define flight system requirements and experiments and to review spacecraft designs as they became available from the study contractors. The Carrier Working Group consisted of representatives from American Satellite, AT&T Long Lines, Bell Telephone Laboratories, Comsat, GTE Satellite, Hughes Communications, ITT, RCA, Satellite Business Systems (MCI), Southern Pacific, and Western Union.

The Advanced Communications Technology Satellite

The efforts of the Carrier Working Group and the industry study contracts led directly to the design and construction of the next NASA experimental communications satellite, known as the Advanced Communications Technology Satellite (ACTS). Its unique feature is that it acts as a "switchboard in the sky." The communications payload incorporates steerable, spot-beam antennas and onboard switching that allows signals to be routed aboard the spacecraft. The Ka-band frequencies used by the ACTS (thirty gigahertz for the uplink and twenty gigahertz for the downlink) were new capabilities for U.S. communications satellites, but the Japanese already had used them on their own satellite. NASA makes the satellite available for experiments by industry, universities, and other government agencies, as well as for tests of new service applications.

Launched 12 September 1993, the ACTS has been perhaps the most successful of NASA's communications satellites. To date, it has operated for two years without failures and has conducted more than 100 experiments and tests. Furthermore, several commercial systems have proposed using ACTS technologies. For example, Motorola, which built the satellite's baseband processor, is incorporating onboard switching in its Iridium system, while Hughes is working on Spaceway, a Ka-band system with spot beams. Despite

21.   Committee on Satellite Communications, Space Applications Board, Assembly of Engineering, National Research Council, *Federal Research and Development for Satellite Communications* (Washington, DC: National Academy of Sciences, 1977), p. 30.

22.   Robert R. Lovell, "The Status of NASA's Communications Program," Pacific Telecommunications Conference, January 1982, p. 4, preprint copy supplied by author.

these technical and commercial successes, the ACTS had a long and tortuous existence during the 1980s, as Congress and the White House debated the philosophy of having NASA develop technology for the U.S. communications satellite industry. While the satellite was still in the design phase, for example, the Reagan administration deleted the satellite from its budget, only to have Congress reinstate the project.

Conclusion

NASA's commercial communications satellite program has produced many significant results over the past 35 years. Although some critics have argued that NASA has overstated its contribution to satellite communications,[23] one can contend that the program has returned far more to the industry than its cost.

One often overlooked key to understanding the value of NASA's experimental communications satellite program is the concept of risk in space design. The high cost of launching spacecraft, coupled with the current impossibility of repairing hardware in synchronous orbit, means that the design of space hardware is driven by the risk of failure. As a result, space hardware designs have been very conservative, using old technology, because of the perceived risk associated with using anything new. Even if a new item can produce greater capability at lower cost, if it increases the risk of failure, it will not be used. For a new technology to fly, its benefits must be overwhelming, thereby precluding incremental improvements in space technology. As a result, space hardware can lag behind the state of the art by more than a decade.

A high risk of failure overshadowed the early years of satellite communications. NASA's launch record in the early 1960s reflected the state of rocket science at the time—namely, rockets failed quite often. Also, the space environment was not well known. The period was a critical time for NASA's involvement in the development of communications satellites. Without NASA, Hughes's Syncom would never have gotten off the ground, and Hughes would not be the world's largest communications satellite maker today.

Despite the benefits that NASA's experimental communications satellite program has brought industry, industry does not look kindly on NASA's development of spacecraft or technology that might undermine a firm's competitive advantage. In other words, a NASA spacecraft must develop new technology that all U.S. companies can use, but the space agency must avoid the construction of spacecraft that might seem to compete with any company. Hughes benefited enormously from NASA's involvement in the early days of communications satellites. Today, as a result, Hughes is the largest builder of commercial communications satellites. In the 1995 Space News "Top 50" list of space companies, Hughes was ranked second.[24]

The 1973 cancellation of NASA's commercial communications satellite research and development program, in retrospect, benefited the space agency. It forced a complete rethinking of the program. To be reinstated in 1978, NASA had to justify the program from scratch and transform it from a public service demonstration into a technology development program. The public service satellite program looked too much like competition to industry, even though the services NASA provided would not have been affordable to public service users.

---

23.   See, for instance, Cunniffe, "Misreading History," *passim.*
24.   Laura K. Browning, "The Top 50 Space Companies," *Space News* 6 (24–30 July 1995): 8.

NASA currently is reorganizing its commercial communications program to prepare for the next generation of research and development following the ACTS, although a need for a followup project is not perceived at the moment. Therefore, NASA plans to develop long-term technology improvements and work on both spectrum management and issues of interoperability between satellite systems. NASA has benefited from the Satellite Industry Task Force, an industry advisory committee chaired by Hughes's Thomas Brackey. At a presentation of its findings on 12 September 1995, attended by Vice President Al Gore, the task force expressed support for the ACTS, but it did not call for NASA undertaking another satellite project.

Chapter 7

From Advent to Milstar: The U.S. Air Force and the Challenges of Military Satellite Communications

by David N. Spires and Rick W. Sturdevant

For more than three decades, the U.S. Air Force has led efforts to expand satellite communications capabilities for military use. From Advent in the early 1960s to Milstar in the 1990s, the Air Force has provided launch vehicles, supporting infrastructure, and most of the communications satellites for the defense community. In so doing, the Air Force has confronted a variety of sometimes related technical, political, and institutional challenges. These focused initially on long-range strategic communications requirements, but increasingly on tactical needs after the first wartime transmission of voice and data from Vietnam to Washington, D.C., via satellite. Air Force engineers often led their commercial counterparts as they probed the boundaries of high-risk technology in an effort to increase payload capabilities. Unlike providers of commercial satellite communications services, however, the Air Force has wrestled with survivability and unique requirements that have increased military satellite costs, even as commercial costs have dropped.

As an example of political challenges, "convergence" proponents have repeatedly criticized military satellite communications; they argue that merging the commercial and military sectors would avoid duplication and save tax dollars. Conversely, civilians who worry about divergent civil and military interests and military people concerned with system security and assured access during conflicts have staunchly defended the need for separate military satellite communications capabilities, even if they must be supplemented by civil and commercial satellite communications to handle total volume. In terms of institutional challenges, the Department of Defense (DoD) traditionally has favored tri-service, or joint force, military satellite communications management for the sake of cost-effectiveness—and to reduce interservice rivalries. The complicated, fragmented nature of military satellite communications management has served historically to render the integration of planning and services more difficult within the Air Force and DoD; furthermore, it has retarded the movement of military satellite communications systems from the realm of research and development into the operational arena.

Origins and Early Efforts

World War II demonstrated the essential requirement of electronically transmitting military information over longer ranges, in greater quantities, and with more reliability and higher security than ever before. A constellation of Earth-orbiting satellites, first proposed by the British science fiction writer Arthur C. Clarke in October 1945, offered a revolutionary way of meeting those requirements. Now a part of early space lore, Clarke's concept of geostationary communications satellites sparked serious military interest, as evidenced by Project RAND's May 1946 report to the Army Air Forces, titled *Preliminary*

Design of an Experimental World-Circling Spaceship.[1] It could not be implemented, however, until the technology for space launch and satellite construction was more advanced.

For the near term, the military services (primarily the Army and Navy) had to satisfy themselves with using the Moon for space communications experiments. Not until 18 December 1958 did an Air Force Atlas B booster successfully carry the Advanced Research Projects Agency's SCORE (Signal Communications by Orbiting Relay Equipment) into low-Earth orbit, from where it delivered President Eisenhower's now-famous Christmas message. The Army followed this achievement in October 1960 with the successful launch of its Courier delayed-repeater communications satellite, which operated in the ultrahigh frequency (UHF) band in a low-altitude orbit (90 to 450 nautical miles, or 167 to 833 kilometers). Meanwhile, under Project West Ford (later Project Needles), the Air Force contracted with Lincoln Laboratory of the Massachusetts Institute of Technology (MIT) to produce 480 million hair-like, copper dipoles, which were launched on 9 May 1963 and reflected radio signals from an orbit nearly 2,000 nautical miles (about 3,700 kilometers) above the Earth.[2]

As the 1960s approached, DoD sought to simultaneously develop a satellite constellation of the sort envisioned by Clarke and create management structures to handle the new communications capabilities. In 1958, the Advanced Research Projects Agency had directed the Army and Air Force to plan for an equatorial synchronous (strategic) satellite communications system, with the Air Force responsible for the booster and spacecraft and the Army in charge of the actual communications elements aboard the satellite, as well as on the ground. The program initially consisted of three projects. Two of these, named Steer and Tackle, used medium-altitude repeater satellites; the third, Decree, called for a synchronous repeater satellite using microwave frequencies. The Pentagon transferred the management of military satellite communications development efforts from the Advanced Research Projects Agency to the Army in September 1959; it soon thereafter combined the three projects into a single program, called Advent.

Once described as a "not quite possible dream," that technologically ambitious undertaking was soon plagued by high costs, inadequate payload capacity, and an excessive satellite-to-booster weight ratio. Those problems caused Secretary of Defense Robert McNamara to cancel Advent on 23 May 1962. Meanwhile, it had become apparent that neither the Army nor any other single service would have overall responsibility for military satellite communications, because in May 1960 the Pentagon combined the strategic

1.   For a discussion of communications satellites in the RAND report, see Douglas Aircraft Company, Inc., *Preliminary Design of an Experimental World-Circling Spaceship*, Report No. SM-11827 (Santa Monica, CA: Douglas Aircraft Company, Engineering Division, 2 May 1946), pp. 14–15. For a probable connection between Arthur Clarke's article and the RAND report, see Donald C. Elder, *Out From Behind the Eight-Ball: A History of Project Echo*, AAS History Series, Vol. 16 (San Diego: American Astronautical Society, 1995), p. 13.

2.   Carl Berger, *The Air Force in Space, Fiscal Year 1961*, Vol. SHO-S-66/142 (Washington, DC: U.S. Air Force Historical Division Liaison Office, April 1966), pp. 84–93; The Aerospace Corporation, *The Aerospace Corporation: Its Work, 1960–1980* (Los Angeles: Times Mirror Press, 1980), pp. 47–49; Maj. Robert E. Lee, *History of the Defense Satellite Communications System (1964–1986)*, ACSC Report No. 87-1545 (Maxwell AFB, AL: Air University Press, 1987), pp. 5–8; Elder, *Out From Behind the Eight-Ball*, pp. 11–13, 51–56. Project West Ford grew out of a 1958 summer study on secure, hardened, reliable communications to overcome problems (such as inadequate power generation and limited real-time communication from low-Earth orbits) associated with the earliest active communications satellites. Although many critics feared that West Ford would produce long-term problems for Earth-based astronomers, most of the dipoles had reentered the atmosphere by early 1966. North American Aerospace Defense Command's Space Surveillance Center, however, did continue tracking remnants as late as 1989. Donald H. Martin, *Communication Satellites, 1958–1992* (El Segundo, CA: Aerospace Corporation, 1991), pp. 8–9; *History of Military Space Operations*, ATC Study Guide S-V95-A-SPV01-SG (Lowry AFB, CO: 3301st Space Training Squadron, Air Training Command, March 1991), pp. 39–40.

communications systems of the three services under a Defense Communications System (DCS) run by the newly created Defense Communications Agency (DCA).[3]

Before authorizing a more realistic military satellite communications project to replace Advent, McNamara opened discussions with Comsat, the corporation that congressional legislation established in early 1963. McNamara questioned why he should fund a separate, costly medium-altitude military satellite system if the military could lease links from Comsat at lesser cost. DoD and Comsat, however, could not agree on costs or the need for separate military repeaters aboard commercial satellites.[4]

Furthermore, the addition of military applications to a civilian system designed for use by other countries created international concerns. On 15 July 1964, after months of fruitless effort, McNamara ended negotiations and opted for the full-scale development of a dedicated military system, which the Air Force had consistently favored to ensure security and reliability. The defense secretary accepted a proposal that Aerospace Corporation had been studying for the Air Force. Although it initially had called for using an Atlas-Agena booster combination to launch a constellation of randomly placed, medium-altitude satellites weighing 100 pounds (forty-five kilograms) each, the development of the more powerful Titan IIIC booster soon resulted in a decision to aim for launching as many as eight satellites at once into a near-synchronous equatorial configuration. The Los Angeles-based Space and Missile Systems Organization (SAMSO), within the Air Force Systems Command, was responsible for developing both the spacecraft and its communications payload; the Army Satellite Communications Agency was designated to manage the ground segment; and DCA had executive management responsibility for this Initial Defense Communications Satellite Program (IDCSP).[5]

The Initial Defense Communications Satellite Program

Originally expected to function as an experimental system, IDCSP rapidly proved its operational worth and became the first phase in a three-phase evolutionary program to provide long-term, survivable communications for both strategic and tactical users. The first seven IDCSP satellites, relatively simple in design to avoid the problems that had plagued Courier and prevented Advent from ever getting off the ground, went aloft on 16 June 1966. Operating in superhigh frequency (SHF), weighing about 100 pounds (forty-five kilograms), and measuring only three feet (one meter) in diameter and nearly three feet (one meter) in height, these spin-stabilized satellites contained no moveable

3. The Aerospace Corporation, *The Aerospace Corporation: Its Work*, pp. 47–49; Lee, *History of the Defense Satellite Communications System*, pp. 5–8; Berger, *The Air Force in Space*, pp. 84–93. When it agreed to terminate the Advent program, DoD received assurances it would have access to NASA's Syncom satellites. On 1 January 1965, NASA transferred Syncom operations to the Pentagon. Over the next two years, the DoD logged extensive time on Syncom 2 and Syncom 3, although use diminished after the IDCSP satellites became operational. See TRW, *Space Log* 4 (Winter 1964–65): 27–28; Heather E. Hudson, *Communication Satellites: Their Development and Impact* (New York: Free Press, 1990), p. 19; Martin, *Communication Satellites*, pp. 12–14.

4. Military and commercial capabilities would not share the same satellite until the Navy contracted with Comsat in 1973 for "gapfiller" service, pending the completion of its Fleet Satellite Communications System. See Michael E. Kinsley, *Outer Space and Inner Sanctums: Government, Business, and Satellite Communication* (New York: John Wiley and Sons, 1976), pp. 199–200; Anthony Michael Tedeschi, *Live Via Satellite: The Story of COMSAT and the Technology that Changed World Communication* (Washington, DC: Acropolis Books, 1989), p. 150; Martin, *Communication Satellites*, pp. 104–05, 183–86.

5. Berger, *The Air Force in Space*, pp. 65–68; The Aerospace Corporation, *The Aerospace Corporation: Its Work*, pp. 48–52; Lee, *History of the Defense Satellite Communications System*, pp. 8–10; House Committee on Government Operations, *Government Operations in Space (Analysis of Civil-Military Roles and Relationships)*, 89th Cong., 1st sess., 4 June 1965, H. Rept. 445, pp. 78–79.

parts, no batteries for electrical power, and only a basic telemetry capability for monitoring purposes. The configuration of each IDCSP platform provided two-way circuit capacity for either eleven tactical-quality voice circuits or five commercial-quality circuits capable of transmitting one million digital or 1,550 teletype data bits per second. The IDCSP satellite's twenty-four-face polyhedral surface accommodated 8,000 solar cells that provided sufficient energy to power a single-channel receiver operating near 8,000 megahertz, a three-watt traveling-wave-tube amplifier transmitting around 7,000 megahertz, and one twenty-megahertz double-conversion repeater. Designed to operate for three years, the actual mean time before failure among IDCSP satellites proved to be six years. DCA declared the system operational and changed its name to the Initial Defense Satellite Communications System (IDSCS) before the launch of the last group of eight satellites on 13 June 1968.[6]

Figure 14

The Mark IV(X) transportable station for satellite communications designed for transmitting and receiving voice, teletype, and facsimile images via NASA's Syncom satellite. During the Vietnam War, the military relied on Syncom for routine administrative and logistical communications between Saigon and Hawaii. (Courtesy of NASA)

6.   Gerald T. Cantwell, *The Air Force in Space, Fiscal Year 1964*, Vol. SHO-S-67/52 (Washington, DC: U.S. Air Force Historical Division Liaison Office, June 1967), pp. 69–76; Gerald T. Cantwell, *The Air Force in Space, Fiscal Year 1965*, Vol. SHO-S-68/186 (Washington, DC: U.S. Air Force Historical Division Liaison Office, April 1968), pp. 42–51; Lee, *History of the Defense Satellite Communications System*, pp. 10–13; The Aerospace Corporation, *The Aerospace Corporation: Its Work*, pp. 48–52; Martin, *Communication Satellites*, pp. 95–96.

Vietnam provided the first opportunity to use satellite communications from a real-world theater of operations. U.S. forces had installed IDCSP ground terminals at Saigon and Nha Trang by July 1967. Under Project Compass Link, IDCSP provided circuits for the transmission of high-resolution photography between Saigon and Washington, D.C. As a result of this revolutionary development, analysts could conduct near-real-time battlefield intelligence from afar. Commercial systems also supplied satellite circuits to support area communications requirements. Even before IDCSP service became available, the military had relied on the NASA-developed Syncom satellite for communications between Saigon and Hawaii. Later, Comsat leased ten circuits between its Bangkok facilities and Hawaii, while the Southeast Asia Coastal Cable System furnished part of the network for satellite terminal access between Bangkok and Saigon. Satellite usage during the Vietnam War established the military practice of relying on commercial space systems for routine administrative and logistical needs while trusting more sensitive command-and-control communications to the dedicated military system.[7]

The Defense Satellite Communications System

While the IDSCS provided good service for nearly ten years and also furnished the basic design for British Skynet and NATO satellites, the first-phase Defense Satellite Communications System (DSCS) satellites remained limited in terms of channel capacity, user access, and coverage. Furthermore, military planners worried about the vulnerability of a command-and-control system that involved a central terminus connected to a number of remote terminals. The subsequent DSCS II design sought to overcome those deficiencies. TRW Systems received a contract from the Air Force Space Systems Division (formerly SAMSO) in March 1969 to develop and produce a qualification model and six flightworthy satellites to be launched in pairs aboard a Titan III. Plans called for a constellation of four active satellites in geosynchronous orbit, supported by two orbiting spares. Each satellite measured nine feet (2.7 meters) in diameter, was thirteen feet (four meters) in height with antennas extended, and weighed 1,300 pounds (590 kilograms).

DSCS II, because it was dual-spun for stability, represented a "giant step" in technical development over its smaller, lighter, and less capable predecessor. A flexible, four-channel configuration provided a variety of communications links for interfacing with various size terminals. It possessed capacity for 1,300 two-way voice channels or 100 million bits of digital data per second, and onboard batteries generated 520 watts of power to complement the satellite's eight solar panels. The five-year design life nearly doubled that of DSCS I, and the new system's redundancy, multichannel and multiple-access features, and increased capability to communicate with smaller, more mobile ground stations especially pleased the Air Force and other users.[8]

---

7. Lt. Col. John J. Lane, Jr., *Command and Control and Communications Structures in Southeast Asia* (Maxwell AFB, AL: Air University, 1981), pp. 113–14. DoD's quest for inexpensive satellite communications during the Vietnam War gave rise to the "30 circuits" episode. When several carriers learned in the summer of 1966 that Comsat intended to lease 30 circuits directly to DoD, in direct violation of the FCC's recently promulgated "Authorized Users" decision, they formally protested. In February 1967, the FCC ordered a so-called composite rate of $7,100 per half-circuit, splitting the traffic evenly three ways among ITT, RCA, and Western Union International. The FCC required Comsat to sell the circuits to the carriers at $3,800. Kinsley, *Outer Space and Inner Sanctums*, pp. 60–62.

8. Thomas Karas, *The New High Ground: Strategies and Weapons of Space-Age War* (New York: Simon and Schuster, 1983), pp. 73–76; Jacob Neufeld, *The Air Force in Space, 1970–1974* (Washington, DC: Office of Air Force History, August 1976), pp. 15–19; Lee, *History of the Defense Satellite Communications System*, pp. 13–21; The Aerospace Corporation, *The Aerospace Corporation: Its Work*, pp. 52–57; Martin, *Communication Satellites*, pp. 100–02.

The orbital history of DSCS II satellites in the 1970s, beginning with the launch of the first pair on 2 November 1971, reveals a somewhat spotty performance record. A Titan IIIC booster failure accounted for the loss of two of the next six satellites; problems with stabilization, antenna pointing, and traveling-wave-tube amplifiers plagued the others. The Air Force responded by contracting with TRW for an additional six satellites of the original design, and later four more with forty-watt traveling-wave-tube amplifiers in place of the twenty-watt amplifiers. Despite another launch failure in March 1978 and continued high-voltage arcing in the power amplifiers, by the early 1980s, the DSCS II constellation not only fulfilled global, strategic communications requirements through forty-six DSCS ground terminals, but also linked the Diplomatic Telecommunications System's fifty-two terminals and the Ground Mobile Forces' thirty-one tactical terminals. Perhaps the best example of the satellite's durability is that DSCS II B4, launched on 13 December 1973, lasted four times longer than its design life. The Air Force did not turn it off until 13 December 1993.[9]

The Air Force had been designing an improved DSCS III satellite since 1974 to meet the military's need for increased communications capacity, especially for mobile terminal users, and for greater survivability. General Electric's DSCS III differed considerably from its phase II predecessor. This third-generation satellite was three-axis stabilized, considerably heavier (2,475 pounds, or 1,123 kilograms, in orbit), and rectangular rather than cylindrical in shape (with dimensions of six by six by ten feet [1.8 by 1.8 by three meters] and a thirty-eight-foot [11.6 meter] span when the solar arrays were deployed). Its 1,000 watts of battery power practically doubled that of DSCS II, as did its ten-year design life. Furthermore, DSCS III signaled a major technological advance by being the first operational satellite to use electronically switched SHF multiple-beam antennas. The sixty-one-beam uplink antenna produced variable gain patterns from Earth coverage, to spot beams, to patterns with nulls in selected directions to counter jamming. The two downlink antennas used nineteen independently switched beams. A gimbaled dish, two horns, and two UHF antennas completed the array. Flexible antenna configurations combined with six transponders, which used forty-watt and ten-watt traveling-wave-tube amplifiers, offered a wide range of services to the growing wideband user community as well as to the ground-mobile force.[10]

The DSCS III satellite also carried a single-channel Air Force satellite UHF transponder with anti-jamming protection for secure voice communications during all levels of conflict. As with most DoD satellites, DSCS III had both an S-band section for use by the Air Force Satellite Control Network and an X-band section that provided redundant command paths and gave Army personnel at eight ground stations worldwide direct control of the transponders and antennas. Beginning with the fourth satellite, the use of improved jam resistance, redundancy, and more powerful amplifiers enabled DSCS III to meet the military's growing requirements for increased capacity and survivability.[11]

9.   Karas, *The New High Ground*, pp. 75–76; Neufeld, *The Air Force in Space*, pp. 15–19; Lee, *History of the Defense Satellite Communications System*, pp. 15–28; The Aerospace Corporation, *The Aerospace Corporation: Its Work*, pp. 55–57; Martin, *Communication Satellites*, pp. 100–02.

10.   Lee, *History of the Defense Satellite Communications System*, pp. 29–32; Martin, *Communication Satellites*, pp. 111–13; Dwayne A. Day, "Capturing the High Ground: The U.S. Military in Space 1987–1995, Part 2," *Countdown*, May/June 1995, p. 18; Lt. Gen. Winston D. Powers and Andrew M. Hartigan, "The Defense Satellite Communications System," *Signal* 39 (11) (July 1985): 53; Giles C. Sinkewiz, "Satellite Communications: Directions and Technology," *Signal* 39 (11) (July 1985): 60–62; A. Nejat Ince, ed., *Digital Satellite Communications Systems and Technologies: Military and Civil Applications* (Boston: Kluwer Academic Publishers, 1992), pp. 368–69.

11.   Karas, *The New High Ground*, pp. 73–76; Neufeld, *The Air Force in Space*, pp. 15–19; Lee, *History of the Defense Satellite Communications System*, pp. 13–21; The Aerospace Corporation, *The Aerospace Corporation: Its Work*, pp. 52–57; Martin, *Communication Satellites*, pp. 100–02.

The orbital history of DSCS III did not include the booster problems that had result-ed in the loss of four DSCS II satellites. On 30 October 1982, a Titan 34D/Inertial Upper Stage vehicle launched the first DSCS III satellite. Another two DSCS III satellites went aloft via the Space Shuttle *Atlantis* in 1985. The *Challenger* disaster and a series of Titan 34D failures then shut down launch operations for two years. Fortunately, the reliability of the DSCS system—and the exceptionally long life span of DSCS III and later DSCS II satel-lites—allowed the constellation to weather the launch crisis better than most other mili-tary satellite systems. The Air Force finally launched a fourth DSCS III satellite via a Titan 34D/Transtage in 1989, after which Atlas II became the preferred launch vehicle. Not until 19 July 1993 did the Air Force complete a full, five-satellite DSCS III constellation. One measure, however, of the confidence that the Air Force has placed in the jam-resistant secure communications capability afforded by DSCS III satellites is that since December 1990, they have been the primary means for transmitting missile warning data from key worldwide sensor sites to correlation-and-command centers at Cheyenne Mountain and elsewhere.[12]

The introduction of new heavy, medium, and light ground terminals beginning in the mid-1970s allowed the military to start phasing out aging equipment first deployed in the 1960s and subsequently modified for use with DSCS II. Although DSCS III continued the practice of using terminals from the earlier system, the 1980s brought new terminals, including eight-foot (2.4-meter) and twenty-foot (six-meter) antennas for the ground-mobile forces. Meanwhile, work continued to convert the entire system from analog to digital transmission by the end of the decade. The DSCS III satellite program also bene-fited from a number of new Air Force acquisition practices. Faced with cost overruns and schedule slips, Air Force officials resorted to milestone billing, and they convinced Congress to approve the practice of multiyear procurement instead of annual buys.[13]

A Fragmented Management System

The fact that DSCS remained a Pentagon-managed system, with DCA, the Army, and the Air Force all playing roles in its day-to-day operation, indicated a much broader insti-tutional problem. Based on DoD Directive 5105.44 of 9 October 1973 and Joint Chiefs of Staff Memorandum of Policy 178 of 17 March 1975, management of military satellite com-munications systems remained fragmented among the Air Force, Navy, Army, and DCA. From the 1980s on, the U.S. military wrestled with that unwieldy situation in an effort to ensure more cost-efficient acquisition and more effective employment of resources through all levels of conflict. The strong tendency, however, since the earliest days of mil-itary satellite communications remained for each service to want its own system. That had been tempered partially during the 1970s by fiscal cutbacks and congressional pressure, which had caused all systems to become more "common user" despite their nomencla-ture. While technology helped achieve a higher degree of efficiency within an expanding

---

12.   Lee, *History of the Defense Satellite Communications System*, pp. 29–32; Martin, *Communication Satellites*, pp. 111–13; Day, "Capturing the High Ground," p. 18; AFSPACECOM/DOF to USSPACECOM/J3M, et al., "Designation of JRSC as Primary Communication System for Missile Warning Data," letter, 28 June 1991, with attachment: AFSPACECOM/DO to USSPACECOM/J3, et al., "Designation of JRSC as Primary Communications System for Missile Warning Data," letter, 17 December 1990, Air Force Space Command, History Office Archives, Colorado Springs, CO.

13.   Lee, *History of the Defense Satellite Communications System*, pp. 32–42. For major decisions associated with the future of DSCS, see *Military Satellite Communications. Opportunity to Save Billions of Dollars*, Report no. NSIAD-93-216 (Washington, DC: Government Accounting Office, 9 July 1993), pp. 9–11.

user community (that is, the cost per capability dropped), the increasing sophistication of military satellite communications still led to dramatic price increases per satellite and required automated tracking, telemetry, and commanding instead of the old, manual methods.[14]

While the creation of the Air Force Space Command in September 1982 signaled the Air Force's commitment to centralizing its space operations, the establishment of the U.S. Space Command three years later clearly offered an opportunity to vest managerial responsibility for all military satellite communications in one organizational chain. The translation of opportunity into reality, however, proved next to impossible. Despite various initiatives, the 1980s brought no definitive answers to the questions of whether the existing military satellite communications management structure should be altered and, moreover, whether the existing acquisition alignment—that is, UHF for the Navy, SHF for DCA, and extremely high frequency (EHF) for the Air Force—should remain binding.[15]

The fussing, fuming, and fumbling over military satellite communications architectural arrangements and their implications continued into late 1990, with the assistant secretary of defense for command, control, communications, and intelligence finally directing DCA to take the lead in developing a suitable architecture. That task included considering the role of smaller and cheaper satellites, the potential for the increased use of commercial satellites, the achievement of U.S. and allied interoperability, and the possibility of cooperative efforts to reduce developmental and operational costs. Almost simultaneously, efforts to develop a new memorandum of policy for military satellite communications management bogged down, after the Air Force Space Command complained that the new version failed to place DSCS executive management within the chain of command of the U.S. Space Command's commander in chief; it left that responsibility in the hands of the Defense Information Systems Agency (the successor to DCA). Systematic

---

14.   DoD Directive 5105.44, 9 October 1973, established the Military Satellite Communications Systems Organization within DCA, provided for coordination of all service and agency efforts, and defined further the military communications satellite roles of the Joint Chiefs of Staff and assistant secretary of defense for telecommunications (later changed to assistant secretary of defense for command, control, communications, and intelligence). The first revision of Joint Chiefs of Staff Memorandum of Policy 178, 1 May 1978, spelled out DCA's, and each service's, executive management responsibilities for military satellite systems. It contained criteria for specifying system interoperability and compatibility. A second revision of Memorandum of Policy 178, dated 4 September 1986, directed that military communications satellite planning be integrated into the same planning process used for other resources. See Lt. Col. Fred Thourot, "MILSATCOM [Military Satellite Communications] Deliberate Planning," *Signal* 41 (June 1987): 63–69; Maj. David J. Fitzgerald and Capt. Timothy G. Learn, "Influencing Satellite Design for Communications Management and Control of MILSATCOM [Military Satellite Communications] Through All Levels of Conflict," 5 March 1987, Air Force Space Command, History Office Archives, Colorado Springs, CO.

15.   Briefing, DCA/MSO, "The Alternative MILSATCOM [Military Satellite Communications] Architectures Study," 23 February 1989, Air Force Space Command, History Office Archives, Colorado Springs, CO. In early 1987, the U.S. Space Command began working with Air Staff, DCA, and the Joint Chiefs of Staff to (1) formulate a policy concept establishing "single chain-of-command from the Joint Chiefs of Staff, through USSPACECOM [U.S. Space Command] and its components, to the MILSATCOM [Military Satellite Communications] operations centers" and (2) provide for three regional space support centers "to facilitate consolidated space operations planning, provide technical and planning assistance to CINCs [Commanders in Chief] and other users, and ensure coordinated employment of space systems." See USSPACECOM, "Concept for MILSATCOM [Military Satellite Communications] Satellite Command and Control—Executive Summary," circa 20 August 1987, Air Force Space Command, History Office Archives, Colorado Springs, CO. Near the end of 1987, the Federal Computer Performance Evaluation and Simulation Center, in response to an initiative from the Space Communications Division at Colorado Springs, awarded a support contract to Booz, Allen & Hamilton for developing a comprehensive military communications satellite information resources architecture "capable of providing a broad, unified framework into which existing and new systems" could evolve. See HQ SPCD/YK to 2 CS/DO, *et al.*, "Award of MILSATCOM [Military Satellite Communications] Information Resources Architecture (IRA) Contract," message, 031830Z February 1988, Air Force Space Command, History Office Archives, Colorado Springs, CO.

efforts to restructure fractionalized military satellite communications management, with the goal of centralizing all responsibilities under the U.S. Space Command and its component commands, failed. Entering the last decade of the twentieth century, the requirements process for military satellite communications seemed bankrupt; attempts at architectural definition bordered on the absurd; and system responsibilities remained split among the Air Force, Navy, Army, and Defense Information Systems Agency.[16]

The Persian Gulf War experience of 1990–1991 made it apparent that the United States and its allies needed a fully integrated communications package for future crises. It could ill afford again to wait until after a crisis arose to assemble such a package. Moreover, an optimum communications network had to emphasize tactical rather than strategic requirements, the need for commercial satellite augmentation, and the importance of a responsive launch capability.[17] In the early 1990s, Air Force planners faced that reality, even as they focused on drastically restructuring the troubled Milstar (Military, Strategic, Tactical and Relay) program to trim costs and save it from cancellation.

Milstar

Milstar had emerged in the late 1970s from an Air Force proposal for a strategic satellite system to be called STRATSAT—a four-satellite constellation designed solely to support nuclear forces. It would avoid potential anti-satellite threats by orbiting at a so-called supersynchronous altitude of about 110,000 miles, and it would operate in the EHF range to provide more band width for spread-spectrum, anti-jam techniques. Considered too ambitious for so limited a mission, STRATSAT gave way to Milstar in 1981.

---

16.   AFSPACECOM/XRFC, "Comments to Draft CJCS MOP 37," staff summary sheet, 25 October 1990; AFSPACECOM/LK to CV/CC, note, 28 November 1990, with ASD/C3I to Director, DCA, "Architecture for Military Satellite Communications," memorandum, 19 November 1990; Joe Mullins and Pravin Jain, Defense Information Systems Agency, "Evolving MILSATCOM [Military Satellite Communications] Architecture and Technology Directions," August 1991; AFSPACECOM/XPFC, "Review of Interim MILSATCOM [Military Satellite Communications] Summit Brief to CSAF," staff summary sheet, 15 November 1991, Air Force Space Command, History Office Archives. At the request of the assistant secretary of the Air Force for space, the Air Force Space Command and the Air Force Systems Command already had begun a communications satellite architectural review in August 1990. See "AFSPACECOM Approach to MILSATCOM [Military Satellite Communications]," briefing, [1991], Air Force Space Command, History Office Archives. The U.S. Space Command's commander in chief agreed in early 1990 to assume from DCA administrative responsibility for the military communications satellite user requirements data base with the intention of delegating it to the Air Force Space Command "for definition and implementation of an automated support system." See USSPACECOM/J4-J6 to Joint Staff/J6, " Administrative Management of the MILSATCOM [Military Satellite Communications] URDB," message, 081245Z May 1990, Air Force Space Command, History Office Archives. Measurable progress did occur during 1990–1992 with respect to aligning military communications satellite operational management under the U.S. Space Command and its component commands. Maj. Gen. Carl G. O'Berry, the U.S. Space Command's director for command control communications and logistics, as well as the Air Force Space Command's deputy chief of staff for systems integration, logistics and support, spearheaded efforts to move "Fleet Satellite Communications" responsibility from the naval communications organization to the Navy Space Command, reassign DSCS tracking, telemetry, and commanding and ground terminal activities to the Army Space Command, and transfer Air Force "Satellite Communications" control from the Strategic Air Command to Air Force Space Command. See Maj. Gen Carl G. O'Berry, interview with Rick W. Sturdevant and Thomas Fuller, U.S. Space Command Headquarters, 7 May 1992, Oral History Interview 92-1, pp. 11–13, Air Force Space Command, History Office Archives.

17.   Alan D. Campen, ed., *The First Information War* (Fairfax, VA: AFCEA International Press, 1992), *passim.* That U.S. forces would have to rely heavily on communications satellites during any Persian Gulf conflict was recognized as early as the 1970s. See Paul B. Stares, *Space and National Security* (Washington, DC: Brookings Institution, 1987), pp. 127–28; Karas, *The New High Ground*, pp. 78–79.

Figure 15

The Milstar (Military, Strategic, Tactical and Relay) program received "Highest National Priority" status in 1983, but it underwent delays, redesigns, and cost overruns that angered a budget-cutting Congress. Consequently, the first Milstar satellite entered orbit in 1994, seven years after its projected launch. (Courtesy of the Air Force Space Command, History Office)

Air Force planners, viewing Milstar as capable of both strategic and tactical operations, proceeded to add numerous requirements to meet more types of missions. President Reagan's assignment of "highest national priority" status to Milstar in 1983 allowed the program to proceed with few funding restrictions. Similar to the Atlas intercontinental ballistic missile program in the 1950s, the development of the necessary "cutting edge" technology for Milstar and procurement, which included fielding the infrastructure for its operational support, proceeded concurrently. In the case of Milstar, unfortunately, those so-called "concurrency procedures" resulted in delays, redesigns, and cost overruns that drew the ire of an increasingly budget-conscious Congress.[18]

Initially designed to provide low-data-rate EHF communication, the eight-satellite Milstar constellation offered cross-link capabilities and extensive hardening against radiation. The EHF range had the advantage of allowing for the use of antennas as small as six

---

18.   Government Accounting Office (GAO), *DoD Acquisition. Case Study of the MILSTAR Satellite Communications System*, Report No. NSIAD 86-455-15 (Washington, DC: GAO, 31 July 1986); GAO, *Military Satellite Communications: Milstar Program Issues and Cost-Saving Opportunities*, Report No. NSIAD-92-121 (Washington, DC: GAO, 26 June 1992); Roger G. Guillemette, "Battlestar America: Milstar Survives A War With Congress," *Countdown*, November–December 1994, p. 22.

inches (about fifteen centimeters) in diameter, which suited highly mobile special operations forces. Titan IV boosters would send four of the satellites into various polar orbits
and the other four into geosynchronous orbits. Because the primary objective was survivability, not high performance, the Milstar design did not include high data rates; each
satellite was to serve no more than fifteen users simultaneously. As a result, it would sup-
plement rather than replace existing satellites such as DSCS and Fleet Satellite
Communications.[19]

Milstar's original strategic orientation seemed anachronistic following the end of the
Cold War. Operation Desert Storm during the Persian Gulf War reinforced interest in promoting Milstar's tactical capabilities, and the program underwent significant downsizing
based on congressional demands and Pentagon reviews. By early 1994, Milstar included
six rather than eight satellites, without the vast array of survivability features and with
fewer ground control stations. The first block of two satellites, designated Milstar I,
retained the limited-use low-data-rate capability, but subsequent Milstar satellites were to
be equipped with a medium-data-rate package to support tactical forces. On 7 February
1994, seven years after its projected launch, the first Milstar satellite went into orbit. The
Air Force anticipated the launch of the first Milstar II satellite in 1999, with transition to
a cheaper, lighter, advanced EHF Milstar III satellite by 2006. By the mid-1990s, despite its
effective use in Haitian operations, it was still not known whether Milstar had the ability
to provide survivable, jam-resistant, global communications to meet the needs of the
national command authorities, battlefield commanders, and operational forces through
all levels of conflict.[20]

Planning and Organizing for the Future

The escalating costs of dedicated military satellite communications such as Milstar,
combined with the inability of the United States during the Persian Gulf War to launch
additional military communications satellites on demand, highlighted the need for using
civil and commercial satellites and launchers in both peacetime and emergency situations.
How to use commercial systems without jeopardizing congressional support for military
satellite communications programs remained a major challenge. Moreover, the need for
an effective integration of military and commercial networks presented yet another challenge. The military sought to save money by finding alternatives to leasing individual communications satellite circuits, which is its historic policy. Although DoD could not readily
identify its total current usage of commercial communications satellites, the consolidation
of needs and procurement of greater overall capacity seemed worthwhile. The creation of
a private military-managed network of commercial communications satellite assets had its
proponents, but the idea foundered because the government could not operate in
nongovernment radio frequency bands. Another alternative involved a "commercially
equivalent" military satellite system that would use military radio frequencies and existing
terminals. Such satellites would cost less and handle more traffic than satellites built to
military specifications. They would offer features commonly found aboard military

19.   James W. Rawles, "Milstar Soars Beyond Budget and Schedule Goals," *Defense Electronics* 21 (February
1989): 66–72; Guillemette, "Battlestar America," p. 19; Day, "Transformation of National Security Space
Programs in the Post–Cold War Era," paper read at 45th Congress of the International Astronautical Federation,
9–14 October 1994, Jerusalem, Israel, pp. 11–12, copy from Air Force Space Command, History Office Archives.
20.   Guillemette, "Battlestar America," pp. 22–23; Day, "Transformation of National Security Programs,"
pp. 3–4; "Satcoms Success Story," *Space Markets* 4 (1991): 11–13.

satellites (for example, steerable spot-beam antennas and secure telemetry and payload control links) but would lack the special survivability features of military models.[21]

By February 1994, when the Government Accounting Office (GAO) reported on the military's use of communications satellites, it was clear that the institutional barriers to efficient, effective military exploitation of commercial communications satellite capabilities remained. GAO recommended "establishing firm policy and procedures for Department of Defense components to coordinate their needs for these services through a central organization."[22] A partial response came later in 1994 with the Pentagon's "Commercial Satellite Communications Initiative," in which the Defense Information Systems Agency set forth cost-saving concepts based on transponder leasing. DoD subsequently drafted a "Military Satellite Communications Master Plan," recommending that commercial satellites be used extensively by the turn of the century for low-security missions. Questions about the necessity of satellite survivability remained, however, and an increasingly vocal group of advocates touted the advantages of fiber optic cables over satellite technology for meeting future military communications needs. Others argued that the real point of discussion was the applicability of satellite and cable technology in particular situations. Despite lingering legislative restrictions on the military's use of commercial systems, the Pentagon relied on commercial carriers to handle most "general purpose" satellite traffic, which amounted to more than 80 percent of all DoD satellite communications requirements. That reliance undoubtedly would increase, because the military could not afford specially designed satellites to handle the five-fold growth in requirements expected between 1995 and 2010. It still seemed appropriate to route critical command-and-control traffic, amounting to less than 20 percent of the total military traffic, over dedicated military communications satellites.[23]

It was apparent by the mid-1990s that rapidly advancing communications technology, made readily available at relatively low cost by the commercial sector, offered the military an attractive alternative to increasingly expensive military satellite communications systems. The greatest roadblocks lay in traditional, service-oriented attitudes and the lingering notion that only military satellite communications could be relied on to be available

21.   Lt. Col. Charles F. Stirling, "Commercial Communication Satellite Application During National Crisis Management," March 1985, Air Force Space Command, History Office Archives; GAO, *Military Satellite Communications: Potential for Greater Use of Commercial Satellite Capabilities*, GAO Report No. T-NSIAD-92-39 (Washington, DC: GAO, 22 May 1992). For background, see C.S. Lorens, DCA, "Concept DCS Commercial Satellite Communications System," 5 June 1981; HQ AFCC/EPPD to HQ USAF/XOKCP, *et al.*, "DCS Commercial Satellite Communications (COMSATCOM) Program Update," letter, 10 March 1983, with two attachments; Commercial Satellite Survivability Task Force, Resource Enhancements Working Group, "Commercial Satellite Communications Survivability Report," 20 May 1983; Donald C. Latham, Deputy Under Secretary of Defense for Command, Control, Communications, and Intelligence, to Chairman, NSDD-97 Steering Group, "Initial Report of the Manager, National Communications System (NCS), on Commercial Satellite Communications Survivability," memorandum, 21 May 1984, with attachment; SPACECMD/KRQS, "Use of Commercial Satellites during National Crisis," staff summary sheet, 2 November 1984, Air Force Space Command, History Office Archives. For a specific example of how ineffective management led to the inefficient use of commercial communications satellites, see "Space Command's Commercial Satellite Communications Program," U.S. Air Force Audit Agency Report, 2 November 1985, Air Force Space Command, History Office Archives.

22.   GAO, *Military Satellite Communications: DoD Needs to Review Requirements and Strengthen Leasing Practices*, Report No. NSIAD-94-48 (Washington, DC: GAO, 24 February 1994), p. 7.

23.   Cheri Privor, "DoD Eyes Commercial Satellites," *Space News* 6 (12–18 June 1995): 3, 37; R.C. Webb, Les Palkuti, Lew Cohn, Lt. Col. Glenn Kweder, and Al Costantine, "The Commercial and Military Satellite Survivability Crisis," *Defense Electronics* 27 (August 1995): 21–25; Pat Cooper and Robert Holzer, "DoD Eyes Satellite Alternative," *Space News* 16 (7–13 August 1995): 1, 28; Donald L. Cromer, "Interoperability Is Key to Viability," *Space News* 6 (25 September–1 October 1995): 15.

with complete certainty during crises. Given this situation, Air Force leaders considered the time ripe to reassert their service's claim to military space leadership—a role also suggested by a number of respected civilian analysts. As Philip Gold, director of defense and aerospace studies at the Seattle-based Discovery Institute and a lecturer at Georgetown University, summarized the situation: "Effective space control demands a revolution on the ground, a revolution in thinking, in procedures, and in relationships." He wrote that the Air Force, with 90 percent of all U.S. military space assets, should be charged with structuring and managing the overall military space program to meet the operational requirements of the other services, the unified commands, and DoD in general. In addition, Gold recommended that the entire relationship among military, civil, and commercial space activities had to reflect that the era of massive technological military-to-civilian spinoffs had passed; technological advances in the commercial arena (namely, the shift to smaller, cheaper boosters and satellites) signaled the emergence of the civilian-to-military "spin-on" era. Despite such enormous obstacles as "governmental over-regulation, from excessive secrecy to surrealistic accounting," the "ponderous culture of weapons development," and concerns about security, Gold asserted that it was time to integrate the military and commercial space efforts.[24]

The Air Force did take the initiative on centralizing military space requirements, of which satellite communications was a part, by focusing on systems acquisition. Arguing that multiple acquisition agencies had led to expensive, less effective capabilities, the Air Force in mid-1994 proposed to the Office of the Secretary of Defense, as well as to the other services, that it be designated executive agent for all space acquisition. The resulting uproar left no doubt that interservice rivalry over space roles and missions continued to haunt the military space program. As Major General Robert S. Dickman, then director of space programs in the Office of the Assistant Secretary of the Air Force for Acquisition, commented, "I don't think anyone anticipated the depth of feelings—animosity may be a more descriptive term—that was evident in the service and joint staff objections."[25]

Although the Air Force initiative withered under fire, it helped crystallize efforts to provide new and effective organizational changes. By the summer of 1995, DoD had created a deputy under secretary of defense for space, established a Joint Space Management Board to coordinate activities between the Pentagon and the Central Intelligence Agency, and designated a DoD space architect. The last became responsible for ensuring compatibility and smooth operations among the different military and commercial systems. Although filled by an Air Force officer, Major General Dickman, the position of space architect remained within the DoD's joint structure.[26]

24.   Philip Gold, "Space Control Blasts Off," *Washington Times*, 20 September 1995, p. 21. For corroboration, see C. Michael Armstrong, "The Paradox of Space Policy," *Space News* 6 (18–24 September 1995): 20; William B. Scott, "Military Space Reengineers," *Aviation Week & Space Technology* 141 (15 August 1994): 20.

25.   Maj. Gen. Robert S. Dickman, "Near Term Issues for the Air Force in Space," prepared remarks presented at the symposium "The USAF in Space: 1945 to the Twenty-First Century," 21–22 September 1995, Washington, DC, p. 2, Air Force Space Command, History Office Archives.

26.   Gen. John H. Tilelli, Jr., Army Vice Chief of Staff, to Dr. John Deutch, Deputy Secretary of Defense, "Organization and Management of Space Activities," memorandum, 26 July 1994, with two attachments; Gen Merrill A. McPeak, USAF Chief of Staff, "Presentation to the Commission on Roles and Missions of the Armed Forces," 14 September 1994, pp. 185–199; Dr. John L. McLucas, "Space Policy," paper presented at the symposium "The USAF in Space: 1945 to the Twenty-First Century," 21–22 September 1995, Washington, DC, p. 18, copy Air Force Space Command, History Office Archives; Andrew Lawler, "U.S. Lawmakers Urge Sole Military Space Chief," *Defense News* 8 (17–23 May 1993): 10, 29; Steve Watkins, "Space Chiefs Assail McPeak Plan," *Air Force Times* 54 (18 April 1994): 3; Robert Holzer and Jason Glashow, "Control of U.S. Space Systems Spurs Service

Conclusion

It is still too soon to assess the impact of the latest changes in the military space organization. Many questions about the future of military satellite communications remain unanswered. Will Milstar provide effective strategic and tactical communications capability? To what extent, and in what way, should the Air Force rely on commercial augmentation? Is a national, combined civil and military network feasible? Or, following the commercial sector's example, should the Air Force consider a constellation of many small, cheap satellites distributed widely in low-Earth orbit to solve the problems of both its growing communications needs and the increasing overcrowding of the geosynchronous orbit? Moreover, how can any satellite communications system, whether military, civil, or commercial, be successful over time without responsive launch capability?

For more than a generation, the fragmented nature of the space community has led to higher system costs, inefficient resource utilization, and the inability to achieve a clear operational (as opposed to research and development) focus for military activities in space. The new Pentagon organization represents a major effort to overcome these glaring deficiencies and, simultaneously, to accommodate the rapidly changing requirements of the 1990s. For the Air Force, the challenge is to work effectively within this joint structure to preserve its capabilities and to provide the best possible space support to the warrior, no matter what the color of the uniform. If successful, the Air Force not only will preserve its leadership role in space, but it will contribute to ensuring a satellite communications architecture that meets the military challenges of the twenty-first century.

Duel," *Space News* 5 (8–14 August 1994): 1; Jason Glashow and Robert Holzer, "U.S. Services Stake Claims to Space Roles," *Space News* 5 (12–18 September 1994): 6; Gen. Merrill A. McPeak, USAF Chief of Staff, "Vying for Military Space Control," *Space News* 5 (26 September–2 October 1994): 15; Steve Weber, "Air Force Defends Plan to Control Pentagon Space Effort," *Space News* 5 (26 September–2 October 1994): 8; Theresa Hitchens, "USAF May Appoint Architect to Rebuild Space Program," *Defense News* 9 (31 October–6 November 1994): 14; Cheri Privor, "NRO Defers Role in Space Architect Office," *Space News* 6 (17–23 April 1995): 4.

Chapter 8

Thirty Years of Space Communications Research and Development at Lincoln Laboratory[1]

by William W. Ward and Franklin W. Floyd

Today, we take for granted the availability of global high-capacity communications circuits. Satellites and cables bring us information and entertainment from almost anywhere. It was not always this way. In the mid-1950s, for example, transatlantic communications relied on several teletype cables, a few dozen voice channels via cables equipped with vacuum-tube repeaters, high-frequency radio (roughly three to thirty megahertz), and the physical transport of messages by plane or ship. The rest of the world was not even that well equipped.

The high-frequency medium always has challenged communications engineers. Under favorable conditions, it provides global point-to-point communications by using relatively small, low-power transmitting and receiving equipment. However, natural phenomena often interfere with high-frequency links, and during war ("cold" or "hot"), they become targets for jamming. Nevertheless, high frequency was the only game in town in the 1950s. As a result, communications for the command and control of U.S. strategic forces worldwide were lacking.

The goal of the space communications program of Lincoln Laboratory at the Massachusetts Institute of Technology (MIT) was and remains the development of reliable, affordable systems of military communications. The program's initial objective simply was to make long-range military communications routinely available—first for large, fixed terminals and then for small, mobile ones. After that objective was reached, the emphasis shifted toward making communications systems capable of functioning, despite the most determined efforts by an adversary to interfere with them by jamming or by physical attack.

Project West Ford

The immediate impetus for Lincoln Laboratory's first work in space communications came from the HARDTACK series of high-altitude nuclear tests carried out in the Pacific Ocean near Johnston Island in August 1958. The first detonation destroyed the ionosphere over a vast area around the test site and interrupted many high-frequency radio communications links, because high-frequency radio signals travel by reflecting off the

1.   The unabridged version of this paper originally appeared as William W. Ward and Franklin W. Floyd, "Thirty Years of Research and Development in Space Communications at Lincoln Laboratory," *The Lincoln Laboratory Journal* 2(1) (1989): 5–34. The authors are indebted to many people inside and outside Lincoln Laboratory for their assistance in the preparation of this history of Lincoln Laboratory's space communications program and are proud to have had the opportunity to chronicle their accomplishments. The authors especially acknowledge the Lincoln Laboratory Library—and the Archives Department in particular.

lower surface of the ionosphere. In particular, the loss of high-frequency radio halted commercial transpacific air transport. The military implications of a high-frequency radio communications failure at a critical time were obvious.

Walter Morrow, at Lincoln Laboratory, and Harold Meyer, then with TRW, considered the problem of high-frequency radio communications failures during the Army's Project Barnstable Summer Study in 1958. They suggested that if natural phenomena, such as solar storms, or thermonuclear detonations disabled the ionosphere as a radio reflector, then an orbiting artificial reflector could replace the ionosphere. Just prior to this, both the Soviet Union and the United States had demonstrated the ability to place satellites in orbit, so the idea seemed feasible.

Morrow and Meyer proposed creating an artificial ionosphere consisting of a pair of belts (one circular polar and one circular equatorial) of resonant scatterers in orbit a few thousand kilometers above the Earth's surface. The scatterers in each belt would be conducting lengths of wire that would resonate at the system's operating wavelength and therefore reflect radio signals—the smaller the wires, the shorter the wavelength, and the easier their distribution from an orbiting dispenser. If the wires were too small, designing adequate transmitters and receivers would become excessively difficult.

Subsequently, Lincoln Laboratory proposed an experiment to demonstrate transcontinental communications by sending full-duplex (that is, simultaneously in both directions) transmissions between terminals in Camp Parks, California, and Westford, Massachusetts. The orbiting wires would act as half-wave dipoles and resonate at about eight gigahertz; communications would be transmitted at 7,750 and 8,350 megahertz. Each scatterer would be a 0.7-inch (about 1.8-centimeter) length of #53 AWG copper wire (0.0007 inch, or about 0.0018 centimeters, in diameter). The experiment required that about 480 million of these forty-microgram dipoles (nineteen kilograms of copper total) be distributed into circular polar orbits at an altitude of about 3,600 kilometers. The average separation between dipoles would be roughly 0.3 kilometer.

Recognizing that a proposal to place vast numbers of anything in orbit would be controversial, Lincoln Laboratory designed the proposed experiment, Project West Ford, to ensure that the experimental belt would not endure. The pressure of incident solar radiation on the orbiting dipoles would change their orbits, so that the perigee of each revolution would move steadily downward. Before long, the orbits would start to dip into the thin upper atmosphere, and atmospheric resistance would slow the dipoles enough that they would fall to the ground. Thus the belt would be removed from orbit within a few years after launch.[2]

Lincoln Laboratory unveiled Project West Ford in 1960 in virtually complete detail, even though the planned experiment was originally secret. It was particularly important to allay the concerns of optical and radio astronomers and other scientists who perceived the experimental belt as potentially harmful, causing interference with scientific observations and auguring worse experiments to come.[3] The experiment originally bore the name Project Needles, and renaming it Project West Ford did little to still the clamor from both sides of the Iron Curtain.[4] Ultimately, reason prevailed, and presidential approval was given for Project West Ford launches, although limited to the bare minimum.

2.   For background on the program, see "Project West Ford" in *Compendium of Communication and Broadcast Satellites, 1958 to 1980* (New York: Institute of Electric and Electronic Engineers (IEEE), 1981), pp. 299–302; Donald H. Martin, "West Ford," in *Communication Satellites 1959–1988* (El Segundo, CA: Aerospace Corporation, 1988), pp. 8–9.

3.   "Project West Ford Issue," *Proceedings of the IEEE* 52 (1964): 449–606.

4.   Petr Beckmann, *Eco-Hysterics and the Technophobes* (Boulder, CO: Golem Press, 1973), pp. 91–92.

On 21 October 1961, the first experiment was launched, piggybacked on another payload, into circular polar orbit from Vandenberg Air Force Base, but it was unsuccessful; the dipoles failed to deploy. Then, on 8 May 1963, a second launch with an improved dispensing arrangement achieved a substantial degree of success. The belt formed and closed over a period of about forty days. Its estimated density was five dipoles per cubic kilometer. The effectiveness of the scatterers proved greater in the early stages of belt formation, when the dipoles were less spread out, and permitted communication data rates of up to 20,000 bits per second. As the months and years passed, the belt became less effective for scatter communications—testimony that it was indeed cleaning itself out of orbit. By early 1966, the removal process was essentially complete.[5]

Project West Ford was an undeniable success, but it had little impact in terms of operational employment. Communication via passive satellites, such as the West Ford dipoles, required users to make large investments in complex terminals and provided only limited capabilities. The success of active communications satellites, beginning with AT&T's Telstar 1 in 1962, swept the field. Now and then, though, the vulnerability of conventional satellite communications to radio-frequency interference, whether intentional or not, is brought forcibly to everyone's attention. Furthermore, most satellites in orbit are fragile and thus vulnerable to physical attack. Therefore, Lincoln Laboratory has focused its work in active-satellite space communications on the development of robust systems that function reliably in the face of formidable levels of interference. However, the lesson of Project West Ford—that point-to-point scatter communications at limited data rates can be extremely survivable—should not be forgotten.

Space Communications at Superhigh Frequency

Following Project West Ford, Lincoln Laboratory embarked on a program to improve the design of active satellites. In most cases, the downlink signal (from a satellite to a surface terminal) is the "weak link" in satellite communications. An uplink can be strengthened by increasing the power of a surface transmitter. In contrast, a satellite downlink must be strengthened by accomplishing a more difficult task: the maximization of the effective isotopically radiated power per unit mass in orbit. To address this downlink problem, Lincoln Laboratory set out to develop high-efficiency spacecraft transmitters in the downlink frequency band. Improved antennas offered an additional benefit. If the spacecraft attitude-control system linked to a high-gain spacecraft antenna could position the antenna within the required beam-pointing range, both downlinks and uplinks would benefit. A series of Lincoln Experimental Satellites (LES) launched between 1965 and 1976 addressed these and other spacecraft-related technological questions.

High-efficiency systems of modulation and demodulation, together with recent advances in encoding and decoding signals for detection and correction of errors, promised significant advantages for communications terminals. Also needed were interference-resistant, multiple-access signaling techniques that permitted the simultaneous use of a satellite by tens or hundreds of users, some of them mobile, without involving elaborate systems for synchronization and centralized control. These and other terminal-related problems were addressed by a series of Lincoln Experimental Terminals (LET) that went hand in hand with the LES.

5.   Irwin I. Shapiro, "Last of the West Ford Dipoles," *Science* 154 (1966): 1445.

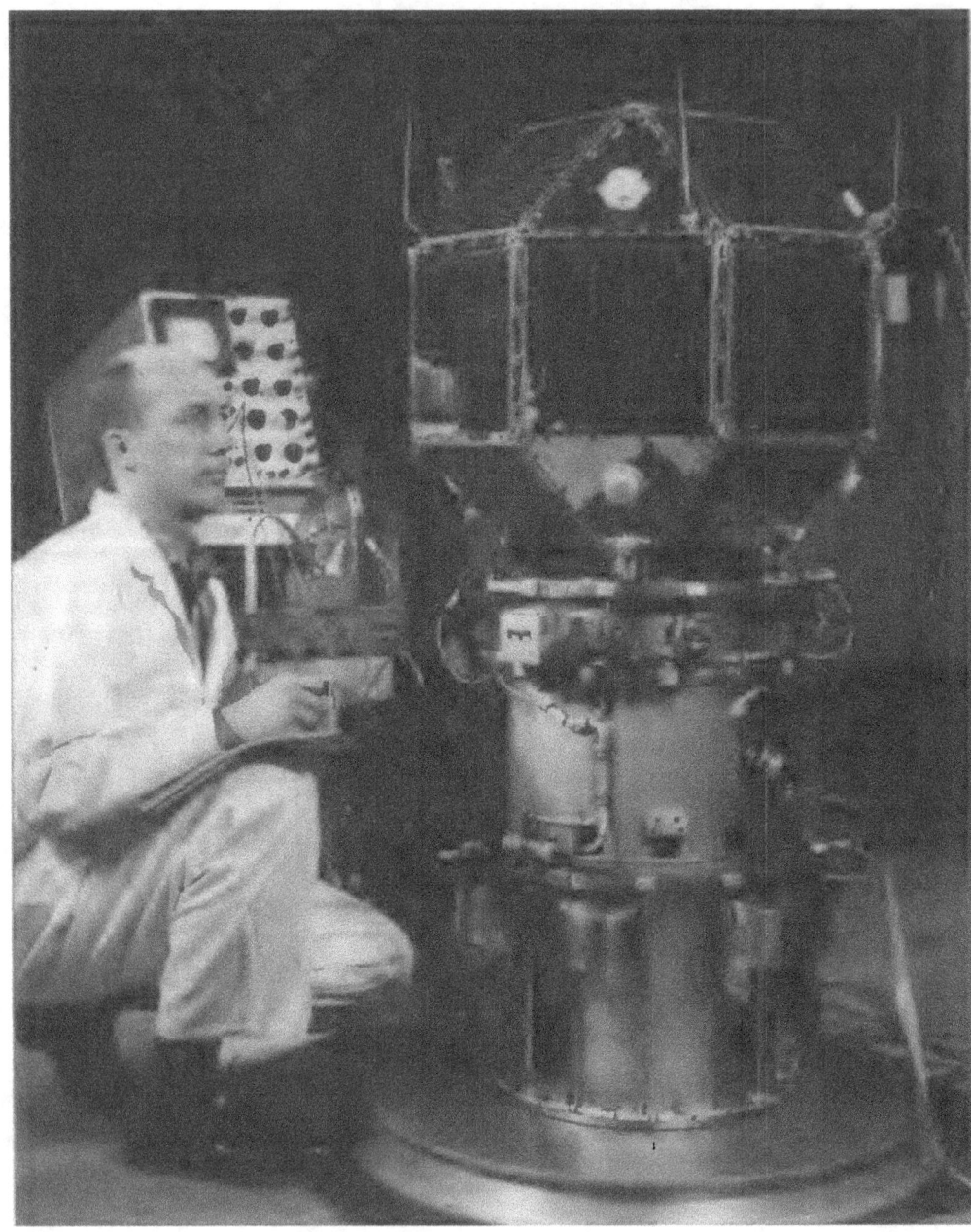

Figure 16

Lincoln Laboratory launched LES-1 in 1965. Compelling the selection of the satellite's frequency band was the availability of superhigh-frequency technology developed for Project West Ford (also known as Project Needles), an attempt to establish jam-proof military communications via a band of orbiting dipoles. (Courtesy of Lincoln Laboratory, photo no. CP202-76)

Lincoln Laboratory's space communications program after Project West Ford began in 1963 with a charter to build and demonstrate military space communications systems.[6] The initial program objective was to build, launch, and field a LES and a LET that would work together as a system and demonstrate practical military satellite communications. The availability of Project West Ford's advanced superhigh-frequency (SHF) technology (at seven to eight gigahertz) contributed to the decision to design LES-1 and LET-1 for that band. The Department of Defense's concurrent procurement of a series of SHF satellites and terminals, commencing with the Initial Defense Communications Satellite Program (IDCSP),[7] meant that lessons learned from LES-1 and LET-1 would find an additional application.

LES-1, launched from Cape Canaveral on 11 February 1965, accomplished only a few of its objectives. Apparently because of miswiring of the ordnance circuitry, the satellite never left circular orbit and ceased transmitting in 1967. LES-2, the twin of LES-1,[8] fared much better; it achieved its planned final orbit on 6 May 1965. The operation with LET-1[9] was successful and commenced the morning after launch.

The next step in Lincoln Laboratory's program in space communications was to park a satellite in geosynchronous orbit. LES-4 was built to fulfill that mission. The satellite was an outgrowth of LES-1 and -2, but it featured more solar cells and a larger array of Sun and Earth sensors.[10] LES-4 carried an instrument for measuring spatial and temporal variations of the energy spectrum of trapped electrons encountered in orbit. This instrument was added to provide information both for scientific interest and to aid the design of future spacecraft.

A Titan IIIC booster was to carry LES-4 and its companion, LES-3 (described in the next section), to a near-geosynchronous altitude and deposit them in circular, near-equatorial orbits. Unfortunately, the booster failed to finish its job, leaving these satellites stranded in their transfer ellipses. This disappointment, however, had its bright side. LES-4's repeated trips between perigee (195 kilometers) and apogee (33,700 kilometers) gave it many opportunities to measure the radiation environment over a wide range of altitudes.[11] Also, its communications system seemed to be working as well as it could under the handicap of being in the wrong orbit. Ultimately, as with the West Ford dipoles, the pressure of solar radiation caused the perigee of LES-4's orbit to descend into the upper atmosphere, and it burned up.

Lincoln Laboratory's accomplishments in SHF space communications opened up a part of the electromagnetic spectrum that remains heavily used today. In fact, succeeding

6. Walter E. Morrow, Jr., "The Lincoln Experimental Communications Satellite and Terminal Program," *AIAA 2nd Communications Satellite Systems Conference, San Francisco, 8–10 April 1968*, AIAA Paper 68-429 (New York: American Institute of Aeronautics and Astronautics, 1968); Herbert Sherman, Donald C. MacLellan, and Philip Waldron, *The Lincoln Satellite Technology Program through 1 January 1968: An Annotated Bibliography*, Lincoln Laboratory Technical Report 450 (Lexington, MA: Lincoln Laboratory, 12 June 1968), DTIC #AD-679559.

7. "IDCSP" in *Compendium of Communication and Broadcast Satellites*, pp. 167–70; Martin, "West Ford," pp. 75–76.

8. Herbert Sherman, Donald C. MacLellan, R.M. Lerner, and Philip Waldron, "Lincoln Experimental Satellite Program (LES-1, -2, -3, -4)," *Journal of Spacecraft and Rockets* 4 (1967): 1448–52; "LES-1 or -2 (Identical)" in *Compendium of Communication and Broadcast Satellites*, pp. 175–77.

9. John W. Craig, *et al.*, *The Lincoln Experimental Terminal*, Lincoln Laboratory Technical Report 431 (Lexington, MA: Lincoln Laboratory, 21 March 1967), DTIC #AD-661577.

10. Sherman, MacLellan, Lerner, and Waldron, "Lincoln Experimental Satellite Program (LES-1, -2, -3, -4)"; "LES-4," in *Compendium of Communication and Broadcast Satellites*, pp. 183–85; Martin, "West Ford," pp. 14–15; J. Bruce Rankin, Mark E. Devane, and Milton L. Rosenthal, "Multi-function Single-Package Antenna System for Spin-Stabilized Near-Synchronous Satellite," *IEEE Transactions on Antennas and Propagation* AP-17 (1969): 435–42.

11. Alan G. Stanley and Jean L. Ryan, *Charged-Particle Radiation Environment in Synchronous Orbit*, Lincoln Laboratory Technical Report 443 (Lexington, MA: Lincoln Laboratory, 15 May 1968), DTIC #AD-677284.

generations of SHF satellites now form the space segment of the Defense Satellite Communications System (DSCS).

Space Communications at Ultrahigh Frequency

Although LES-1, -2, and -4 showed the capabilities of SHF for reliable communications between fairly massive terminals, these technologies were not immediately available to small tactical units, such as vehicles, ships, aircraft, and specialized ground troops, all of which needed direct, dependable communications. Only a large command-post airplane or a sizable ship could be equipped with an SHF terminal capable of working with the DSCS satellites in orbit or those planned for the immediate future.

Moreover, because high levels of SHF radio power could not be generated in the satellites, the downlink continued to limit system performance. Each terminal's antenna aperture had to be large enough to capture the downlink signal, and the price paid for a large antenna aperture was a narrow antenna beam that had to be pointed directly toward the satellite. Small tactical units, especially those in motion, could not accommodate such an antenna system.

Communications links at much lower frequencies in the military UHF (ultrahigh-frequency) band (225 to 400 megahertz) solved the downlink problem. Solid-state circuits could generate substantial amounts of UHF power in a satellite.[12] A relatively uncomplicated low-gain terminal antenna could provide a broad beam, thereby simplifying the task of pointing an antenna in the direction of the satellite, as well as a sizable aperture. Such antennas were particularly appealing for aircraft.[13] UHF terminals also promised to be comparatively simple and inexpensive, and they could be produced in large numbers.

The feasibility of satellite communications at longer wavelengths was demonstrated at VHF by Hughes Aircraft Company on 8 May 1964. The company used teletype-rate signaling (60 words per minute) through the Syncom 2 satellite from one ground terminal to another nearby.[14] On 27 January 1965, teletype-rate satellite communication to and from an airplane in flight was demonstrated by using the Syncom 3 satellite, operating in the same mode as Syncom 2, and a ground terminal at Camp Parks, California.[15] NASA's ATS-1 satellite, launched in December 1966, also participated in experiments of this sort.

In 1965, the Department of Defense established Tri-Service Program 591 (Tactical Satellite Communications) to enable the Army, Navy, and Air Force to evaluate the potential usefulness of satellite communications in the military UHF band. Lincoln Laboratory was chosen to provide the satellites essential to the test program. LES-5 was to be built and launched as soon as possible; LES-6 would incorporate improvements on LES-5 and would be launched a year later. The three military services would procure test terminals that would work with LES-5 and -6 and would arrange for their installation in ships and aircraft.

12.   David M. Snider, "A Theoretical Analysis and Experimental Confirmation of the Optimally Loaded and Over-Driven RF Power Amplifier," *IEEE Transactions on Electron Devices* ED-14 (1967): 851–57; Alvise Braga-Illa and David M. Snider, "Transmitted-Power Maximization in Communication Satellites," *Journal of Spacecraft and Rockets* 6 (1969): 173–77.

13.   C.A. Lindberg, "A Shallow-Cavity UHF Crossed-Slot Antenna," *IEEE Transactions on Antennas and Propagation* AP-17 (1969): 558–63.

14.   L.A. Greenbaum and R.A. Boucher, "VHF Teletype Demonstrations," Hughes Aircraft Co., Ref. 2230.3/152 (23 June 1964), NASA-CR-57989 (contractor's report).

15.   Allen L. Johnson, "Two Hundred Years of Airborne Communications," *Aerospace Historian* 31 (1984): 185–93; *VHF Aircraft Satellite Relay—Final Report of Flight Test*, Report No. 481-1016-958 (Baltimore: Bendix Radio Division, April 1965).

Lincoln Laboratory carried out two programs to measure the characteristics of the UHF environment. In the first program, receiving equipment was installed in aircraft and flown over representative cities and varied terrains, and the ambient radio noise was measured.[16] In the second program, propagation phenomena between satellites and airborne terminals were examined. For this second program, LES-3 was built in haste, using technology from LES-1, -2, and -4, and was launched along with LES-4 on 21 December 1965.

LES-3[17] was essentially an orbiting signal generator, emitting a signal near 233 megahertz.[18] Given the degree of smoothness of the planet's surface relative to the one-meter wavelength of 300 megahertz (the middle of the military UHF band), much of its surface is mirrorlike. As a result, electromagnetic waves could be propagated between the satellite and the airborne terminal by more than one path. By knowing the likely parameters of the signal delays, the communications system designers were able to construct modulation and demodulation circuits that would not be confounded by multipath-propagation effects.

As mentioned, booster problems trapped LES-3 and LES-4 in transfer ellipses, instead of equatorial near-geosynchronous orbits. The actual orbit of LES-3 did permit the gathering of multipath-propagation data over a wide variety of terrains, and it gave the Lincoln Laboratory test team a reason to fly to exotic destinations to receive LES-3 signals reflected by representative types of terrain. As with LES-4, LES-3 ultimately reentered the atmosphere and disintegrated. LES-5, launched on 1 July 1967, and LES-6, launched on 26 September 1968, shared a strong family resemblance.[19] To sum up, satellite communications in the military UHF band worked well. The tri-service terminals aboard aircraft and ships and in the field communicated readily through the orbiting LES-5.[20]

LES-6 and the Hughes-built UHF/SHF TACSAT[21] (launched 9 February 1969) placed substantial communications resources in geostationary orbit, and the Department of Defense procured large quantities of UHF terminals. Because more than two satellites were needed for worldwide coverage, a series of satellites, including Gapfiller

16.   George Ploussios, "Electromagnetic Noise Environment in the 200 MHz to 400 MHz Band on Board Aircraft," *Proceedings of the IEEE* 54 (1966): 2017–19; George Ploussios, "City Noise and Its Effect upon Airborne Antenna Noise Temperatures at UHF," *IEEE Transactions on Aerospace and Electronic Systems* AES-4 (1968): 41–51; George Ploussios, *Noise Temperatures of Airborne Antennas at UHF*, Lincoln Laboratory Technical Note 1966-59 (Lexington, MA: Lincoln Laboratory, 6 December 1966), DTIC #AD-644829.

17.   Sherman, MacLellan, Lerner, and Waldron, "Lincoln Experimental Satellite Program (LES-1, -2, -3, -4)"; "LES 3" in *Compendium of Communication and Broadcast Satellites*, pp. 179–81.

18.   Kenneth L. Jordan, Jr., "Measurement of Multipath Effects in a Satellite/Aircraft UHF Link," *Proceedings of the IEEE* 55 (1967): 1117–18.

19.   William W. Ward and Burt E. Nichols, *Lincoln Experimental Satellite-5* (LES-5) *Transponder Performance in Orbit*, Lincoln Laboratory Technical Note 1968-18 (Lexington, MA: Lincoln Laboratory, 1 November 1968), DTIC #AD-686421; Donald C. MacLellan, Hugh A. MacDonald, Philip Waldron, and Herbert Sherman, "Lincoln Experimental Satellites 5 and 6," in Nathaniel E. Feldman and Charles M. Kelly, eds., *Progress in Astronautics and Aeronautics*, Vol. 26, *Communication Satellites for the Seventies: Systems* (Cambridge, MA: MIT Press, 1971), pp. 375–98; "LES-5" and "LES-6" in *Compendium of Communication and Broadcast Satellites*, pp. 187–93; William L. Black, Bradford Howland, and Edward A. Vrablik, "An Electromagnetic Attitude-Control System for a Synchronous Satellite," *Journal of Spacecraft and Rockets* 6 (1969): 795–98; Milton L. Rosenthal, Mark E. Devane, and Bernard F. LaPage, "VHF Antenna Systems for Spin-Stabilized Satellites," *IEEE Transactions on Antennas and Propagation* AP-17 (1969): 443–51.

20.   Irwin L. Lebow, Kenneth L. Jordan, Jr., and Paul R. Drouilhet, Jr., "Satellite Communications to Mobile Platforms," *Proceedings of the IEEE* 59 (1971): 139–59.

21.   "TACSAT" in *Compendium of Communication and Broadcast Satellites*, pp. 225–27; Martin, "West Ford," pp. 77–78.

(MARISAT),[22] FLTSAT (U.S. Air Force satellite communications),[23] and LEASAT,[24] were launched. A new series, UHF Follow-On (UFO) satellites, is now under development. Use of the UHF spectrum for military satellite communications is by no means limited to the Pentagon. The Soviet Union has announced a series of Volna ("Wave") satellites that incorporate UHF uplinks and downlinks,[25] and the United Kingdom has included UHF provisions in its series of Skynet 4 satellites, the first of which was launched on 10 December 1988.[26] Although it is very difficult to defend a communications satellite with a UHF uplink against a determined jamming attack, the relative simplicity and comparative cheapness of UHF military satellite communications terminals make this part of the spectrum highly attractive, and it is likely to remain in use for a long time.

Multiple-Beam Antennas

LES-1, -2, and -4 showed that the SHF band could provide reliable communications within certain limitations. The antenna systems on these satellites were small in terms of wavelength, and their beams were much larger than Earth coverage (which is about 18 degrees from synchronous altitude). The next level of sophistication in SHF space communications was a satellite antenna system with a mechanically pointable, less-than-Earth-coverage beam. This advancement was achieved through the governmental procurement of communications satellites, such as the second generation of the DSCS satellites (the TRW DSCS II series).[27] Lincoln Laboratory undertook the task of developing and demonstrating, in orbit, an antenna system that could allow satellite operators to aim the transmitting (downlink) power to receivers and simultaneously reduce the receiving (uplink) sensitivity in directions that might include sources of jamming or other interference.

Such an antenna system can be built in two ways. In the phased-array approach, many separate transmit and/or receive modules (each of which has a beam width much larger than Earth coverage) are controlled individually in amplitude and phase, so that the sum of their signals—a result of constructive and destructive interference—approximates the desired transmitting or receiving antenna pattern covering the Earth. In the multiple-beam-antenna approach, many separate antenna feeds form a dense set of narrow pencil beams covering the Earth. The signals from this collection of beams are adjusted in amplitude and phase ("weighted") and combined to approximate the desired antenna pattern. Each approach has its merits and shortcomings, and the appropriate choice depends on the application.

Lincoln Laboratory began a program to demonstrate, in orbit, a nineteen-beam (multiple-beam) antenna for reception at SHF. An Earth-coverage horn was to be used for the transmission. The thirty-inch-diameter (seventy-six-centimeter-diameter) aperture of the receiving antenna yielded a nominal three-degree resolution throughout the cone subtended by the Earth from geosynchronous-satellite altitude. The ground control terminal

22.   "MARISAT" in *Compendium of Communication and Broadcast Satellites*, pp. 51–53; Martin, "West Ford," p. 84.

23.   "FLTSATCOM" in *Compendium of Communication and Broadcast Satellites*, pp. 159–61; Martin, "West Ford," pp. 87–90.

24.   "LEASAT" in *Compendium of Communication and Broadcast Satellites*, pp. 171–73; Martin, "West Ford," pp. 93–95.

25.   "Volna 1 through 7 Series" in *Compendium of Communication and Broadcast Satellites*, pp. 87–89; Martin, "West Ford," pp. 104–05.

26.   "Skynet-4" in Martin, "West Ford," p. 95.

27.   "DSCS-II" in *Compendium of Communication and Broadcast Satellites*, pp. 151–53; Martin, "West Ford," pp. 80–82.

was to calculate the weights for the individual beams to approximate the desired antenna pattern and to transmit the weights to the satellite by telecommand.

The multiple-beam antenna, which was kept facing the Earth by the attitude control system, dominated the configuration of LES-7. Solar cell arrays followed the Sun to collect energy as LES-7 revolved in orbit. Work got under way to develop the satellite bus, which consisted of structure and housekeeping systems, power, propulsion, attitude control, thermal control, telemetry, and telecommand, in parallel with the development of the multiple-beam antenna and associated communications systems.[28]

By early 1970, it became apparent that LES-7 was ahead of its time. Because there was not enough Pentagon support for the mission, the funding required for the satellite's development, launch, and evaluation in orbit was not available. With considerable regret, Lincoln Laboratory put aside the LES-7 flight program, but it developed the critical multiple-beam-antenna technology "on the bench" and at an antenna test range.[29] In time, the multiple-beam-antenna concept was applied to the third generation of the DSCS satellites. Each General Electric-built DSCS III carries three multiple-beam antennas: two nineteen-beam SHF antennas for transmission and one sixty-one-beam SHF antenna for reception.

Space Communications at Extremely High Frequency

Lincoln Laboratory developed and built for the Department of Defense a pair of experimental communications satellites, LES-8 and -9, designed to operate in coplanar, inclined, circular, geosynchronous orbits and to communicate with each other via intersatellite links (cross-links), as well as with fixed and mobile terminals.[30] Uplinks, downlinks, and intersatellite links at extremely high frequency (EHF) augmented the military UHF band. The EHF band held out the promise of abundant band width to accommodate many simultaneous users and spread-spectrum systems of modulation and demodulation for anti-jamming communications links. For reasons of convenience, operating frequencies in the Ka-band (thirty-six to thirty-eight gigahertz) were selected.

Intersatellite links in the range of fifty-five to sixty-five gigahertz would have been desirable, because the absorption by oxygen molecules would attenuate signals passing through the atmosphere. However, it soon became apparent that the technology of 1971, on which LES-8 and -9 had to be based, would not support such an enterprise. Very little test equipment was commercially available for frequencies above forty gigahertz. Therefore, attention focused on the development of the needed components and subsystems at the Ka-band.[31]

One of the strengths of Lincoln Laboratory's space communications program is that it encompasses the development of both terminals and satellites under one roof. The transmission and reception for satellite lines providing substantial anti-jamming capability, such as those links through LES-8 and -9, are complex when compared to links that rely

---

28.   "LES 7" in *Compendium of Communication and Broadcast Satellites*, pp. 195–97; Martin, "West Ford," p. 17.

29.   Andre R. Dion and Leon J. Ricardi, "A Variable-Coverage Satellite Antenna System," *Proceedings of the IEEE* 59 (1971): 252–62.

30.   Frederick W. Sarles, Jr., Leonard W. Bowles, L.P. Farnsworth, and Philip Waldron, "The Lincoln Experimental Satellites LES-8 and LES-9," IEEE Electronics and Aerospace Systems Convention (EASCON), Arlington, Va., 1977, *Eascon-77 Record*, p. 21-1A, IEEE no. 77CH1255-9; "LES 8 and 9" in *Compendium of Communication and Broadcast Satellites*, pp. 199–206; Martin, "West Ford," pp. 31–33.

31.   F. John Solman, Carl D. Berglund, Richard W. Chick, and Brian J. Clifton, "The K Systems of the Lincoln Experimental Satellites LES-8 and LES-9," AIAA Paper 78-562, *A Collection of Technical Papers: AIAA 7th Communications Satellite Systems Conference, San Diego, California, April 24–27, 1978* (New York: AIAA, 1978), p. 208–15. This paper contains a bibliography of other papers on K-band technology for LES-8 and -9. It also appeared in *Journal of Spacecraft and Rockets* 16 (1979): 181–86.

on unprotected transponders, such as the links through the earlier LES. It would be very difficult if the space and terrestrial segments of a modern military satellite communications system were developed separately and their first operating encounter took place after launch. Lincoln Laboratory conducted extensive end-to-end testing of communications lines before launch, including the terminals that Lincoln Laboratory developed and those developed by the Air Force and the Navy. The generally smooth course of testing in orbit owed much to the prelaunch testing at Lincoln Laboratory.[32]

The LES-8 and -9 intersatellite links successfully addressed the key technical problems that confront the implementation of satellite-to-satellite communications.[33] LES-8 and -9 were launched together on 14 March 1978 aboard a Titan IIIC booster. Their launch was tailored to suit the companion Naval Research Laboratory Solrad IIa and Solrad IIb satellites and to facilitate the evaluation of specialized gyroscopes designed by the Charles Stark Draper Laboratory that LES-8 and -9 carried as flight experiments.[34] LES-8 and -9 are powered by radioisotope thermoelectric generators (RTGs) and have no solar cells or batteries. The design of the RTGs had to assure their physical integrity in the event of a launch failure, so that the potential environmental hazard would be acceptable. During compatibility tests and preparations for launch, Lincoln Laboratory had to develop special procedures and pay scrupulous attention to health factors to ensure that workers would not be overexposed to particle radiation.[35] The RTGs were well worth the effort. The compatibility of these rugged power sources with complex signal-processing circuitry has been well established. These RTGs are similar to the ones that power the Voyager spacecraft, which have been exploring the outer planets and beyond since 1977.

LES-8 and -9 represent the high-water mark to date of Lincoln Laboratory's program in space communications. In addition to their complex communications system, these satellites have systems and subsystems for housekeeping functions, including attitude control,[36] onboard (secondary) propulsion,[37] telemetry,[38] and telecommand. Lincoln

---

32.   L.J. Collins, L.R. Jones, David R. McElroy, David A. Siegel, William W. Ward, and David K. Willim, "LES-8/9 Communications-System Test Results," AIAA Paper 78-599, *A Collection of Technical Papers: AIAA 7th Communications Satellite Systems Conference, San Diego, California, April 24–27, 1978* (New York: AIAA, 1978), p. 471–78.

33.   David M. Snider and David B. Coomber, "Satellite-to-Satellite Data Transfer and Control," AIAA Paper 78-596, *A Collection of Technical Papers: AIAA 7th Communications Satellite Systems Conference, San Diego, California, April 24–27, 1978* (New York: AIAA, 1978), p. 457; William W. Ward, David M. Snider, and Ronald F. Bauer, "A Review of Seven Years of Orbital Service by the LES-8/9 EHF Intersatellite Links," *IEEE International Conference on Communications: Integrating Communication for World Progress, Boston, MA, 19–22 June 1983* (Piscataway, NJ: IEEE, 1983), pp. E1.1.1–10.

34.   Lawrence J. Freier, R.N. Masters, J.A. DeFrancesco, R.A. Harris, G.J. Blakemore, and R.A. Jeffery, *Third-Generation Gyro System Final Report*, Report R-843 I, Charles Stark Draper Laboratory, Inc., February 1975.

35.   C.E. Kelly and C.W. Whitmore, "Performance of the Multi-Hundred-Watt Radioisotope Thermoelectric Generator (MHW/RTG) Power Source for the LES-8/9 Satellites," AIAA Paper 78-534, *A Collection of Technical Papers: AIAA 7th Communications Satellite Systems Conference, San Diego, California, April 24-27, 1978* (New York: AIAA, 1978), p. 728–35.

36.   Franklin W. Floyd, Carl H. Much, Norval P. Smith, Joseph R. Vernau, and J.P. Woods, "Flight Performance of the LES-8/9 Three-Axis Attitude-Control System," AIAA Paper 80-0501, *A Collection of Technical Papers: AIAA 8th Communications Satellite Systems Conference, Orlando, FL, 20–24 April 1980* (New York: AIAA, 1980), p. 159.

37.   Carl K. Much and Franklin W. Floyd, Jr., "The Cold-Ammonia Propulsion System for the Lincoln Experimental Satellites, LES-8 and -9," *Proceedings of the JANNAF Propulsion Meeting, Atlanta, GA, 7–9 December 1976*, Vol. 3 (Laurel, MD: Chemical Propulsion Information Agency and Johns Hopkins University, Applied Physics Laboratory, 1976), p. 491.

38.   Frederick W. Sarles, Jr., John H. Helfrich, Paul F. McKenzie, and James K. Roberge, "The LES-8/9 Telemetry System: Part I, Flight-System Design and Performance," *Record of the International Telemetry Conference 78*, Los Angeles, 14–16 November 1978, p. 645; John H. Helfrich, Atle M. Gjelsvik, Charles M. Rader, Daniel C. Rogers, and C.E. Small, "The LES-8/9 Telemetry System: Part II, Ground-Terminal Design and Performance," *Ibid.*, p. 657.

Laboratory continues to be responsible for the upkeep of LES-8 and -9. The Lincoln Experimental Satellite Operations Center operates and maintains these satellites, and it will continue to serve them as long as they remain useful.[39] The technologies of onboard signal processing and of EHF transmission and reception, successfully demonstrated in the LES-8 and -9 Joint Test Program, have been incorporated in subsequent military satellite communications procurements. The single-channel transponders on the DSCS III satellites and the Milstar communications system itself (see below) have flowed directly from LES-8 and -9.

Switchboards in the Sky

Following the launch of LES-8 and -9 in 1976, Lincoln Laboratory addressed the problem of providing affordable anti-jamming communications to many small, mobile users. LES-8 and -9 and their associated terminals demonstrated that UHF and EHF communications systems could be cross-banded in signal-processing satellites and serve the needs of a limited number of small, mobile terminals. It was tempting to try to extend the approach to meet the needs of a large number of users, because the relative simplicity and cheapness of UHF terminals made that part of the spectrum attractive. Unfortunately, the military UHF band (225 to 400 megahertz) does not have enough available band width to provide the required levels of anti-jamming protection, and the application of adaptive nulling, an anti-jamming technique, was not enough to overcome this handicap.[40] Therefore, under Lincoln Laboratory's new approach, all anti-jamming space communications links were assigned to the EHF band.

The arguments for and against EHF are well known.[41] The use of the EHF bands can overcome the frequency congestion that affects both civilian and military systems at lower frequencies. However, the major advantage to military users is that EHF also supplies the band widths necessary to implement robust jam-proof systems based on spread-spectrum technologies. By using advanced spread-spectrum and other techniques, including extensive onboard signal processing, a satellite of modest size can serve large numbers of small mobile users simultaneously with highly jam-resistant communications channels. The probability that covert transmissions from terminals wishing to remain unnoticed will be intercepted is also reduced at EHF. On the negative side, the effects of rain attenuation on EHF links require that the minimum elevation angle of the satellite relative to the terminal must be significantly higher than for lower frequency systems.

In consultation with its sponsors, Lincoln Laboratory conceived a "strawman" EHF system and built a test-bed satellite with terminal hardware that served as the focus for the in-house technology development program. The groundwork for this EHF system concept

39.   William W. Ward, *Developing, Testing, and Operating Lincoln Experimental Satellites 8 and 9 (LES-8/9)*, Lincoln Laboratory Technical Note 1979-3 (Lexington, MA: Lincoln Laboratory, 16 January 1979), DTIC #AD-AO69095/8.

40.   Joseph T. Mayhan, "Adaptive Antenna Design Considerations for Satellite Communication Antennas," *IEE Proceedings: Microwaves, Optics and Antennas* 130, Part H (1983): 98–108; Joseph T. Mayhan and Franklin W. Floyd, *Factors Affecting the Performance of Adaptive Antenna Systems and Some Evaluation Techniques*, Lincoln Laboratory Technical Note 1979-14 (Lexington, MA: Lincoln Laboratory, 9 August 1979), DTIC #AD-AO81711/4.

41.   William C. Cummings, Pravin C. Jain, and Leon J. Ricardi, "Fundamental Performance Characteristics That Influence EHF MILSATCOM Systems," *IEEE Transactions on Communications* COM-27 (1979): 1423–35; C.J. Waylan and G.M. Yowell, "Considerations for Future Navy Satellite Communications," IEEE Electronics and Aerospace Systems Convention (EASCON), Arlington, VA, 1979, *EASCON-79 Record*, p. 623–27, IEEE No. 79CH1476-1.

was laid during the successful development and demonstration of the Ka-band components, subsystems, and systems in LES-8 and -9. The essential features of the strawman EHF system were demonstrated on the bench at Lincoln Laboratory in 1980 and 1981 in the combined operation of the test-bed satellite with its test-bed terminal.

Subsequently, the EHF system concept and the associated technologies in development at Lincoln Laboratory served as a point of departure for thinking about EHF systems within the Department of Defense military satellite communications community. In December 1981, the Pentagon decided to go ahead with a new enterprise, Milstar (the Military Strategic Tactical and Relay satellite), which incorporated many technical features of Lincoln Laboratory's strawman EHF system in its own system and in its Common Transmission Format.

Lincoln Laboratory was asked to support Milstar development by building two Air Force communications satellite (FLTSAT) EHF Packages (FEPs). The communications capabilities of an FEP, when installed on TRW's Air Force UHF/SHF communications satellite, are a subset of those of a full Milstar satellite payload.[42] The first FEP was integrated with FLTSAT-7, which was launched on 4 December 1986; the second was launched in 1989 as part of FLTSAT-8. The electronics and antenna assemblies of each FEP were built by Lincoln Laboratory under very tight power (305 watts) and mass (111 kilograms) constraints, so that they would be compatible with the existing Air Force satellite design. The FEP also has facilitated the early operational test and evaluation of the Milstar EHF/SHF terminals being developed by the Army, Navy, and Air Force. The complexity of an FEP communications system is far greater than that of the LES-8 or -9 satellites, even though the FEP has fewer parts. Integrated circuits in the early 1980s, when the FEP design choices were made, were more sophisticated than when LES-8 and -9 were designed in the early 1970s.

The major innovation in the FEP was a computer-based resource controller that establishes data channels operating at different data rates, via different antenna beams and other means, to support the communications needs of individual users. The onboard access controller receives requests for user service from each user terminal's computer. The controller in turn sets up the requested services and informs the user terminals' computers of its actions via a downlink order wire. Once a channel has been set up, the FEP converts uplink message formats to downlink message formats and retransmits user data via either or both of the FEP's two antenna beams. Although the computer-to-computer dialogs between the FEP and the user terminals are complex, the required interactions between human and machine are user friendly and can be performed easily by terminal operators.[43]

The experience gained in operating the Lincoln Experimental Satellite Operations Center for the control of LES-8 and -9 in orbit was directly applicable to the task of controlling an FEP in orbit. The greater sophistication of the FEP (compared to the LES-8 or -9) has resulted in a much lower workload in the FEP Operations Center than in the Lincoln Experimental Satellite Operations Center, where any change in the configuration of the satellite's communications system requires human intervention. The resource controller in the orbiting FEP carries out most of its computer-to-computer transactions with users and would-be users without supervisory intervention. Two FEP Operations Centers

42.   David R. McElroy, "The FEP Communications System," AIAA Paper 88-0824, *A Collection of Technical Papers: AIAA 12th International Communications Satellite Systems Conference, Arlington, VA, 13–17 March 1988* (New York: AIAA, 1988), p. 395.

43.   Marilyn D. Semprucci, "The First Switchboard in the Sky: An Autonomous Satellite-Based Access/Resource Controller," *Lincoln Laboratory Journal* 1 (1988): 5–18.

have been built. One is installed permanently at Lincoln Laboratory; the other, transportable but by no means mobile, has been installed at a Navy facility in Maine. (The Navy is the operational manager of the FEP Communications System.)

During the FEP program, Lincoln Laboratory concentrated on the challenging technologies required for the FEP, taking advantage of the satellite-bus technologies already developed and proven in space by TRW's series of Air Force satellites. It has been gratifying that the FEP aboard FLTSAT-7 arrived safely in orbit and has worked well since launch. While the success of the FEP program speaks well for Lincoln Laboratory's approach to implementation and quality assurance in building reliable spacecraft, some of the success of Lincoln Laboratory's program in space communications has to be attributed to plain luck. Consider the case of the FEP carried by FLTSAT-7.

Although FLTSAT-6 carried no FEP, FLTSAT-7 and -8 each carried one. The sponsor decided to juggle the launch schedule and interchange FLTSAT-7 and -6 to get an EHF package into orbit as early as possible. As a result, FLTSAT-7 was launched on 4 December 1986, and FLTSAT-6 followed on 26 March 1987 during a storm. Lightning struck the Atlas/Centaur booster. After the rocket engine noise ceased, and a series of muffled explosions occurred aloft, one of the authors of this chapter, who was at Cape Canaveral at the time, said to himself: "There but for the grace of God and the United States Navy went the first FEP."

Advanced EHF/SHF Terminals

In another Milstar-related activity, Lincoln Laboratory designed and built the Single-Channel Objective Tactical Terminal (SCOTT), which was the advanced model of the Army's Milstar EHF/SHF terminal. In 1983, Army personnel successfully tested this terminal, mounted in a tracked military vehicle, against a satellite simulator in the field. The Army's production version of SCOTT has many of the features that were first demonstrated in Lincoln Laboratory's development model.

As an outgrowth of the SCOTT work, Lincoln Laboratory conducted a feasibility study in 1983 that resulted in a conceptual design for a human-portable, Milstar-compatible EHF/SHF terminal. The development of the Single-Channel Advanced Milstar Portable (SCAMP) terminal was completed shortly after the launch of the first FEP, and it has operated successfully with the FEP. There are numerous diverse needs for limited-capability terminals of this class, which offer most of the advantages of Milstar communications without the full range of options.

Optical Space Communications

The success of optical communications for some terrestrial applications is undeniable. Laser and fiber optic technologies have brought about cables that seriously compete with communications satellites. However, optical space communications has been the "wave of the future" for many years. The advent of the laser, with its promise of coherent radiation across the transmitting apertures and correspondingly fine, high-gain antenna beams, has led to very encouraging link-performance calculations. To the best of the authors' knowledge, however, no one has yet demonstrated a nontrivial optical intersatellite link in space.

Lincoln Laboratory once considered putting optical intersatellite links on LES-8 and -9, in addition to the millimeter-wave links. The optical feature was dropped from the satellite's configuration in late 1971 when it became clear that the current state of the art

Figure 17

The Single-Channel Advanced Milstar Portable terminal (SCAMP), shown above with Lincoln Laboratory engineer David M. Snider, illustrates the degree of reduction achieved in the size of satellite communications equipment over the decades. Compare SCAMP, for example, with the Army's Mark IV(X) mobile unit shown in Chapter 7, Figure 14. (Courtesy of Lincoln Laboratory, photo no. 5978-87-5C)

in solid-state laser-diode technology was inadequate for a flight experiment and that the project's resources could not support an optical link. Progress in available components, coupled with new insights in system design, has since made it attractive to resume work in this area, often called "LASERCOM."

Lincoln Laboratory is now developing a technology base for high-data-rate intersatellite links that could be realized with small-aperture, lightweight, low-power optoelectronic packages. The approach taken uses solid-state laser diodes and silicon-diode detectors operating in a heterodyne mode.[44] Modulated continuous-wave transmission and heterodyne detection will be combined in the system design to provide communications significantly superior to the more commonly used systems based on pulsed transmission and direct energy detection (commonly known as the photon-bucket approach).

Although Lincoln Laboratory began to prepare in 1985 for a demonstration of heterodyne LASERCOM technology in orbit, program constraints have ruled it out for the present. Lincoln Laboratory is now building an engineering model of a complete heterodyne LASERCOM system, which will address all critical technological areas and issues. When the world is ready for LASERCOM, the technology will be available.

---

44.   Vincent W.S. Chan, "Space Coherent Optical Communication Systems: An Introduction," *Journal of Lightwave Technology* LT-5 (1987): 633–37; Vincent W.S. Chan, "Intersatellite Optical Heterodyne Communications Systems," *Lincoln Laboratory Journal* 1 (1988): 169–86.

Future Developments in Space Communications

The next goal for reliable military satellite communications systems is the extension to high-data-rate applications of the robust, jam-resistant technologies for low-data-rate applications that FEP has demonstrated. Considering the large band widths that will be required, these new systems will most likely be implemented at EHF, at least for the uplinks and downlinks. The effects of bad weather, even clouds, on optical links between satellites and ground terminals seem certain to rule out LASERCOM for applications in which consistent link availability is important. However, optical links between satellites and airborne platforms flying above the weather may meet specific military needs. The technology of radio intersatellite links has been amply demonstrated in orbit by LES-8 and -9 and by NASA's Tracking and Data Relay Satellite System.[45] It is only a matter of time— and of continued support—until LASERCOM intersatellite links are similarly demonstrated.

Intersatellite link technologies have not yet found civilian application. The Intelsat 6 series of communications satellites, the first of which was launched in late 1991, was designed well after the LES-8 and -9 intersatellite links had been demonstrated in orbit. However, it was not found economically justifiable to include intersatellite links in the Intelsat 6 satellites, nor have such links appeared in successive generations of these birds. Nevertheless, the time for civilian intersatellite links will come.

Conclusion

In the more than 30 years since the launch of Sputnik, the field of space communications has reached a high level of maturity. The mission failures that occasionally besmirch the record of each spacefaring nation cannot obscure the numerous remarkable—and useful—achievements that have taken place. Notable among them are the contributions of space communications, both economically and in terms of increased international stability. The field of space communications allows national leaders to stay in touch with one another, and it gives them more control over their military resources, thus reducing the possibility of accidental war. The field has changed the ways in which societies function and interact, and it promises to do far more. To quote the science fiction writer Arthur C. Clarke (who first suggested the geostationary communications satellite): "What we are building now is the nervous system of mankind, which will link together the whole human race, for better or worse, in a unity which no earlier age could have imagined."[46]

---

45. "TDRSS/Advanced WESTAR" in *Compendium of Communication and Broadcast Satellites*, pp. 139–47; Martin, "West Ford," pp. 152–56.

46. Arthur C. Clarke, "The Social Consequences of Communications Satellites," paper presented at the 12th International Astronautical Congress, Washington, DC, 1961, and published in Arthur C. Clarke, *Voices from the Sky: Previews of the Coming Space Age* (New York: Harper & Row, 1965), p. 139.

Chapter 9

Billion Dollar Technology: A Short Historical Overview of the Origins of Communications Satellite Technology, 1945–1965

by David J. Whalen

In the four decades since the launch of Sputnik, communications satellites remain the only truly commercial space technology. Less than eight years after Sputnik, a revenue-producing satellite, Early Bird, emerged. Thus far in the 1990s, the average number of communications satellites launched annually has been twenty, at an average cost of more than $50 million for the satellite and another $50 million (or more) for the launch vehicle. The average spent annually on communications satellites is in excess of $2 billion.

In the face of this overwhelming commercial success, most analysts of the early development of communications satellites have concentrated on politics and policy studies, almost excluding the technological and economic origins of this industry. These analysts tend to discuss the passage of the Communications Satellite Act of 1962 and the ups and downs of the NASA communications satellite program, with little reference to technology or economics. As a consequence of this emphasis on political studies, conventional wisdom asserts that the government developed communications satellite technology because industry was unwilling or unable to face the high costs and high risks associated with communications satellite research and development. This conventional wisdom assumes that industry would not develop new communications satellite technology because it was too risky, there was a lack of funds, and industry would not be able to appropriate the benefits.[1]

While some communications satellite technology flows from one manufacturer to another, much is protected by patents, and even more is protected by the difficulty of learning new technology. Technology transfer, even when facilitated by cooperation, is often difficult. Many early geosynchronous satellites used techniques pioneered by Hughes on Syncom in 1963. The Hughes-Williams patent was the subject of litigation for years, but it proved to be quite valuable to Hughes. Eventually, most other manufacturers and the U.S. government had to pay royalties to Hughes. Perhaps more important, Hughes has dominated the manufacture of communications satellites since the first Syncom in 1963. The risk of competitors appropriating technology is greatly overstated.

Capital has rarely been the most important issue when investors looked at communications satellites; access to the market has been far more important. Telecommunications has been a multibillion-dollar business for most of this century. AT&T, in particular, certainly had the financial resources to invest in communications satellites.

---

1. See, for example, Marcia S. Smith, "Civilian Space Applications: The Privatization Battleground," in Radford Byerly, Jr., ed., *Space Policy Reconsidered* (Boulder, CO: Westview, 1989), pp. 105–16; Linda R. Cohen and Roger G. Noll, "The Applications Technology Satellite Program," in Linda R. Cohen and Roger G. Noll, eds., *The Technology Pork Barrel* (Washington, DC: Brookings Institution, 1991), pp. 149–77.

In 1955, John R. Pierce of AT&T compared the estimated cost of satellite communications with the cost of the first transatlantic telephone cable (TAT-1) then being laid between the United States and Europe. The cable would provide approximately thirty-six voice circuits at a cost of around $35 million. Pierce then asked his readers: "Would a channel 30 times as wide [as TAT-1], which would accommodate 1080 [30x36] phone conversations or one television signal, be worth 30x35 million dollars: that is, a billion dollars?"[2] AT&T was in a position to fund a billion-dollar project. In 1959, with net sales of over $7 billion, AT&T was in a better position to fund communications satellite research and development than NASA, whose entire budget was only a few hundred million dollars.

In late 1964, after the successful launch of Syncom 3 and prior to the launch of Early Bird, NASA Administrator James E. Webb asked his staff: "How did we get so much communications satellite technology for so little money?"[3] His question has never been adequately answered, because most analysts have concentrated on political, rather than technological, issues. This chapter addresses the history of the development of that technology and, in the process, attempts to answer Webb's question.

Postwar Technology and Resources

The roots of nearly all of the technologies associated with communications satellites extend back to World War II. Although billions were spent on the Manhattan Project, radar research and development, and the development of the B-29, the program that most stimulated a longing to exploit space was the relatively inexpensive German V2 rocket. The V2 made it clear that rockets were practical and, with a little more development, could place a satellite in orbit.

Although Hermann Oberth in 1923 and George O. Smith in 1942 had speculated about geosynchronous or, in Smith's case, Earth-Venus relay communications satellites, the acknowledged "father"—or, in his own words, "godfather"—of communications satellites was Arthur C. Clarke, a Royal Air Force officer at the time who hoped to revitalize the British Interplanetary Society after the war. Clarke was the first, in his October 1945 *Wireless World* article, to develop rather fully the concept of a geosynchronous satellite. Three of these satellites, each fixed over a specific longitude on the equator, would provide complete global coverage. Clarke envisioned their use for television programming, and he assumed that the satellite would be occupied by humans to change the vacuum tubes on a regular basis.[4]

The *Wireless World* article did not receive wide publicity. Clarke's real contribution to satellite communications was to continue pressing for a geosynchronous system in other publications. His book *The Exploration of Space*[5] included a system of three geosynchronous satellites, and many readers viewed it as a blueprint for the entire space program.[6] At the same time as the publication of Clarke's *Wireless World* article, moreover, U.S. Navy researchers began developing launch vehicles and satellites,[7] and, failing to find higher

2.   John R. Pierce, "Orbital Radio Relays," *Jet Propulsion* 25 (April 1955): 153–57.

3.   Paraphrase of Webb's comments at the 22 September 1964 Program Review, referred to in W.A. Radius to ADA/Shapley, 10 December 1965, Thompson papers, NASA History Office, Washington, DC.

4.   Arthur C. Clarke, "Extra-Terrestrial Relays: Can Rocket Stations Give World-Wide Radio Coverage?," *Wireless World* 51 (October 1945): 305–08.

5.   Arthur C. Clarke, *The Exploration of Space* (New York: Harper & Row, 1952).

6.   For example, former NASA Associate Administrator Homer E. Newell credited Clarke as the champion of communications satellite applications. Homer E. Newell, *Beyond the Atmosphere: Early Years of Space Science* (Washington, DC: NASA SP-4211, 1980), p. 106.

7.   Harvey Hall, *Early History and Background on Earth Satellites*, report no. 405:HH:dr (Washington, DC: Office of Naval Research, 29 November 1957).

level support, they approached the Army Air Forces with a proposal for a joint program. Ultimately, the Army Air Forces was as unwilling as the Navy to fund the project.

Despite this failure, one positive outcome was an independent study of the feasibility of Earth satellites by Project RAND (then a part of the Douglas Aircraft Corporation and not yet an independent, nonprofit research and development organization) on behalf of the Army Air Forces.[8] Completed on 12 May 1946, the RAND report observed, among other points, that a satellite launch would have a dramatic effect on world opinion and that the satellite might have notable use as a communications relay.[9] In spite of a curtailment of military research and development expenditures under the Truman administration, subsequent RAND studies of geosynchronous communications satellites and the political and psychological aspects of launch vehicles appeared in 1949 and 1950, respectively.[10] These RAND studies emphasized that whichever nation first launched a satellite would score a tremendous psychological victory and that the satellite could be used for reconnaissance. In the words of historian Walter A. McDougall, the 1950 RAND report "more than any other, deserves to be considered the birth certificate of American Space Policy."[11] Over the following years, RAND, the Air Force, North American Aviation, and other defense companies worked on reconnaissance, rather than on other satellite applications. Instead, communications satellites were to be born from civilian commercial telecommunications developments.

Following World War II, telephone use underwent a dramatic increase. Communications in general grew at more than three times the gross national product (GNP) rate of growth. A large component of that growth was in both domestic and international long-distance traffic. However, few improvements had been made since the laying of the last transatlantic telegraph cable in 1928. Radiotelegraphy and radiotelephony, while rapidly overtaking underwater cable telegraphy, suffered from noise that was inherent in the technology. The first transatlantic telephone cable (TAT-1) was not laid until 1956. TAT-1 was mainly an AT&T venture in collaboration with such foreign partners as the British Post Office and Canada's Overseas Telecommunications Corporation. Indicative of the cable's success was the immediate inclusion of radio companies (RCA, ITT, and Western Union) in the TAT-1 arrangements. They recognized the superiority of the new technology and wanted to take part in its exploitation. Next came TAT-2 in 1959, owned by AT&T, the French Postal and Telecommunications Administration (Direction des Télécommunications Sous-Marines), and the Deutsche Bundespost, and TAT-3, owned by AT&T and the British Post Office, in 1963.[12] Behind the laying of these transatlantic cables was a dramatic growth in international telephone traffic throughout the postwar period, generally at the rate of 20 percent per year (Figure 18).

---

8. Douglas Aircraft Corporation formed Project RAND in late 1945 to advise the Army Air Forces. It became an independent organization in 1948.

9. Project RAND, *Preliminary Design of an Experimental World-Circling Spaceship*, Report No. SM-11827 (Santa Monica, CA: Project RAND, May 1946).

10. Project RAND, *Satellite to Surface Communication—Equatorial Orbit*, RAND RM-603 (Santa Monica, CA: Project RAND, July 1949); Paul Kecskemeti, *The Satellite Rocket Vehicle: Political and Psychological Problems*, RAND RM-567, (Santa Monica, CA: Project RAND, 4 October 1950).

11. Walter A. McDougall, . . . *The Heavens and the Earth: A Political History of the Space Age* (New York: Basic Books, 1985), p. 108.

12. U.S. Department of Commerce, *1984 World's Submarine Telephone Cable Systems* (Washington, DC: U.S. Government Printing Office, 1984), pp. 90–91, 101, 127. TAT-4 was laid in 1965 amid some discussion of its cost-effectiveness relative to satellites. TAT-5, laid in 1970, provided a dramatic increase in capability to 720 voice circuits for a relatively modest $79 million.

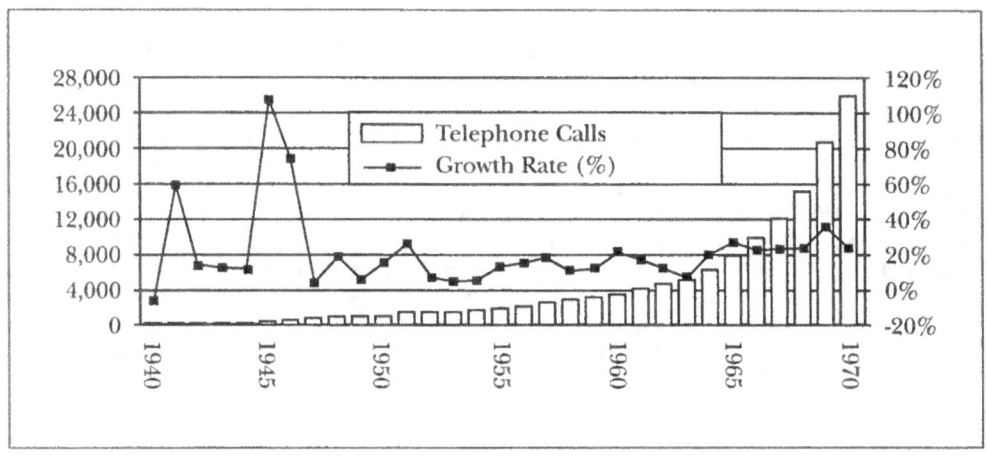

Figure 18
International telephone calls (1940–1970). (Source: *Historical Statistics of the United States*)

Although RCA developed much of the radio and television technology, AT&T, specifically its Bell Telephone Laboratories, made many breakthroughs in basic electronic devices, the most famous of which was the transistor. AT&T also was deeply involved in developing the maser and the traveling-wave tube. The maser is a low-noise amplifier, an important component of ground stations receiving weak satellite signals; the traveling-wave tube is a high-gain linear amplifier used in satellite transmitters. While masers originated at Columbia University, Bell Telephone Laboratories worked hard at adapting the technology to underwater telephone cables. Bell also developed the traveling-wave tube for use in microwave relay stations. Furthermore, AT&T performed extensive research and development work for the U.S. government, including nuclear weapons research (Sandia), missile studies (Safeguard), and Apollo program support (BellComm). While other companies also had research and development laboratories, AT&T stood at the top in terms of the caliber of the research and development performed, as well as the company's willingness to invest vast sums.

Similar to AT&T, most of the firms that later became involved in satellite communications had been around for some time. General Electric was the oldest, having been formed by the merger of Thompson-Houston and Edison General Electric in 1892. AT&T was formed in 1895. Both ITT, created to manage a variety of telephone and telegraph companies initially in Latin America, and RCA, organized to pool radio patents, were formed in 1920. GTE was a telephone competitor of AT&T. Hughes, which started in the 1930s as a "hobby-shop" for Howard Hughes, grew dramatically in the postwar period. All of these companies were major electronics manufacturers, and all, except Hughes and General Electric, had significant stakes as communications system operators.

These firms generally were quite large. In 1959, all of these companies combined spent more than NASA. AT&T and General Electric alone accounted for more than 2 percent of the GNP. The following are the 1959 U.S. GNP, federal outlays, the budgets of NASA and the Department of Defense (DoD), and the net sales of the eight largest American companies (industry figures are net sales):

U.S. GNP	$483.7 billion
Federal outlays	$92.1 billion
DoD	$44.6 billion
NASA	$0.3 billion
AT&T	$7.4 billion
General Electric	$4.3 billion
RCA	$1.4 billion
Lockheed	$1.3 billion
ITT	$0.7 billion
Bendix	$0.7 billion
GTE	$0.4 billion
Western Union International	$0.3 billion

(Source: *Historical Statistics of the United States*)

AT&T dominated the telecommunications industry; consequently, any improvement in telecommunications technology ultimately was to the advantage of AT&T. By 1957, when Sputnik began orbiting the Earth, all the technologies necessary for satellite communications had been invented. All that remained was to demonstrate these technologies and to compare the different innovations that had been discussed in the technical literature.

Sputnik: The Catalyst

The media panic resulting from the "surprise" launch of Sputnik 1 on 4 October 1957 did not evidence itself immediately in public policy, but it soon began to have an effect. A flurry of military rocket development ensued. The following month, both the Thor and Jupiter rockets were ordered into production, and by the end of the year, more than a half-dozen Thors had been fired. Thor would become the workhorse of the space program. Then, on 31 January 1958, the United States launched its first satellite, Explorer 1, followed on 17 March by Vanguard 1. All together, the United States had seven successful launches in 1958; the Soviet Union had only one.[13] American missile technology now was available to launch communications experiments.

Meanwhile, John R. Pierce and Rudolf Kompfner of AT&T, independent inventors of the traveling-wave tube, saw an opportunity for AT&T to launch an experimental communications satellite. Sometime in early 1958, Pierce and Kompfner saw a picture of the shiny 100-foot (thirty-one-meter) sphere that William J. O'Sullivan of the National Advisory Committee for Aeronautics (NACA) Langley Research Center had proposed for launch to undertake atmospheric research. It reminded Pierce of the 100-foot (thirty-one-meter) communications reflector he had envisioned orbiting around Earth in 1954. That summer, while Pierce and Kompfner were attending an Air Force-sponsored meeting on communications at Woods Hole, Massachusetts, Pierce met William H. Pickering, director of the Jet Propulsion Laboratory (JPL), and the three engineers discussed the possibility of using a sphere such as O'Sullivan's for communications experiments. Pickering volunteered the support of JPL. To support their plan, Kompfner and Pierce presented a paper at George Washington University in Washington, D.C., on 6–7 October 1958.[14]

---

13.   Dated events, unless specifically footnoted, are from House Committee on Space and Astronautics, *A Chronology of Missile and Astronautic Events [1915–1960]*, 87th Cong., 1st sess., 1961, HR-67, and subsequent annual NASA publications.

14.   John R. Pierce, "Transoceanic Communication by Means of Satellites," *Proceedings of the IRE* 47 (March 1959): 372–80; John R. Pierce, *The Beginnings of Satellite Communications* (San Francisco: San Francisco Press, 1968), pp. 9–12.

The Pentagon had its own space program. In fact, before the advent of NASA, DoD had been the primary space agency and remained dominant for some time thereafter. DoD's initial responsibility for space activities was with the Advanced Research Projects Agency (ARPA). In the period between the announcement of the intention to form NASA in early 1958 and the actual formation of the space agency in late 1959, ARPA was responsible for managing all U.S. space programs and later for apportioning responsibility for those programs among the three armed services and the newly formed NASA.

In November 1958, representatives of NASA, ARPA, and the Bureau of the Budget met with the president's science advisor to discuss satellite communications. They agreed that ARPA would concentrate on active satellites, while NASA would develop passive satellites.[15] NASA already was committed to launching balloons to study atmospheric density, and it was the only agency interested in Pierce's plan to bounce signals off balloons in orbit.

At the same time, Hughes was inching into the field of communications satellites. In 1959, when the Air Force canceled the F-108 long-range interceptor in response to the cancellation of the Soviet advanced intercontinental bomber program, the Hughes Aircraft Company lost its F-108 contracts and laid off 20 percent of its employees. Frank Carver, manager of the F-108 fire-control system design group, saw the layoff coming and asked Harold A. Rosen, an electrical engineering Ph.D. from California Institute of Technology (Caltech),[16] to explore potential markets for the skills of the Advanced Development Laboratory personnel. Later in 1959, Donald D. Williams, a Harvard physics major, joined Rosen, and over the following months, they worked on the design of a lightweight geosynchronous communications satellite.[17]

Congressional hearings held 3 and 4 March 1959 on "satellites for world communication"[18] illustrated the disparate lines along which military, civilian, and commercial satellite communications were developing. Of the six organizations making presentations, four of them, NASA, ARPA, AT&T, and ITT, were actively engaged in communications satellite research. AT&T and ITT even expected to make a profit in this endeavor. John Pierce, in making the AT&T presentation, mentioned that about two dozen professionals and various technicians were working on satellite communications at AT&T, a half-dozen of whom were full time. Although Pierce did not mention the development of the transistor by Bell Telephone Laboratories or his own invention of the traveling-wave tube (both necessary to satellite communications), the message was obvious: AT&T could do the job and already had started.[19]

Following Pierce was R.P. Haviland of General Electric. Haviland, based on his experience with the early Navy space effort (1945–1947) was General Electric's in-house "satellite expert." He described two possible communications satellites. One, for the future, was a huge manned television broadcast space station with living quarters and enormous antennas à la Clarke and costing under $2 billion. The second was a less ambitious project intended for the present. Sixteen satellites, each with four transmitters and capable of

15.   T. Keith Glennan to James H. Douglas, 18 August 1960, NASA History Office.

16.   William Pickering, John Pierce, and Harold Rosen all had doctorates in electrical engineering from Caltech.

17.   Harold A. Rosen, "Harold Rosen on Satellite Technology Then and Now," *Via Satellite* 8 (July 1993): 40–43; GM-Hughes Electronics, "History and Accomplishments of the Hughes Aircraft Company," undated, p. 12, Hughes Aircraft Company Archives, El Segundo, CA; Edgar W. Morse, "Preliminary History of Project SYNCOM," 1 September 1964, pp. 32–34, History Notes HHN-14, NASA History Office.

18.   House Committee on Science and Astronautics, *Satellites for World Communication*, 86th Cong., 1st sess., 1959, (86) H1733-3, *passim*.

19.   *Ibid.*, pp. 35–46.

200 simultaneous teletype messages, would provide global coverage. The cost of the whole system was between $100 and $150 million to establish and perhaps $50 million annually to operate. General Electric thus seemed prepared to enter the field of communications satellites, too.[20]

The next presenter was Henri G. Busignies, President of ITT Laboratories. He discussed ongoing research in satellite communications at ITT, which recently had been concentrating on twenty-four-hour satellite systems. ITT could handle the necessary satellite and ground station communications hardware, and it had joined with Curtis-Wright and Aerojet for assistance with the satellite vehicle itself. As with AT&T and General Electric, ITT was interested in communications satellites, but it seemed to be looking for a partner, especially one with a knowledge of satellites.[21]

The main message of the congressional hearings was that satellite communications were being taken seriously. ARPA, at least in the opinion of the AT&T engineers, was too ambitious—and perhaps insufficiently knowledgeable. AT&T, JPL, and NASA Langley were proceeding with a simple passive experiment (Project Echo), which, if successful, would lead to more complex satellites. AT&T clearly saw itself as the leader in passive satellite programs. The most important ground facilities would be those of Bell Telephone Laboratories in New Jersey and JPL in California. NASA would build and launch the balloon. AT&T treated JPL as a knowledgeable junior partner, but it considered the rest of NASA almost incompetent. Their worst comments were reserved for ARPA engineers and managers, though, whom they saw as unrealistically grandiose in their approach.[22]

Hughes and AT&T Push Forward

Meanwhile, industry pushed forward their communications satellite plans. Rosen and Williams, joined by Tom Hudspeth, were busy at Hughes Aircraft Company. By the summer of 1959, they had designed, at least conceptually, a lightweight communications satellite, and they were ready to make presentations to upper management.[23] Rosen's satellite proposal then underwent evaluation in an in-house review. The study recommended that Hughes seriously consider a commercial venture in satellite communications. Even though certain aspects of the proposal were overly optimistic, communications satellites could generate prestige for the company. The major obstacle would be commercial: finding enough traffic to make the system profitable. The review recommended, therefore, a thorough study of the business aspects of satellite communications and talks with General Telephone, the largest Bell competitor.[24]

The Hughes Aircraft Company Task Force on Commercial Satellite Communication[25] met for the first time a few weeks later on 12 October 1959. The group determined that the most critical aspect of satellite communications was not the technology, but the commercial prospects. The AT&T monopoly would be difficult, but not necessarily impossible, to circumvent. A satellite's value need not be determined necessarily by its ability to carry

20.   *Ibid.*, pp. 47–56.
21.   *Ibid.*, pp. 56–62.
22.   John R. Pierce, "BTL Tracking Proposal," April 2, 1959; W.C. Jakes, "Visit to Washington on March 31, 1959," 7 April 1959; J.R. Pierce to J.A. Morton, 14 April 1959, AT&T Archives, Warren, NJ.
23.   A.S. Jerrems to F.R. Carver, 17 September 1959, Hughes Aircraft Company Archives, El Segundo, CA.
24.   S.G. Lutz to A.V. Haeff, "Evaluation of H.A. Rosen's Commercial Satellite Communication Proposal," 1 October 1959, Hughes Aircraft Company Archives.
25.   The task force's working members were E.D. Felkel, S.G. Lutz, D.E. Miller, H.A. Rosen, and J.H. Striebel.

telephone traffic; that value could come from prestige, a claim on a section of the stationary orbit, a wide bandwidth service (such as television), or military communications. Many questions were left to be answered, however. What about competition from AT&T and RCA? Should government support be sought?[26]

The recommendations of the Hughes task force strongly endorsed the Rosen satellite proposal. Hughes had a cost advantage that derived, according to the task force, from the "Hughes brand of system engineering." NASA, RCA, Space Electronics, and the Army had proposed solving the problems of geosynchronous satellites with large complex spacecraft, while Hughes alone (successfully) had attempted to design a cheap lightweight spacecraft. The task force recommended starting immediately to capture the prestige of having the first satellite in orbit, funding the development of the traveling-wave tube separately as a commercial product, and exploring cooperation with General Telephone.[27]

Following a briefing a week later on 26 October 1959 by Dr. A.V. Haeff of the task force, L.A. "Pat" Hyland, Hughes General Manager, suggested determining whether new technologies developed for satellite communications could be patented and coordinating efforts with NASA as other firms had done on Atomic Energy Commission projects. On those projects, the government financed the development of a type or class of reactor but allowed industry to develop it commercially and to retain commercial rights over the patents that it generated. Hyland also urged that dialogue with potential partners not be engaged until after both these issues were resolved. Finally, he decided, the development of the traveling-wave tube should proceed only if its value were comparable to other research projects.[28]

On 5 November 1959, Donald Williams of Hughes traveled to NASA headquarters for a meeting with Homer J. Stewart, who headed a small planning group under NASA Administrator T. Keith Glennan. Although most Hughes senior executives were not too upset at the thought of NASA taking over their patent rights, Williams, the inventor of the orbit and altitude control system, prefaced his discussions with a statement that Hughes did not want to lose their proprietary rights by talking to NASA. Stewart assured him that this would not happen. Williams emphasized the Hughes interest in proceeding with their satellite program as a commercial venture. Stewart saw the Hughes-funded commercial communications satellite program as being in line with the traditional U.S. policy of letting the private sector run telecommunications, but, Stewart warned with some prescience, a faction in Congress would oppose privately owned commercial communications satellites.[29]

As with Hughes, AT&T was proceeding with its own communications satellite program. During most of 1959, AT&T concentrated on building antennas, transmitters, and receivers for the Echo program. By November, successful Moon-bounce experiments had been conducted between AT&T's Bell Telephone Laboratories facilities at Crawford Hill, New Jersey, and JPL's Goldstone facilities.[30] AT&T and several of the other companies

    26.   S.G. Lutz to A.V. Haeff, "Economic Aspects of Satellite Communication," 13 October 1959, Hughes Aircraft Company Archives.

    27.   S.G. Lutz to A.V. Haeff, "Commercial Satellite Communication Project: Preliminary Report of Study Task Force," 22 October 1959, Hughes Aircraft Company Archives.

    28.   L.A. Hyland to A.V. Haeff and C.G. Murphy, "Communications Satellite," 26 October 1959, Hughes Aircraft Company Archives.

    29.   D.D. Williams to D.F. Doody, "Discussions with Dr. Homer J. Stewart, NASA," 23 November 1959, Hughes Aircraft Company Archives.

    30.   AT&T/BTL, "PROJECT ECHO, Monthly Report No. 3, Contract NASW-110, December 1959," AT&T Archives.

interested in satellite communications (notably ITT, but also Hughes) viewed the ground facilities as the most important component of a communications satellite system. In general, from this early period to the present, more money has been spent on ground stations than on the communications satellites themselves. The prevalence of the Bell Telephone Laboratories technology in its own systems suggested to AT&T's researchers that the company had a technological advantage over other companies.[31]

In addition to work on passive satellites, such as Echo, AT&T was not neglecting active satellite design. By August 1959, Leroy C. Tillotson, a senior engineer at Bell Telephone Laboratories, had described a satellite design quite similar to Telstar in a memorandum, and another Bell memorandum described a satellite traveling-wave tube. By the end of 1959, studies of spacecraft power systems (such as solar cells, Ni-Cd batteries, and DC-DC converters), structures, space environment, thermal control, and attitude control also had been completed. Perhaps more important was a growing commitment to the development of active satellites. In the words of Pierce: "by the end of 1959 our thoughts were directed toward a simple, low-altitude active satellite as the next step."[32]

NASA Joins the Game

By the end of 1959, the various commercial communications satellite programs were moving along. Hughes had designed the basic Syncom satellite, and it was considering developing that satellite in partnership with NASA.[33] AT&T and NASA were finishing preparations for Echo, and the firm was proceeding with an active satellite design. NASA, the fledgling space agency, still had no communications satellite policy despite its participation in the Echo program.

Aside from Echo, the first reference in the diaries of NASA Administrator T. Keith Glennan to communications satellites is an entry dated 27 July 1960, in which he refers to a meeting with George B. Kistiakowsky, the president's science advisor, to talk about communications satellites. The two were concerned that public policy had not been developed in this area, yet the "pressures generated by AT&T and by the military as well as by other industrial suppliers are building up quite a fire."[34]

The next day, Glennan assigned Robert G. Nunn, Jr., NASA's Assistant General Counsel, the task of preparing an outline of a position paper to be delivered at a cabinet meeting in which NASA would request responsibility for preparing administration policy on communications satellites. A week later, Nunn and John A. Johnson, NASA's General Counsel, met with Glennan to discuss what was now called "the communications satellite problem." The following days' diary entries were filled with references to communications satellites, including an 11 August meeting with an AT&T delegation that had been exploring the possibility of a joint communications satellite program with partners in Britain, France, and West Germany. In the space of a few days, NASA, at least as evidenced by

---

31.   Pierce, *The Beginnings of Satellite Communications*, pp. 15–22; A.C. Dickieson, "TELSTAR: The Management Story," unpublished monograph, Bell Telephone Laboratories, July 1970, pp. 29–32, AT&T Archives.

32.   Pierce, *The Beginnings of Satellite Communications*, pp. 22–23; Dickieson, "TELSTAR," pp. 32–34.

33.   Robert K. Roney to A.E. Puckett, "Communications Satellite Review Analysis," 27 January 1960; R.L. Corbo to T.W. Oswald, "Revision of the Communication Satellite Structure and General Arrangement," 29 January 1960; Rosen and Williams, "Commercial Communications Satellite," October 1959, Appendix B, Hughes Aircraft Company Archives.

34.   T. Keith Glennan, *The Birth of NASA: The Diary of T. Keith Glennan* (Washington, DC: NASA SP-4105, 1993), p. 189.

Glennan's diary entries, had gone from a position of almost no interest in communications satellites to one of very high interest—even "owning the problem."[35]

NASA had concentrated on passive satellites, as specified in the 1959 agreement with DoD, and had assumed that the Pentagon would be responsible for active satellites. That policy was about to change. After consulting with Budget Director Maurice H. Stans and Deputy Secretary of Defense James H. Douglas, Jr., Glennan developed a new agreement with DoD that allowed NASA to proceed with an active communications satellite program. Discussions with General Counsel John Johnson made it clear to Glennan that many other policy problems had to be solved as well.

AT&T remained the leading firm with a communications satellite program, although on 19 August 1959, George Metcalf of General Electric, contractor for the military communications satellite Advent, communicated to Glennan his firm's interest in developing a commercial launch and data-acquisition service, including a Pacific Ocean launch site.[36] AT&T, in contrast, was willing to put its own money into satellite communications. On 15 September 1959, George Best and William Baker of AT&T met with Glennan to provide more background on AT&T's interest in communications satellites. They informed him that AT&T was prepared to spend $30 million for three satellite flights—more if they had any success. This was the first proposal Glennan had received from industry in which company, rather than government, funds were to be committed.

Almost a week later, on 21 September, as Glennan discussed his upcoming trip to Bell Telephone Laboratories for an Echo demonstration, his lawyers and deputy argued that his presence would give the appearance of NASA support for AT&T. Glennan was not amused. He wrote in his diary: "AT&T is going to be in the business and if we are going to take leadership in getting this program off the ground, it seems to me that we have to take a positive rather than a negative viewpoint in matters of this kind." Glennan attended the demonstration the next day. A picture of Glennan and the six Federal Communications Commission (FCC) commissioners in attendance was transmitted by facsimile to the Naval Research Laboratory, then returned to Bell in Holmdel, New Jersey, via the Echo balloon. Glennan was impressed.[37]

Nonetheless, the development of administration policy on communications satellites did not move forward during the final days of the Eisenhower administration. November came and went with no sign of the cabinet paper Glennan had proposed reading since July and had promised for the 11 November 1960 cabinet meeting. Glennan seems to have been surprised that 11 November was a holiday. Generals Wilton B. Persons (Eisenhower's chief of staff) and Andrew J. Goodpaster (White House staff secretary) chided Glennan, telling him that the president wanted to mention communications satellites in his State of the Union message. On 7 December 1959, Glennan, Nunn, and Johnson met with representatives of AT&T in their Manhattan offices. Glennan suggested that AT&T might be better off by minimizing its role in satellite communications to avoid monopoly problems. The provision of ground stations for the upcoming satellite program at no cost to the government would be a good start, he suggested.

The next day at a NASA meeting on the subject, Abe Silverstein, NASA's director of space flight programs, objected strongly to the presence of private companies in the communications satellite business. Glennan was amazed. Another meeting with AT&T, this time in Washington, included the details of their program. The program impressed Glennan,

35.   *Ibid.*, pp. 189–207.
36.   *Ibid.*, pp. 207–10.
37.   *Ibid.*, pp. 232–34.

but he was beginning to believe that the Washington establishment was too anti-business to tolerate an AT&T monopoly of satellite communications. After giving Eisenhower a copy of the communications satellite briefing paper on 19 December, Glennan presented the paper to the whole cabinet on 20 December 1959. Eisenhower released the details of the paper at the end of the month as his communications satellite policy. It emphasized the traditional private nature of the U.S. telecommunications industry.[38]

On 23 December 1959, as most members of the Eisenhower administration were packing their bags, and while Keith Glennan was taking his Christmas holiday, Robert Nunn and John Johnson met with Attorney General William P. Rogers to discuss NASA's future relationship with AT&T. Nunn, the special assistant for communications satellite policy, was not a political appointee; he and General Counsel Johnson continued at NASA under the Kennedy presidency. They outlined key elements of future NASA policy. Nunn believed that "AT&T [was] realistically the only company capable of doing the job." On the other hand, the monopoly power of AT&T and its attempt to "preempt" the role of communications satellite builder and operator "would in effect select AT&T as the 'chosen instrument' of the United States." Rogers argued that the government must not act to put AT&T in a preemptive position, nor must it appear to do so. Moreover, Rogers advised, "the Executive Branch probably should obtain at least the acquiescence of Congress." Nunn and Johnson showed Rogers the Glennan position paper, which the White House had released. Although he had no overall problem with the paper, Rogers objected to two sentences that specified "private enterprise" and rejected government operation of a communications satellite system. The sentences remained in the statement.[39]

The Kennedy Administration Takes Control

Before his election, John F. Kennedy's attitudes toward the space program are not clear, but his transition team developed some strong feelings about the space program in general and about satellite communications in particular. Jerome B. Wiesner, assigned the task of examining the U.S. space program and advising the president, was critical of the space program. In particular, he believed that developing a satellite communications system was beyond the investment capabilities of industry.[40] Wiesner, though, had proposed a joint U.S.-U.S.S.R. communications satellite program in 1959.[41] Lying beneath Wiesner's evaluation of communications satellites was a strong belief that this new technology should not become an AT&T monopoly. The Kennedy administration was prepared to overturn the Eisenhower view that private industry was the logical candidate to develop commercial satellite communications. They were not alone inside Washington.

Despite the positive temper of talks between NASA and AT&T, NASA civil servants made it clear that they were against AT&T involvement in satellite communications. As Glennan wrote in his diary: "[Leonard] Jaffe and Silverstein seem[ed] determined that anything short of having someone other than AT&T win the competition will be tantamount to following a 'chosen instrument' policy."[42] Glennan had done his best, but he left behind civil servants who seemed strongly biased against private industry, or at least against AT&T.

38.   *Ibid.*, pp. 278–92.

39.   Robert G. Nunn, "Memorandum for the Record," 23 December 1960, Nunn Papers, NASA History Office.

40.   House Committee on Science and Astronautics, *Defense Space Interests*, 87th Cong., 1st sess., 1961, (87) H1857-2, pp. 22–23.

41.   George B. Kistiakowsky, *A Scientist at the White House* (Cambridge, MA: Harvard University Press, 1976), p. 52.

42.   Glennan, *The Birth of NASA*, p. 303.

That bias permeated NASA policy on satellite ground stations. Both AT&T and ITT had offered to provide stations for active communications satellite experiments at no cost to the government. NASA accepted AT&T's offer for ground stations in the United States, but the space agency felt that the State Department, rather than AT&T, should make arrangements for foreign stations. Although AT&T already had cleared the way, NASA reached formal agreements with Britain (14 February 1961) and France (16 February 1961) to participate in the testing of Relay, a medium-altitude repeater, and Rebound, a low-altitude balloon reflector.

Part of defining satellite communications policy involved determining the separate responsibilities of NASA and the FCC. Those agencies defined their respective roles in a memorandum of understanding signed 27 February 1961. The FCC remained responsible for spectrum allocation, while NASA looked after technology and policy.[43] However, the question of public versus private ownership remained to be answered. On 27 February 1961, new NASA Administrator James E. Webb met briefly with Robert Nunn and asked for a briefing on communications satellite developments. The following day, Nunn obliged. He emphasized that NASA had a policy formation role and suggested that Webb read both the United Research report favoring government ownership of communications satellites and the Glennan cabinet paper favoring private ownership.

The question of public versus private satellite ownership also entered the drafting of NASA's budget. In March 1961, NASA was preparing its fiscal year 1963 budget estimates. The budget rested on several assumptions—namely, no funding for operational systems (only demonstrations would receive support), ground support, or passive satellite development.[44] The NASA fiscal year 1962 budget proposed by the outgoing Eisenhower administration included $34.6 million for communications satellite development. Some controversy lingered as to how much industry should reimburse NASA for communications satellite development. In December 1960, Glennan had asked Budget Director Maurice Stans to include the $10 million industry contribution in the NASA budget, so that the government would not be dependent on industry. Stans refused. Then, during a 14 January 1961 press conference on the NASA budget, NASA Deputy Administrator Hugh L. Dryden was asked why industry, given their obvious interest, was paying only $10 million. Dryden gave an uncommitted response. The following month, Webb, NASA Associate Administrator Robert C. Seamans, Jr., and Senator Robert S. Kerr (D-OK) discussed the $10 million. Kerr and Webb agreed that the figure should appear in the supplemental budget, and the $10 million finally showed up in the March amendment to the budget.[45]

Meanwhile, AT&T was losing ground at NASA headquarters. In April, AT&T President Fred Kappel and NASA Administrator James Webb exchanged a curious series of letters. In a letter of 5 April 1960, Kappel complained that Webb had stated publicly that "NASA [had] yet to receive a firm proposal from any company" to form a communications satellite development partnership with NASA. Kappel recapped AT&T's communications with NASA over the past year and pointed out that, in its Relay proposal, AT&T

43. NASA, *Aeronautical and Astronautical Events of 1961* (Washington, DC: U.S. Government Printing Office, 1962), pp. 4–8; Minutes: Administrator's Staff Meeting, 18 January, 26 January, and 2 February 1961, NASA History Office.

44. Abe Silverstein to Assistant Directors, *et al.,* "Fiscal Year 1963 Preliminary Budget Estimates; additional information concerning," 1 March 1961; A. Silverstein to H. Goett, "Designations for Missions and Payloads," 1 March 1961, NASA History Office.

45. Robert C. Seamans, NASA Exit Interview, Seamans Folder, NASA History Office, pp. 13, 25–26; Hugh L. Dryden, Press Conference transcript, 14 January 1961, NASA History Office, p. 21; James E. Webb to The Director, Bureau of the Budget, 13 March 1961, NASA History Office; Robert L. Rosholt, *An Administrative History of NASA, 1958–1963* (Washington, DC: U.S. Government Printing Office, 1966), p. 195.

had volunteered to share costs, even to contract privately, for rockets and launch facilities.[46] Webb replied in a rather unfriendly tone: "I am told that your letter of December
14th was delivered by a number of your associates [James Fisk and George Best], that an
extended conference ensued, and that it was made clear that NASA would not permit
your company, or any other, to pre-empt the program of the United States in this area."[47]
In contrast, Glennan's response to the 14 December meeting was that AT&T had proposed "a rather good program."[48] The battle lines were being drawn, and AT&T was on
the wrong side.

Drawing the Battle Lines

On 12 April 1961, Russian Major Yuri A. Gagarin achieved the first manned orbital
flight. A few days later, the Cuban Bay of Pigs debacle ran its ill-fated course. Among the
many U.S. reactions to these events was an urge to do something spectacular in space
before the Soviet Union. Some suggested that space was the realm of the military, not
NASA, the civilian space agency.[49] On 20 April 1961, President Kennedy asked Vice
President Lyndon B. Johnson, in his role as chairman of the Space Council, to recommend a program for the United States to beat the Soviet Union in space. Johnson replied
that the United States could probably win a race to the Moon.

As the United States embarked on this race to the Moon, the FCC became embroiled
in communications satellite policy. On 19 January 1961, the regulatory agency authorized
AT&T to launch an experimental communications satellite system. Its only previous significant communications satellite action had been to reopen docket 11866 (known as
"above 890," because it related to the general allocation of frequencies above 890 megahertz) in May 1960 (and modified in December 1960 as docket 13522, to address only frequencies for satellite communications), which sought views on the frequencies required
for space communications. In response to what seemed to be a rush to decision in the
satellite communications arena, the FCC opened docket 14024 in March 1961 to solicit
opinions on the "administrative and regulatory problems" associated with commercial
satellite communications systems.

The responses to FCC dockets 11866, 13522, and 14024 figured in the testimony given
before the House Committee on Science and Astronautics during hearings on communications satellites held 8–10 May 1961. Those hearings again demonstrated the willingness
of private industry to invest and build in the field of satellite communications, but the
relationship between government and industry had yet to be defined. In his opening
remarks, committee chair Overton Brooks (D-LA) set the tone: "The proper relationship
between Government and Industry must be defined . . . the most desirable business
arrangements should be determined at the earliest possible time."[50]

The response of General Electric to FCC docket 14024 mirrored that firm's eagerness
to join the communications satellite field. The company proposed that the aerospace and
communications industries jointly raise $250 million to put the General Electric medium-

46.   Fred R. Kappel to James E. Webb, 5 April 1961, copy in John M. Logsdon, gen ed., *et al.*, *Exploring
the Unknown: Selected Documents in the History of the U.S. Civil Space Program, Volume III: Space Applications*
(Washington, DC: NASA, forthcoming).
47.   James E. Webb to Fred R. Kappel, 8 April 1961, in Logsdon, *Exploring the Unknown, Volume III: Space
Applications*, forthcoming.
48.   Glennan, *The Birth of NASA*, p. 290.
49.   This suggestion even came from within NASA. See Don Ostrander to Dr. Seamans, "Reflections on
the American Posture in Space," 21 April 1961, NASA History Office.
50.   House Committee on Science and Astronautics, *Communications Satellites, Part 1*, 87th Cong., 1st
sess., 1961, (87) H1898-1-A, p. 1.

altitude equatorial system of ten satellites in place using the Atlas-Agena launch vehicle. Another $250 million would be needed to build the ground system. General Electric pointed out that the communications industry had sufficient expertise to build the ground stations, and the aerospace industry had the expertise to build the satellites. General Electric was ready to invest $25 to $50 million of its own money in satellite communications; government support, but not government funding, would be required. As did ITT, General Electric called attention to the need for foreign participation and the apparent capability of the United States to support only *one* commercial communications satellite system. Moreover, as had all the other presenters except AT&T, General Electric emphasized the importance of avoiding a monopoly.

Lockheed's response to docket 14024 was a proposal for the creation of a new organization, Telesat, to operate a global system of geosynchronous communications satellites, two over the Atlantic Ocean and two over the Pacific Ocean. Communications carriers, other companies, and the general public would own Telesat. Telesat would not have any foreign owners, but foreign organizations would have their own ground stations and might receive an undivided ownership interest in the satellites, but not in Telesat itself. Lockheed foresaw that government subsidy would be advisable during Telesat's early years to reap the prestige benefits of inaugurating a global communications satellite system; moreover, the $200–$315 million system would not be self-supporting until sometime in the mid- to late 1970s.

Meanwhile, NASA attempted to sort out satellite communications policy and the fulfillment of Kennedy's wish to see the United States beat the Soviet Union in space. On 12 May 1961, shortly after the premature closing of the House hearings, NASA Associate Administrator Robert Seamans presented NASA Administrator James Webb and Deputy Administrator Hugh Dryden with a summary of the accelerated NASA program proposed in response to the president's desire to beat the Soviets. It included communications satellites, as well as a manned lunar landing and meteorology projects.[51] Memoranda between Seamans and NASA counsel Robert Nunn raised the question of why NASA should develop an operational communications satellite system. As Nunn pointed out, "the communications industry continues to affirm its own clear intent and obvious ability to achieve the same objective."[52]

In any case, NASA was proceeding with its own experimental communications satellite program and, in carrying it out, relied on the expertise of private industry. On 18 May 1961, NASA awarded RCA the contract to build Relay, a medium-altitude repeater satellite. Ironically, RCA had stated publicly its preference for a twenty-four-hour satellite. Although NASA did not announce the standings, the ranking apparently was (1) RCA, (2) Hughes, (3) Philco, and (4) AT&T. Some NASA participants in the evaluation process expressed surprise; they had expected the AT&T proposal to be better, if not the best. The deciding factor was apparently the RCA ten-watt traveling-wave tube. AT&T was not entirely out of the game, however.

The relationship between government (NASA) and industry still had not been set forth when President Kennedy made his speech of 25 May 1961 that challenged the nation to land an American on the Moon before the end of the decade. That speech included satellite communications as a NASA goal.[53] On 15 June 1961, Kennedy directed

---

51.   Robert Seamans to James Webb and Hugh Dryden, "Status of planning for an accelerated NASA program," 12 May 1961, NASA History Office.

52.   Robert Nunn, "Memorandum for the Associate Administrator," 16 May 1961, Nunn Papers, NASA History Office.

53.   NASA News Release 61-112, "Statement by James E. Webb, Administrator," and NASA News Release 61-115, "Budget Briefing," 25 May 1961; Minutes, Administrator's Staff Meeting, Thursday, 25 May 1961, NASA History Office.

the Space Council to study communications satellite policy. The FCC was delaying its decision on government-industry joint ventures until the first week of July, after the Space Council study. Discussions between NASA and AT&T over patent policy reached a conclusion that pleased the FCC commissioner and the assistant attorney general. This policy not only gave the government royalty-free use of AT&T patents, but also licensing rights.[54] NASA Administrator Webb had a clear vision of the appropriate spheres of responsibility of NASA and the FCC, with which NASA had a memorandum of understanding. The FCC was "to take proper action on the problem of organizing the resources of private industry in such a manner as to meet governmental requirements and conform to public policy," while NASA had "the job of developing the space technology which any private organization authorized by the FCC will be able to utilize to provide communication services to the public."[55]

The House communications satellite hearings, suspended in May at the request of NASA, resumed on 13 July 1961. The purpose of the hearings, in the words of committee chair Overton Brooks, was, among others, "to determine the extent that private industry should participate in the space communication program." The testimony of Webb is illustrative of his attitude toward industry involvement in communications satellites. When questioned about the financial contributions of private industry, Webb responded that "there are certain things no private industry can undertake on its own at this particular stage of the game."[56] This is a strange comment. As we have seen, AT&T was willing to fund communications satellite research and development by itself, and General Electric and Lockheed had formed "joint-venture corporations" to develop and operate a commercial communications satellite system.

Webb seems to have been committed to private ownership of the satellite communications system, preferably by a joint venture of international communications carriers as proposed by the FCC. Webb also seems to have been committed to NASA control of space policy and space technology issues. His stand on offering launch services to industry, furthermore, seemed to have shifted. In May 1961, he was willing to launch an AT&T satellite, but he had not been willing the month before when his acrimonious correspondence with Kappel took place.

Webb told Fred Kappel of AT&T that NASA would sell him launch services. Shortly thereafter, NASA announced the award of the Relay contract to RCA; then on 27 July 1961, NASA and AT&T entered into agreements for the reimbursable launch of Telstar. On 11 August 1961, NASA signed a sole-source contract with Hughes to build Syncom, its first geosynchronous satellite.[57]

By the end of 1961, two passive experiments (NASA/AT&T's Echo balloon and DoD's Project West Ford), two medium-altitude active experiments (NASA/RCA's Relay and AT&T's Telstar), and two geosynchronous experiments DoD/General Electric's Advent and NASA/Hughes's Syncom) were under way. All technological options were in play. General William M. Thames, then commander of the Army Advent Management Agency, testified before Congress that the military Advent satellite system could handle all traffic needs and would be ready in 1965. Webb and Dr. Edward C. Welsh, Executive Director of the Space Council, both testified at the same hearing that only one system would be viable. Webb stated that "you simply cannot start two or three communication satellite

54.   Minutes, Administrator's Staff Meeting, Thursday, 22 June 1961, NASA History Office.
55.   House Committee on Science and Astronautics, *Communications Satellites, Part 1*, 87th Cong., 1st sess., 1961, (87) H1898-1-A, p. 464.
56.   *Ibid.*, p. 461.
57.   In a 1960 conversation with Seamans, Glennan had suggested just such a "sole-source" procurement of the Hughes system after the policy had been worked out. Seamans NASA Exit Interview, p. 28.

systems. . . . Therefore, the Government policy has been to say we will create the conditions under which one system will be established."[58]

The situation called for some policy direction from the White House. That policy was shaped by Edward Welsh, former legislative assistant to Senator Stuart Symington and, since 1951, a member of the Senate Aeronautics and Space Sciences Committee who had helped draft the Space Act of 1958. Welsh's experience made him somewhat a political partisan and a strong advocate of "trust-busting." Soon after his appointment by Kennedy to the Space Council, Welsh suggested that the communications satellite policy was a natural for the Space Council. In the next year, he dedicated more staff time to this issue than any other. Welsh influenced Kennedy's 25 May 1961 speech that committed the United States to building a global communications satellite system.

Kennedy wanted the system to be global, be sensitive to the needs of the developing world, and serve the public interest. In response to a request from the president to prepare a policy recommendation aimed at accelerating the creation of an operational communications satellite, Welsh drafted a statement favoring private ownership and control and circulated it among staff members from DoD, NASA, the State Department, the Atomic Energy Commission, the FCC, the Justice Department, the Office of Civil Defense Management, the Bureau of the Budget, and the Office of the Science Advisor. Even before the paper was published, Welsh publicly stated that the Space Council did not favor government operation of the satellite system.[59]

After Welsh presented the paper at the Space Council meeting of 5 July 1961, President Kennedy released the paper on 24 July. The paper proposed placing responsibility for the communications satellite system in private hands. It assigned government the roles of regulation, negotiation with foreign countries and organizations, research and development, and launch services. Subsequently, several liberal members of Congress sent Kennedy a letter suggesting that the government avoid any decision that might result in a satellite communications monopoly.[60]

In November 1961, President Kennedy asked the Space Council to prepare a plan for implementing the program outlined in the 24 July 1961 statement. Welsh decided that implementation would require legislation, so in September, he hired Dr. Charles S. Sheldon from the Congressional Research Service as a technical expert. Sheldon, the son of an engineer, had a Ph.D. in economics from Harvard. He and Welsh were the primary writers of the proposed legislation. Welsh felt, as Glennan had before him, that a policy vacuum on communications satellites existed. Welsh also was concerned that, if the proposed system were to be privately owned and operated, it must be competitive, not a monopoly. The final version of the plan went to the White House on 30 November 1961.

Comsat

At this point, Congress began to play a role in shaping communications satellite policy. The House had studied the issue as early as 1959, when its Committee on Science and Astronautics held hearings on "Satellites for World Communications."[61] In early 1962,

58.   House Committee on Science and Astronautics, *Communications Satellites, Part 2*, 87th Cong., 1st sess., 1961, (87) H1898-1-B, p. 739.
59.   Hal Taylor, "Council Favors Private Ownership," *Missiles and Rockets* 9 (3 July 1961): 11, 40.
60.   Letter, Hubert Humphrey, Estes Kefauver, Wayne Morse, *et al.*, to Kennedy, 24 August 1961, reprinted in Senate Committee on Foreign Relations, *Communications Satellite Act of 1962*, 87th Cong., 2d sess., 1962, (87) S1539-4, pp. 51–54.
61.   House Committee on Science and Astronautics, *Satellites for World Communications*, 86th Cong., 1st sess., 1959, (86) H1733-3.

three communications satellite bills were introduced into the Senate: the Kerr bill (S. 2650, introduced 11 January 1962); the administration bill (S. 2814, 27 January 1962); and the Kefauver bill (S. 2890, 26 February 1962).[62]

The Kerr bill, which resembled the FCC position, favored ownership by a consortium of existing communications carriers. The bill would mandate the creation of a new corporation capitalized at $500 million. Shares, at $100,000 each, would be sold in minimum lots of five to U.S. common carriers authorized by the FCC. The Kerr bill would prohibit AT&T from going its own way, but would do little to minimize AT&T's domination of international telephony, as the firm probably would buy a plurality, if not a majority, of the shares. Senator Robert Kerr also was a cosponsor of the administration bill, so it is unclear how committed he was to his own bill. Delbert D. Smith, a lawyer who studied these events, has suggested that Kerr might have been trying to make the Administration bill look like the middle ground between carrier ownership (his bill) and government ownership (the Kefauver bill).[63]

The Kefauver bill, inspired by that Senator Estes Kefauver's distrust of industry's— especially AT&T's—monopolistic tendencies, called for a communications satellite system owned and operated by the government. Kefauver supported his proposal with three main arguments. First, private ownership would evolve into a monopoly. Second, the initial system would of necessity consist of low-altitude satellites inferior to geosynchronous satellites; investment in this initial inferior system would delay the creation of a superior geosynchronous system. Third, because satellites were developed at government expense, the benefits should accrue to the public, not to profit-making private corporations.

The administration bill steered a middle course between the Kefauver and Kerr bills. It assumed that private ownership maximized efficiency. As did the Kerr bill, the Administration bill called for the formation of a new corporation, but one with a broad ownership base. Not only international carriers, but other corporations and private citizens could purchase shares. The administration bill also placed limits on the number of shares any single entity could own. Foreign organizations could own shares, too, as well as ground stations.[64]

Despite Kefauver's arguments, it is obvious in hindsight that the bill did not delay geosynchronous systems; Early Bird, Comsat's first satellite, launched in March 1965, had a geosynchronous orbit. It is also obvious that the costs of satellite development were borne by *both* government and industry. AT&T paid for the development of Telstar and reimbursed NASA for the launch services. Hughes paid the development costs of the protoflight Syncom satellite, although NASA underwrote the construction of the actual flight models. Only Relay was entirely a government-funded satellite, but the remarkably short time between contract award and launch suggests that RCA had been spending its own money for some time. Only launch vehicles were completely funded by the government, and, given AT&T's willingness to pay for launches, this did not have to be the case.

The introduction of these three bills in the Senate made it clear that AT&T's investment in satellite manufacturing was wasted. The operating system was to be either sponsored or entirely operated by the government, with no place for AT&T satellites. AT&T, though, had spent more money on ground stations than on satellites, because the large number of satellites in low orbit proposed for AT&T's Telstar system favored the building

---

62.   Delbert D. Smith, *Communication via Satellite: A Vision in Retrospect* (Boston: A.W. Sijthoff, 1976), pp. 93–103.

63.   *Ibid.*, p. 104.

64.   Senate Committee on Foreign Relations, *Communications Satellite Act of 1962*, 87th Cong., 2d sess., 1962, (87) S1539-4, pp. 13–21.

of simple satellites and complex ground stations. The horn-and-maser AT&T ground station design, however, lost out to cheaper designs using parabolic dish antennas and parametric amplifiers. Telstar would not bring home any profits, and AT&T could not fall back on just pride in its performance. Problems discovered in testing caused AT&T to request a launch delay.[65] AT&T's pride was taking a lot of hits.

AT&T also suffered from the restrictions placed on it as a prerequisite for participating in the communications satellite program. NASA had rights, including licensing rights, to all AT&T communications satellite inventions after May 1961. NASA, not AT&T, would undertake all negotiations with foreign government telecommunications administrations. NASA would coordinate all tests and all publicity. AT&T found none of these restrictions acceptable, but the firm was more interested in building satellites and ground stations than in debating terms.

As AT&T's star fell, Hughes's rose. That firm had redesigned the Syncom satellite to accommodate its launch on a Thor-Delta from Cape Canaveral. In the process, Syncom grew from thirty to sixty pounds. The added weight permitted the addition of new design features. Syncom was a joint program of NASA and DoD. Although the Syncom communications capability was often disparaged as "only one voice-channel," it had one advantage over Relay and Telstar: the Syncom transponder could be used continuously, not just when the batteries were fully charged. The lightweight, low-power Hughes traveling-wave tube was a key part of its success.[66] Harold Rosen of Hughes and Leonard Jaffe of NASA began discussions concerning potential improvements to Syncom, and in February 1962, Hughes presented its plans to NASA for the Syncom Mark II.[67] The Hughes Syncom held great promise. It could be sold to the congressionally approved satellite system operating organization, Comsat, the U.S. military, foreign customers, and perhaps NASA, too.[68]

In the faceoff between medium-altitude and geosynchronous satellites, the successful launch of AT&T's Telstar on 10 July 1962 demonstrated that medium-altitude communications satellites were eminently practical. However, this did nothing to improve AT&T's chances of operating such a system, for by then it was clear that Congress, out of fear of creating a monopoly, would not allow AT&T to do so.

Following the introduction of the three communications satellite bills in the Senate, Congressman Oren Harris introduced H.R. 11040, which was identical to the Kerr bill, in the House on 2 April 1962. The bill passed the House 354 to 9 on 3 May.[69] Meanwhile, the Senate Committee on Aeronautical and Space Sciences reported favorably on Kerr's bill on 2 April 1962.[70] The Senate Commerce Committee then reported favorably on the Kerr bill, and it was brought before the full Senate on 14 June 1962.

The movement toward private operation by a consortium of communications carriers seemed to be gathering momentum, although Senator Kefauver and his allies attacked the bill for several days. On 21 June 1962, the bill was withdrawn to allow other business to be completed. When the debate on the bill resumed a month later on 26 July, a different climate prevailed. On 10 July 1962, AT&T had succeeded in launching the first Telstar satellite.

---

65.   Dickieson, "TELSTAR," pp. 85–130.

66.   *Ibid.,* pp. 157–91.

67.   Described in the brochure of Hughes Aircraft Company titled *SYNCOM,* NASA History Office.

68.   Anonymous (C.G. Murphy?), "Policy Statement for Exploitation of HAC [Hughes Aircraft Company] Communications Satellites," circa early 1962, Hughes Aircraft Company Archives; Dickieson, "TELSTAR," pp. 85–127, 129–55. Ironically, the Telstar uplink (6,390 megahertz) and downlink (4,170 megahertz) frequencies were those land microwave frequencies that AT&T earlier had argued, in the "Above 890" decision, would be unsuitable for satellite communications.

69.   "House Votes 354-to-9 Approval of Space Communications Firm," *Washington Post,* 4 May 1962, p. A1.

70.   Senate Committee on Foreign Relations, *Communications Satellite Act of 1962,* 87th Cong., 2d sess., 1962, (87) S1539-4, *passim.*

After receiving the approval of the Foreign Relations Committee on 10 August 1962, the Kerr bill came before the full Senate, where it immediately became the subject (again) of a filibuster by Senators Estes Kefauver (D-TN), Albert Gore (D-TN), Wayne Morse (D-OR), Russell Long (D-LA), Ralph Yarborough (D-TX), Maurine Neuberger (D-NY), Ernest Gruening (D-AK), and Paul Douglas (D-IL). In an effort to end the filibuster, the Senate passed a historic cloture motion on 14 August 1962 by a vote of sixty-three to twenty-seven. The Senate passed the amended House bill by a vote of sixty-six to eleven on 17 August; ten days later, the House passed the amended bill. President Kennedy signed it into law as the Communications Satellite Act on 31 August 1962.[71]

Figure 19

President Kennedy played a key and lasting role in shaping the development of satellite communications. His administration blocked the extension of AT&T's monopoly of terrestrial telecommunications into space. In his famous 25 May 1961 speech, which declared the United States would land a human on the Moon by the end of the decade, Kennedy also committed the United States to building a global communications satellite system. Depicted above: On 31 August 1962, the President signed into law the Communications Satellite Act, which called for government creation of a private corporation, the Communications Satellite Corporation (Comsat). (Courtesy of NASA)

The Communications Satellite Act mandated that the government create a private corporation, the Communications Satellite Corporation, now commonly known as Comsat. On 4 October 1962, President Kennedy named the thirteen members (called

---

71. The political actions leading to the passage of the Communications Satellite Act of 1962 have been studied in great detail in books, articles, and dissertations. Much of the debate is detailed in the various published hearings. The description above is compiled from various sources, especially Smith, "Civilian Space Applications"; Jonathan F. Galloway, *The Politics and Technology of Satellite Communications* (Lexington, MA: Lexington Books, 1972); Michael E. Kinsley, *Outer Space and Inner Sanctums* (New York: John Wiley & Sons, 1976).

Incorporators) of the temporary Comsat board of directors: six lawyers, three financiers, one labor representative, and one engineer. Beardsley Graham, the engineer, had worked at the Stanford Research Institute and Lockheed, where he had been active in satellite communications studies carried out in partnership with GTE and RCA. The conclusion of the Lockheed studies was that the main issues were business, regulatory, and international relations, not technical. Graham also was convinced that geosynchronous systems would be in place in the near future.[72]

The first task of Comsat was to incorporate, which it did in the District of Columbia on 1 February 1963. Its second task was to issue stock, perhaps worth as much as $500 million.[73] The stock would have to be sold before any election of the board of directors. Before any stock was sold, however, Comsat obtained FCC authorization to borrow up to $5 million from banks, including Continental Illinois, of which David Kennedy, an Incorporator, was an officer.

Meanwhile, the board of directors was undergoing some key changes in personnel. During January 1963, Chairman Philip Graham resigned. The following month, the corporation announced that Leo D. Welch of Standard Oil Company (New Jersey) had been named chairman and that Under Secretary of the Air Force Joseph V. Charyk had been named president of Comsat. Welch's background was in international finance; Charyk's background was technical and included significant experience in reconnaissance satellites.[74]

In addition to the standard problems associated with getting a company off the ground, Comsat had three major concerns: (1) satisfying Congress; (2) maintaining good relations with the Europeans; and (3) deciding on what the eventual operational system would look like. Congress was unhappy with the high salaries paid to Welch and Charyk ($125,000 and $80,000 per year, respectively), their luxurious offices in a Washington mansion (Tregaron), and the general uncertainty involved in the enterprise.[75] The Europeans were not certain whether they wanted to be part of any American system. In December 1962, the Conference of European Postal and Telecommunications Administrations formed a committee to study the issue of joining an American-led global communications system. While recognizing that the majority of international telecommunications traffic originated or terminated in the United States, the Europeans were anxious to gain maximum control and make equipment sales.[76] Deciding on Comsat's operational system—geosynchronous versus medium-altitude—was a more complicated problem.

---

72.   Beardsley Graham, "Satellite Communication," *International Science and Technology* 13 (January 1963): 69–74.

73.   Jack Raymond, "President Names Satellite Board," *New York Times*, 5 October 1962, p. 1.

74.   "Satellite Job is Resigned by Graham," *Washington Evening Star*, 26 January 1963, p. A9; "Satellite Firm Selects Top Officers, Hints at Stock Sale in a Year," *Wall Street Journal*, 28 February 1963, p. 32; Associated Press, "Welch, Charyk Picked to Head Satellite Firm," *Washington Post*, 1 March 1963, p. A3.

75.   John W. Finney, "Space Radio Plan Creating Doubts," *New York Times*, 24 April 1963, p. 2.

76.   Cecil Brownlow, "U.S.-Europe Comsat Agreement Predicted," *Aviation Week & Space Technology* 78 (22 April 1963): 74–75.

Geosynchronous or Medium-Altitude Orbit?

The choice of a Comsat system was becoming ever more urgent. In July 1963, the FCC made public its concern that Comsat "no longer has definite plans for an early issue of stock." The appointed Comsat board was in the position of making decisions about the future of the company, which should have been made by the shareholders. Comsat responded by suggesting that decisions had to be made about the system configuration before issuing stock. Until a basic program was outlined, it was unclear how much equity capital should be acquired, and it was uncertain how to categorize risks, which was a legal requirement.[77]

Support for a geosynchronous Comsat operating system had been growing for some time. In September and October 1962, the Applications Subcommittee of the House Committee on Science and Astronautics had held hearings on "Commercial Communications Satellites" for the purpose, in the words of Congressman Ken Hechler (D-WV), of determining "the most effective and least expensive system for commercial development"[78]—specifically, whether the system should be in geosynchronous or medium-altitude orbit. Representatives of Hughes, NASA, AT&T, the office of the director of Defense Research and Engineering (to which ARPA reported), the State Department, and the U.S. Information Agency gave testimony; by far the longest presentations were those by Hughes and NASA.

Making the case for a geosynchronous system were Fred Adler, Hughes Space Systems Division Manager, and Gordon Murphy, Syncom Program Manager. It seems that Murphy startled the committee members when he stated that in addition to the NASA contract for three Syncom satellites, a NASA study contract for advanced satellites also was in place. Furthermore, Murphy declared, "we expect that the later contract will lead to an initial operational communication satellite demonstration in the first half of 1964. We call the advanced satellite Syncom Mark II."

Murphy outlined the work conducted at Hughes from 1959 to 1961 on satellite design and testing, all with company funds. As a result, Hughes could launch its first Syncom only 17 months after signing a contract. This was less than RCA's 19 months from contract startup to launch Relay, but more than AT&T's 14 months from contract startup to launch Telstar. However, Murphy told the committee that Hughes needed a NASA commitment by March 1963 to build flight vehicles, as well as a NASA commitment to build ground stations. The Telstar, Relay, and Syncom ground stations were all owned by others. Murphy's major argument for a geosynchronous system was that it could be installed sooner than a medium-altitude system because fewer satellites would have to be launched, and the simplicity of the ground stations allowed them to be installed much more quickly.[79]

Actual satellite launch experience, however, seemed to favor the medium-altitude satellites, such as Telstar and Relay. The Telstar and Relay satellites had been launched relatively successfully, but Syncom failed after its injection into geosynchronous orbit.[80] Most

---

77.   Robert C. Toth, "F.C.C. Sees Delay in Space Stock," *New York Times*, 27 July 1963, p. 3; "Satellite Corp., Prodded by FCC, May Tell More This Week About Plans for Stock Sale," *Wall Street Journal*, 29 July 1963, p. 7.

78.   House Committee on Science and Astronautics, *Commercial Communications Satellites*, 87th Cong., 2d sess., 1962, (87) H1898-4, p. 1.

79.   *Ibid.*, pp. 1–4.

80.   "S. African Observatory Spots Missing Syncom," *Washington Post*, 1 March 1963, p. A8; "Development of the Relay Communications Satellite," *Interavia* 17 (June 1962): 758–59; Sidney Metzger and Robert H. Pickard, "RELAY," *Astronautics and Aerospace Engineering* 1 (September 1963): 64–67; NASA, *Final Report on the Relay 1 Program* (Washington, DC: U.S. Government Printing Office, 1965), pp. 63–90; Robert E. Warren, "Syncom 1 Progress Report No. 4," 4 March 1963, NASA History Office.

analysts felt that geosynchronous satellites held the most promise. Syncom subsequently showed that geosynchronous orbit was attainable, but the satellite's light weight constrained the communications payload to a single telephone circuit, which was significantly less than the wideband capabilities of Relay and Telstar.

Support for a geosynchronous system also began to appear in the literature. For example, in September 1963, Siegfried Reiger and Joseph Charyk of Comsat published articles in the journal of the American Institute of Aeronautics and Astronautics that outlined the system's technical choices. In summarizing the problem of orbit selection, Reiger admitted that most engineers agreed that a medium-altitude system could be made operational sooner than a geostationary system, "[b]ut there also appears to be general agreement that in the long run the [geo-]stationary-satellite concept offers the greatest promise and growth potential." Reiger made the point that insufficient geostationary satellite experience existed to evaluate the problems associated with this orbit.[81]

Meanwhile, NASA continued to underwrite Hughes's Syncom experimental geosynchronous satellite program. Within a year of receiving a sole-source contract from NASA to build three Syncom satellites, Hughes Aircraft Company engineers began studies of an Advanced Syncom, a geosynchronous satellite whose payload would support thousands of voice circuits rather than the "single voice circuit" of its predecessor Syncoms. The traveling-wave tubes would be more powerful; the antenna would direct most of its energy toward the Earth's surface; and the satellite transponders would allow more than one ground station to use the same transponder at the same time.[82]

NASA's funding of Syncom did not go unnoticed, however. Congressional objections to the spending of taxpayer dollars for the benefit of a private corporation, Comsat, made NASA's job particularly difficult. In an effort to allow more time for Syncom 2 to prove itself, and to persuade Congress of the benefits of Advanced Syncom, NASA extended the Hughes study contract another two months in June 1963.[83] The successful launch of Syncom on 26 July 1963 made it clear that geosynchronous satellites were practical. Complex ground station problems were solved with apparent ease; the on-board jets successfully corrected the satellite's drift and positioned it over the Atlantic Ocean, ready to carry traffic.[84]

Hughes was extremely active in attempting to find additional markets for Syncom. One idea, resurrected from a suggestion made by Hughes's Donald D. Williams in 1959, before Harold Rosen convinced him communications was a better application, was to use Syncom as a navigational system.[85] More immediate was the unsuccessful attempt to persuade DoD that the Hughes geosynchronous satellite system performed better than its medium-altitude competitor.[86] More productive were Hughes's discussions with Robert R. Gilruth, Apollo Program Manager and Director of the Marshall Space Flight Center, to

81.   Siegfried H. Reiger, "Commercial Satellite Systems," *Astronautics and Aerospace Engineering* 1 (September 1963): 26–30; Joseph V. Charyk, "Communications Satellite Corporation: Objectives and Problems," *Astronautics and Aerospace Engineering* 1 (September 1963): 45–47.

82.   Paul E. Norsell, "SYNCOM," *Astronautics and Aerospace Engineering* 1 (September 1963): 76–78; NASA Press Release, "Technical Background Briefing Project SYNCOM," 29 January 1963, NASA History Office.

83.   R. Darcey to H. Goett, 14 June 1963, quoted in Weitzel, "The Origins of ATS," NASA HHN-83, 1968 (unpublished), p. 50, NASA History Office.

84.   R.W. Cole to D.D. Williams, "Summary of Orbital Data for Syncom 2 (A-26)," 18 August 1963, Hughes Aircraft Company Archives.

85.   L.M. Field to L.A. Hyland, "Possible use of Syncom as a Navigational System—Microwave Loran," 9 August 1963, Hughes Aircraft Company Archives.

86.   C.G. Murphy to L.A. Hyland, "Synchronous Altitude Communication Satellite System," 16 September 1963, Hughes Aircraft Company Archives.

use Syncom for communications links among Project Apollo's widespread tracking stations.[87] Attempts to find additional markets accelerated in the wake of the news that NASA would not pursue the Advanced Syncom flight program. As NASA Administrator James Webb thought, Congress would not allow NASA to fund hardware development because of the perception that NASA was subsidizing Comsat.[88]

The Early Bird Hatched

Meanwhile, Comsat still was trying to choose the base satellite system orbit. Among the announcements the firm made at the end of 1963 was a request for proposals for a satellite design, either medium-altitude orbit or geosynchronous orbit, which would constitute Comsat's basic system. No decision on the basic system would be made until after the designs had been evaluated.[89]

The choice of communications satellite system orbit was not easy. AT&T and the military were convinced that medium-altitude satellites were best, while many others believed that high-altitude, geosynchronous satellites were preferable. NASA seemed to be leaning toward both geosynchronous and medium-altitude satellites. For example, in April 1962, Milton Stoller, director of NASA's Office of Applications, stated that "when all the technology is in hand, it will be the synchronous-orbit satellites which will be the most attractive to us."[90] However, in September 1962, Leonard Jaffe, director of the communications group under Stoller, suggested that a medium-altitude system similar to Telstar would be best.[91] Then, on 25 February 1963, NASA announced its intention to concentrate on geostationary satellites. All of these opinions and decisions complicated Comsat's task of selecting a satellite operating system.

The Comsat request for proposals received four responses in early February 1964. One design group consisted of AT&T teamed with RCA, while another had TRW teamed with ITT. Hughes and Philco individually submitted the other two proposals.[92] Neither the Philco, the AT&T-RCA, nor the TRW-ITT proposal was for a geosynchronous system. Hughes had the only geosynchronous satellite proposal.[93]

Many experts assumed that the initial commercial system would be placed in a medium-altitude orbit. In December 1963, Hughes proposed a commercial version of Syncom, which would be ready for launch in early 1965 and which could serve both experimental and operational needs. Without committing to a geosynchronous orbit for the basic system, whose characteristics were to be determined later, Comsat accepted the Hughes proposal and launched an "early bird" prior to making a final decision on system type. This "early bird" would consist of "a synchronous satellite orbited on an experimental-

87.   L.A. Hyland to Robert Gilruth, 21 October 1963, Hughes Aircraft Company Archives.
88.   R.E. Warren to R. Garbarini, 10 September 1963; and R. Garbarini to H. Goett, 24 September 1963, both quoted in Weitzel, "The Origins of ATS," pp. 55–56.
89.   Comsat Press Release, "Commercial Communications Satellite Engineering Design Proposals Requested of Industry; Corporation Regards Move As Major Step," 22 December 1963, NASA History Office.
90.   "NASA's ComSat Funding to Climb," *Missiles and Rockets* 10 (2 April 1962): 17.
91.   "NASA sees TELSTAR-Type Satellite as Best for World-Wide System," *Aviation Week & Space Technology* 77 (24 September 1962): 40.
92.   There were several problems associated with these teams. Philco and TRW had teamed on medium-altitude communications satellites. ITT was an adviser to DoD on medium-altitude communications satellites. It was quite possible that medium-altitude communications satellites would be canceled, and Comsat would provide the capability required by the military.
93.   "Communications Satellite Corp. Gets Bids for Designing System From Six Companies," *Wall Street Journal*, 12 February 1964, p. 1; Cecil Brownlow, "International Comsat Agency Considered," *Aviation Week & Space Technology* 80 (17 February 1964): 34.

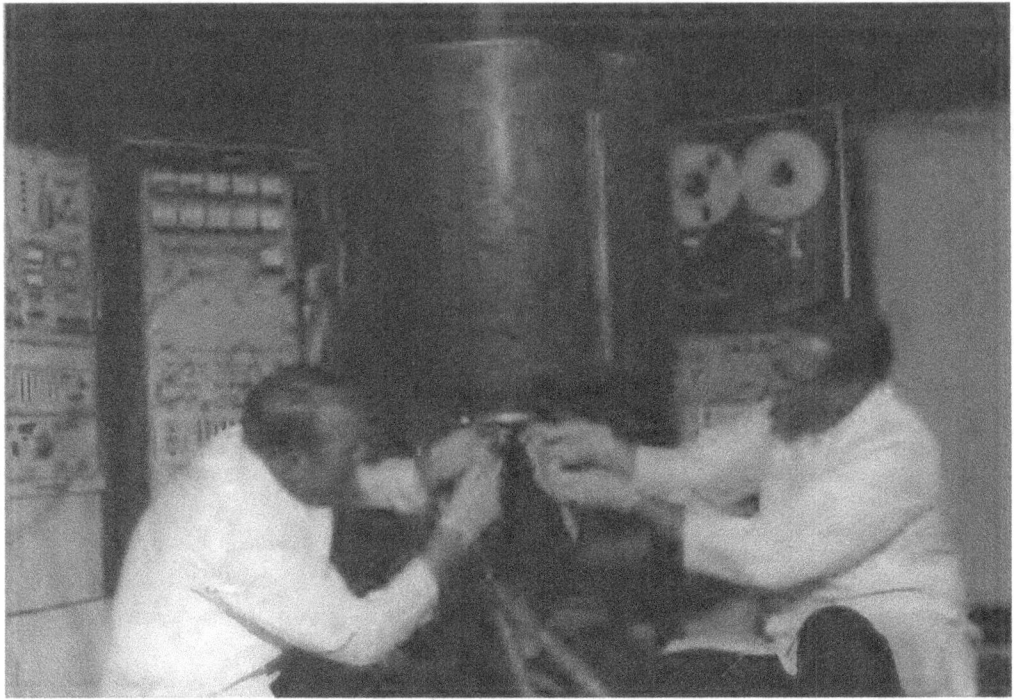

Figure 20

Early Bird, shown here, was Comsat's first satellite. Launched in March 1965 into a geosynchronous orbit, Early Bird (known also as Intelsat I) suggested the commercial potential of communications satellites. (Courtesy of NASA, photo no. 66-H-150)

operational basis in 1965, with a bandwidth and power which can provide a capability for television or, alternatively, for facsimile, data, or telegraphic message traffic or for up to 240 two-way telephone channels."[94]

Comsat considered Early Bird, as the satellite came to be called, as both experimental and commercial, but not the final system choice. Comsat proposed to make the satellite available for commercial use after conducting tests with it. Part of the Early Bird rationale was the success of Syncom (launched 26 July 1963) and the resulting desire to experiment with a similar satellite operating at so-called commercial frequencies (at four and six gigahertz).

Before Early Bird construction could begin, however, Comsat needed FCC approval. The FCC was not comfortable with Comsat's intent to use the AT&T Andover ground station. It preferred that Comsat own and operate its own ground stations.[95] On 4 March 1964, Comsat requested permission from the FCC to launch Early Bird in early 1965 into geostationary orbit over the Atlantic Ocean. The eighty-five-pound (38.6-kilogram) satellite was to provide 240 voice circuits or one television channel. AT&T planned to lease

94.   Comsat Press Release, "Commercial Communications Satellite Engineering Design Proposals."

95.   John J. Kelleher to Leonard Jaffe, Memorandum, "FCC Action on Early Bird," 24 January 1964, Jaffe papers, NASA History Office.

100 circuits on Early Bird to handle peak telephone loads and for cable replacement. In April, the FCC approved the launch of Early Bird.[96]

Before the launch of its Early Bird satellite, Comsat received a strong vote of confidence from the business world. The first five million shares of Comsat stock (at $20 each) were sold exclusively to communications common carriers. By the 23 March 1964 deadline, more than 200 carriers had notified the FCC of their interest in purchasing Comsat shares, and by 26 May 1964 (the official deadline for bidding on shares), AT&T had offered to purchase $85 million of the $100 million worth of shares reserved for communications carriers. Because the stock was oversubscribed, AT&T was permitted to buy stock worth only $57.9 million, or 29 percent of all shares. ITT bought stock worth $21 million (11 percent); GTE took $7 million (4 percent); and RCA got $5 million (3 percent). Other carriers bought the remaining 5 percent. The Comsat charter stipulated that the communications carriers elect six members of the board of directors and that each holder of 8 percent of total shares earned a director's seat. Consequently, AT&T had at least three seats guaranteed, and ITT had at least two seats.[97]

The second set of five million shares were sold to the public, with the FCC responsible for apportioning sales. When these shares were offered on 2 June 1964, the public snapped them up. Comsat stock was very popular—so popular that the size of purchases was limited and, as a result, Comsat stock ownership was dispersed. The average public shareholder held only twenty-seven shares. Of the 130,000 shareholders, about 120,000 held less than 100 shares. As a result, there was upward pressure on the stock price (from $20 to $48 per share by mid-August 1964).[98]

Entangling Alliances

As Comsat prepared its experimental Early Bird program and undertook a search for the final satellite system design, questions regarding the use of the commercial system by the military and foreign organizations had to be resolved. As early as 1960, well before the passage of the Communications Satellite Act, international arrangements started when AT&T suggested the possibility of a satellite program to its foreign cable partners. Both AT&T and NASA continued those discussions throughout 1961 and 1962. While the Communications Satellite Act of 1962 was under consideration by Congress, the Europeans deliberated their options.

---

96.   "Ocean Satellite by 1965 is Sought," *New York Times*, 5 March 1964, p. 4; "Comsat Files with FCC," *Missiles and Rockets* 14 (9 March 1964): 8; "Communications Satellite Firm Negotiates Spacecraft Contract with Hughes Aircraft," *Wall Street Journal*, 17 March 1964, p. 6; "FCC Gives ComSat Go-Ahead," *Missiles and Rockets* 14 (20 April 1964): 10.

97.   John P. MacKenzie, "200 Carriers File for Comsat Shares," *Washington Post*, 24 March 1964, p. 4; "Satellite Corp. to Market Stock at $20 a Share," *Wall Street Journal*, 7 May 1964, p. 32; Philip J. Klass, "Comsat Firm Files $200-Million Fund Plan," *Aviation Week & Space Technology* 80 (11 May 1964): 25–26; Robert Hotz, "Selling Shares in Space," *Aviation Week & Space Technology* 80 (11 May 1964): 17; John P. MacKenzie, "Industry Snaps Up Its Half of Comsat Stock," *Washington Post*, 28 March 1964, p. 2; "Satellite Corp. Picks 6 Candidates for New Board," *New York Times*, 9 July 1964, p. 43C; "Comsat Stock Widely Distributed," *Space Daily*, 14 August 1964, p. 227; S. Oliver Goodman, "130,000 Owners Listed in Initial Report of Comsat," *Washington Post*, 14 August 1964, p. D7; John W. Finney, "Comsat Takes On Private Role As Holders Stage First Meeting," *New York Times*, 18 September 1964, p. 45; Suzanne Montgomery, "Comsat Board Awaits Senate Nod," *Missiles and Rockets* 15 (28 September 1964): 34.

98.   Alexander R. Hammer, "Stock of Comsat Shows a Big Gain," *New York Times*, 20 August 1964, p. 37; "Comsat stock Widely Distributed," *Space Daily*, 14 August 1964, p. 227; S. Oliver Goodman, "130,000 Owners Listed in Initial Report of Comsat," *Washington Post*, p. D6.

They decided that they could maximize their benefits from the U.S. satellite program by negotiating as a bloc. In December 1962, the European Conference of Postal and Telecommunications Administrations began formal studies to establish a basis for discussions with the United States. After a series of formal and informal discussions among members of the European Conference on Satellite Communications, the Europeans agreed that a consortium, rather than the series of bilateral agreements preferred by the U.S. State Department, was the best approach, with Comsat serving as the consortium manager. During meetings held in Washington, 21–25 July 1964, agreements were reached that created the International Telecommunications Satellite Organization (known as Intelsat), thereby establishing a framework for the international use of communications satellites.[99]

While Intelsat furnished a framework for international satellite communications, the question of military use remained unanswered. The partial ownership and operation of Intelsat by foreign agencies determined that Intelsat could not provide communications services for the U.S. military. Therefore, at a press conference on 15 July 1964, Secretary of Defense Robert S. McNamara announced that DoD would build its own satellite communications system. Both Leo Welch and Joseph Charyk of Comsat attempted, in letters to McNamara, to "keep the door open" for the military, but Comsat, at least for the moment, was not committed to the medium-altitude communications satellites preferred by the Pentagon.[100]

On 25 January 1965, Comsat—not Congress—reopened the issue of providing communications satellite services to the military. Comsat proposed launching twenty-four Hughes satellites, similar to Early Bird, eight at a time. Comsat blamed the Pentagon for the previous debacle, and they wanted to try again. Philco, the winner of the 1963 DoD satellite contract competition, had not been able to proceed for more than a year because of Comsat's efforts. Philco was annoyed that the whole process had started over again. Comsat proposed a sole-source contract with Hughes, leaving Philco out. Philco protested to the FCC on the basis that the Communications Satellite Act required competitive bidding. Eventually, Philco built the Pentagon's communications satellites, but Comsat's political moves delayed launch from 1964 to 1966.[101]

As Comsat carried out this attempt to capture military customers, Hughes was beginning to gain momentum in the competition to be the primary provider of communications satellites. The year 1964 had begun with a contract for two geosynchronous satellites (model HS-303, the Early Bird) for Comsat. In March, NASA had awarded Hughes a contract for five Applications Technology Satellites, and in August, Syncom 3 was launched into geostationary orbit. Syncom 3 relayed television images from the 1964 Tokyo Olympic Games to the United States via the Navy ground station at Point Mugu, California, and a new ground station at Kashima, northeast of Tokyo. Although the transmission was not up to "commercial quality," no one really noticed. Many wondered then and later if space was the right place to spend human resources, and many critics could not discriminate between the practical and the prestigious.[102]

99.   House Committee on Government Operations, Military Operations Subcommittee, *Satellite Communications: Military-Civil Roles and Relationships*, 88th Cong., 2d sess., 1964, (88) H2086-1-A, pp. 89–96.

100.   *Ibid.*, pp. 11–12, 51–53, 105–13.

101.   Comsat Press Release, 25 January 1965, NASA History Office; "Comsat's Defense Bid Challenged by Philco," *The [Washington] Evening Star,* 2 February 1965, p. 3; "FCC Bars Comsat Pact," *Missiles and Rockets* 16 (8 February 1965): 9; "Comsat to Seek Bids on DoD Comsat," *Space Daily,* 17 February 1965, pp. 239–40; Larry Weekley, "Comsat Bows to FCC, Invites General," *Washington Post,* 17 February 1965, p. C6.

102.   Anthony Michael Tedeschi, *Live via Satellite* (Washington, DC: Acropolis Books, 1989), p. 31; "Leveled on Comsat Olympic Coverage," *Space Daily,* 14 September 1964, p. 237; "5 Technological Satellites will be Developed by Hughes," *New York Times,* 4 March 1964, p. 7; "Hughes Gets ATS Pact," *Missiles and Rockets* 14 (9 March 1964): 8; A.S. Jerrems to H.A. Rosen, "Syncom Publicity," 21 September 1964, Hughes Aircraft Company Archives.

Comsat was "bullish" on Hughes, but it still was not clear whether the basic Comsat system would be geosynchronous. Syncom had been successful, and orbital and attitude control seemed to be much simpler than originally thought. However, telephone companies (generally) experimented only with the medium-altitude Telstar and Relay satellites because Syncom used military, not commercial, frequencies. Comsat stock, selling for almost three times the initial offering price, was constantly moving; no one knew what was going to happen in the end. Early Bird was eagerly anticipated.

Early Bird Flies

U.S. and European demand for Early Bird circuits was building. AT&T wanted to use 100 of the satellite's 240 circuits, while Canada, Britain, France, and West Germany also were anxious to participate.[103] NASA was responsible for the launch, but except for basic tracking services, the space agency had no responsibilities after placing the spacecraft in orbit. Comsat personnel performed all the orbital and control functions. Thus, as prescribed by the Communications Satellite Act, NASA provided launch services, but Comsat was in charge.[104]

Early Bird was launched on 6 April 1965. The spacecraft was almost identical to Syncom 3, but it used commercial (six and four gigahertz) rather than military (eight and two gigahertz) frequencies for uplink and downlink communications. Similar to Syncom 3, Early Bird's orbit was geostationary, not just geosynchronous—that is, its orbit was in the plane of the equator, not inclined to it.[105] The eighty-five-pound (38.6-kilogram) Early Bird was a more sophisticated spacecraft than the original twenty-five-pound (11.3-kilogram) "commercial communications satellite" that Harold Rosen, Donald Williams, and Tom Hudspeth had envisioned in 1959, but it clearly originated in the modified design they had developed in 1960.[106]

Formal Early Bird communications experiments began on 10 April 1965; commercial service began on 1 June. Early Bird's 240 voice-channel capacity almost equaled the 317-channel capacity of all existing Atlantic telephone cables—and it cost much less! The economics of Early Bird were astounding. The most up-to-date underwater telephone cable carried fewer channels and cost about ten times as much the satellite. Early Bird and its launch vehicle cost Comsat around $7 million, a small fraction of its $200 million capitalization.

Despite the demand for Early Bird telephone channels, an article in *U.S. News & World Report* emphasized transatlantic television as the most "visible" capability of the new satellite.[107] The linking of Europe and America in a global television extravaganza on 2 May 1965 perhaps fed that image of the satellite. Before the launch of Early Bird, though, Hughes had announced that the technology to build a television broadcast satellite was available. Harold Rosen of Hughes suggested that Arthur C. Clarke's 1945 dream of a television broadcast from space could be made a reality with the NASA Applications Technology Satellite program.

---

103. Robert C. Cowen, "Commercial Relay Satellite has Date for Spring," *Christian Science Monitor,* 15 December 1964, p. 3; Eric Wentworth, "Comsat Gyrations Scrutinized by Federal Agencies in Case a Sudden Price Plunge Sparks Public Furor," *Wall Street Journal,* 22 December 1964, p. 3; "NASA Signs Agreement to Launch Comsat's Early Bird," *Space Daily,* 28 December 1964, p. 269.

104. W.J. Weber, Memorandum for the Record, "Summary of HS-303, Early Bird Communications Satellite Program," 15 January 1965, NASA History Office.

105. NASA Mission Operations Reports, NASA History Office.

106. Comsat Press Release, "Early Bird Fact Sheet III: Early Bird," 30 March 1965, NASA Mission Operations Reports, NASA History Office.

107. "As Comsat Gets Down to Business," *U.S. News & World Report,* 12 April 1965, p. 7.

On 13 May 1965, ABC television filed with the FCC for permission to launch a television relay satellite—the first domestic communications satellite. Comsat responded, however, that Congress had granted it a monopoly on satellite communications, but it would be glad to provide a relay service for ABC. The potential market for a satellite television relay was huge. Estimates of AT&T revenues for relaying television on terrestrial circuits was about $50 million per year. Nonetheless, telephone traffic was to dominate Comsat and Intelsat satellites for many years. Even so, it was clear that television, not telephony, captured the public's attention and began to give rise to the notion of the "global village."[108]

Meanwhile, the scramble for Early Bird telephone circuits was under way. In June 1965, AT&T filed with the FCC for permission to lease 100 voice channels from Comsat. AT&T was still negotiating with European telecommunications administrations, but it expected to have thirty-six links with Britain, twelve with West Germany, ten with France, and more with other countries. RCA filed for thirty circuits, ITT wanted forty-one, and Western Union International sought fifty-five. In all, Comsat had requests for 226 out of Early Bird's 240 circuits.

Commercial service officially was not to start until 27 June 1965, but the FCC granted emergency permission to begin commercial service after the failure of one of the transatlantic cables days earlier. The FCC then allocated voice circuits to AT&T (seventy-five circuits), ITT (ten), Western Union International (ten), RCA (ten), and Canada (six). AT&T managed to find partners for sixty of its circuits, but only one other circuit was in operation: an RCA circuit to Germany. European stations were not yet capable of handling the full traffic load, so the allocation of all circuits was held in abeyance. A long article in the *Wall Street Journal* bemoaned the slowness of the Europeans.[109] The Europeans were making it clear that international telecommunications required two equal partners. They were not going to tolerate a global satellite communications system dominated by the United States.[110]

European developments aside, Comsat was about to acquire a major domestic customer. Negotiations with NASA would provide Project Apollo with high-quality, wideband, global communications systems linking NASA's tracking stations in real time with the Houston Mission Control Center. This led to Comsat filing an application with the FCC on 30 September 1965 to build and launch four geosynchronous communications satellites to provide NASA those services. Comsat proposed to buy those four satellites from Hughes at a total cost of $11.7 million, while NASA assumed a total liability (monies the agency would pay if it did not use the system) of $10.5 million for the entire network. At

108. Barry Miller, "Hughes Proposes TV Broadcast Satellite," *Aviation Week & Space Technology* 82 (1 February 1965): 75–77; Howard Simon, "Test of Early Bird's TV Heightens Optimism," *Washington Post*, 8 April 1965, p. A2; "Early Bird Satellite Relays a TV Signal in a Surprise Test," *Wall Street Journal*, 8 April 1965, p. 2; "Early Bird Orbit is Nearly Perfect," *New York Times*, 10 April 1965, p. 11; Suzanne Montgomery, "Early Bird Operation May be Speeded," *Missiles and Rockets* 16 (12 April 1965): 12; Val Adams, "2 Continents See Global TV Today," *New York Times*, 2 May 1965, p. 35; Jack Gould, "Comsat to Assay ABC Satellite," *New York Times*, 15 May 1965, p. 63.

109. Jerry E. Bishop, "Lagging Early Bird," *Wall Street Journal*, 2 August 1965, pp. 1, 14.

110. "First Formal Bid to be Comsat Customer Filed by AT&T With FCC," *Wall Street Journal*, 3 June 1965, p. 4; "RCA Follows AT&T in Asking to Become Customer of Comsat," *Wall Street Journal*, 4 June 1965, p. 6; Larry Weekley, "Early Bird Line Demand Grows," *Washington Post*, 8 June 1965, p. D8; "Comsat Gets Go-Ahead to Put Early Bird Into Commercial Use," *Wall Street Journal*, 21 June 1965, p. 3; "Early Bird Given Leasing Go-Ahead," *New York Times*, 24 June 1965, p. 57.

a time when only one-fourth of the Early Bird capacity was being used, Comsat had acquired a major new customer.[111]

On the last day of 1965, Comsat released its first quarterly report, which listed revenues of $966,000 from Early Bird operations. Given that these revenues were gathered during the first half of the life of an investment in excess of $7 million, those figures were quite disappointing. On the other hand, Early Bird was primarily an experiment, and it had been an extremely successful experiment. Comsat still had almost $188 million in cash out of its initial $200 million capitalization, and it had contracted to provide Atlantic and Pacific Ocean service to support the Apollo program.[112]

A Choice Is Made

Still, no decision on the basic Comsat system had been made. Comsat had pushed forward with Early Bird (Intelsat I), an experimental operational system, and Intelsat II, the special-purpose system launched to provide NASA communications services. When Comsat granted the Early Bird geosynchronous contract to Hughes, the communications company also awarded two study contracts for medium-altitude-orbit satellites to two teams, AT&T-RCA and TRW-ITT. The assumption beneath those study contracts was that a medium-altitude experimental system might follow Early Bird some time in 1966. However, events were unfolding too swiftly for that kind of experimentation; furthermore, experts generally assumed that Telstar and Relay had proven the medium-altitude case. It was time to choose.

Comsat had a variety of system designs from which to choose. Hughes, the builder of Early Bird and Intelsat II, as well as Comsat's partner in the venture to supply military communications satellite services, proposed an "advanced Early Bird," the geosynchronous HS-304. AT&T and RCA, the builders of Telstar and Relay, proposed a system of eighteen satellites in random polar orbits. These two firms held one-third of Comsat's shares, and AT&T was the dominant telecommunications company. TRW and ITT proposed a controlled, or "phased," system of twelve satellites in similar orbits. In May 1965, Comsat's Joseph Charyk announced that the corporation was no longer considering the "random" approach.[113]

Then, on 17 August 1965, Comsat, on behalf of the Interim Communications Satellite Committee, issued a request for proposals for an "advanced satellite" (Intelsat III), the basic satellite system design. The request for proposals stipulated that the satellite should be suitable for use in either a geosynchronous orbit *or* in a phased system at altitudes between 6,000 and 12,000 miles (9,656 and 19,312 kilometers), although preferably applicable to *both* orbital types. The request for proposals further specified a capacity of 1,000 two-way

111. Comsat Press Release, "Comsat Files Application for FCC Authorization for Proposed Satellites for Apollo Service," 30 September 1965; Robert C. Seamans, Jr., to Dr. Joseph V. Charyk, letter, 23 November 1965; Robert C. Seamans, Jr., to Dr. Joseph V. Charyk, letter, 30 December 1965, all in NASA History Office. Also see "Comsat Seeks to Buy 4 Super Early Birds," *New York Times,* 20 October 1965, p. 11; "Comsat Files Apollo Satellite Contract," *Space Daily,* 21 October 1965, p. 262; "First Commercial Satellite to be Placed Over Pacific," *New York Times,* 4 November 1965, p. 28C; "4 Satellite System Ordered by Comsat," *New York Times,* 25 November 1965, p. 3.

112. Comsat Press Release, "Comsat Gives Highlights of Corporation's Progress in an Interim Report to Shareholders," 31 December 1965, NASA History Office.

113. "Comsat Studies," *Aviation Week & Space Technology* 80 (6 April 1964): 19; "The Big Four in Comsat Competition," *Space Daily,* 9 April 1965, p. 223; "Early Bird Follow-On," *Aviation Week & Space Technology* 82 (19 April 1965): 28; Lyle Denniston, "Comsat Satellite Decision Hints of Earlier Profits," *Washington Evening Star,* 11 May 1965, p. 72.

Figure 21

The Intelsat V series of geosynchronous communications satellites illustrates the great growth in satellite size that had been achieved since the first Intelsat, Early Bird, launched in 1965 (Figure 20). Its greater size (and weight) compared to its predecessors suggests the larger number and variety of functions it could perform—and had to perform—as the number of geostationary slots began to fill up. (Courtesy of NASA, photo no. 85-H-402)

voice circuits, a five-year lifetime, a weight of approximately 240 pounds (109 kilograms), two repeaters, a directional antenna, and multiple access capability. RCA, TRW, and Hughes submitted bids. On 16 December 1965, Comsat announced that it was negotiating with TRW for at least six Intelsat III satellites at a cost of approximately $20 million. TRW had offered a design capable of operating at both medium and geosynchronous altitudes. Comsat still had not decided which orbit to use.[114]

On 29 December 1965, again on behalf of Intelsat, Comsat issued a new request for proposals for design studies for its fourth-generation satellites (Intelsat IV). These would have a capacity of 6,000 voice circuits (or ten TV channels) and a five-year lifetime, and they would weigh less than 2,300 pounds (1,043 kilograms). These satellites would be *geosynchronous*. Comsat had finally chosen a system![115]

Arthur C. Clarke could be proud, but so could John R. Pierce, Harold A. Rosen, Donald D. Williams, Tom Hudspeth, Sid Metzger, Siegfried Reiger, Leonard Jaffe, and the hundreds of other engineers who had helped to make commercial satellite communications a reality. Political forces may have determined the form of the final system, and economic forces had provided much of the impetus, but none of these forces would have produced the global communications satellite system without the efforts of the engineers and technicians who envisioned, developed, demonstrated, and deployed the "billion dollar technology."

---

114. Comsat Press Release, "Comsat Asks Manufacturers to Submit proposals on Advanced Satellite for Global Communications," 17 August 1965, NASA History Office; "Early Bird Follow-On," *Aviation Week & Space Technology* 83 (23 August 1965): 27; Larry Weekley, "Comsat Negotiates With TRW on Large Satellite Contract," *Washington Post,* 17 December 1965, p. C8; "Comsat is Negotiating for New Satellites With TRW; Users' Charges May be Cut," *Wall Street Journal,* 17 December 1965, p. 24.

115. John Noble Wilford, "Comsat Seeking Bigger Satellite," *New York Times,* 30 December 1965, p. 21; "Comsat Asks Designs for Big, New Satellite," *Washington Evening Star,* 30 December 1965, p. A9.

Conclusion

All of the technologies for communications satellites had existed in some form since at least the end of World War II when Arthur C. Clarke wrote his *Wireless World* article. Most, especially the electronics technologies, were relatively mature by 1955 when John R. Pierce wrote his article for *Jet Propulsion*. Missiles, the critical enabling technology, arrived on stage on 4 October 1957. In the words of Pierce: "The necessary spurs to concrete action [on communications satellites] came with the successful launching of Sputnik I by the USSR."[116]

Equally important, the international telecommunications market was booming in the late 1950s and could provide sufficient revenues to underwrite a "billion dollar technology." AT&T was ready to finance a global system out of its own funds. General Electric and Lockheed, probably in pursuit of satellite sales, were anxious to form a consortium to share the risk. Hughes, GTE, and ITT were anxious to discover a way around AT&T's dominant position in the telecommunications market.

Meanwhile, the U.S. government seemed to have many minds during the Eisenhower administration. While the president himself seemed to have been wholly in favor of private (AT&T) development and ownership of a global satellite communications system, others within the administration and civil service were less enthusiastic. Some felt that AT&T was monopolistic; others believed that satellite communications was something government should do. The Kennedy administration was less committed to private ownership and more committed to the potential prestige of a government-sponsored program. The Bay of Pigs and the flight of Gagarin in the spring of 1961 finally tilted government away from private development of communications satellites, thereby guaranteeing that AT&T would not be able to dominate satellite communications.

Prior to the government takeover of communications satellite development in 1961, private industry had undertaken extensive development using its own funds. Hughes had built and tested a Syncom prototype. AT&T had built large antennas for the Echo program, had designed Telstar, and had built a few prototype Telstar components. If other companies were not spending the millions of dollars invested by Hughes and AT&T, they nonetheless were pursuing communications satellite and ground station technology with their own resources.

In spite of the Pentagon's lack of direct involvement in commercial communications satellite development, the military market allowed commercial manufacturers to develop the expertise that would have allowed them to compete with AT&T in the early days and with Hughes later on. General Electric based its attempt to enter the commercial communications satellite business in the early 1960s on its experience with the military's Advent program. Despite this early work, General Electric did not launch a communications satellite until 1978. The Japanese Broadcast Satellite built by General Electric was similar to Advent in many ways, albeit a far more sophisticated version. General Electric later built the DSCS III series for DoD, and after purchasing RCA, the company allowed RCA's Astro Division to continue building communications satellites, while General Electric's Space Division concentrated on military and Earth-observing markets. RCA, after being out of the market from the last Relay (1964) until the launch of Satcom F-1 (1975), was very successful in the U.S. domestic satellite market.

---

116. John R. Pierce, *The Beginnings of Satellite Communications* (San Francisco: San Francisco Press, 1968), p. 9.

TRW (through its subsidiary Space Technology Laboratories) entered the communications satellite arena as the systems engineers for the Pentagon's Advent and NASA's Relay programs. In 1965, the company won the Intelsat III series contract. Hughes was favored in that competition, but that firm refused to design a satellite capable of functioning in either a geosynchronous or a medium-altitude orbit. TRW built additional satellites for DoD and NASA, but it was never strong in the commercial market. Lockheed, for its part, lobbied for a place for aerospace companies within the organizational structure that became Comsat. Later, the company attempted to enter the domestic communications satellite market as an operator with CML Satellite Corporation (later Satellite Business Systems), but not until late 1995 did Lockheed begin building communications satellites (for the Iridium low-Earth orbit system).

The most successful manufacturer of communications satellites has been the Hughes Aircraft Company. From early 1959, when the Syncom design began to come together, to 1961, when NASA gave Hughes a sole-source contract for Syncom development, Hughes devised various designs and strategies to enter the communications satellite business as manufacturer, operator, or some combination of the two. One strategy involved allowing NASA to fund satellite development much as the Atomic Energy Commission had funded the development of nuclear power plants. Although only one of many strategies, top management supported it, and it is a fair representation of what took place.

NASA considered the Hughes design faulty, but the agency chose it as the next most promising program when the Kennedy administration made available increased communications satellite research and development funding in mid-1961. Problems with the military Advent program, which had been in trouble from the start, contributed to the joint NASA-DoD decision to proceed with Syncom. Development work on Syncom then led to the commissioning of the Advanced Syncom study program and, somewhat later, to the Intermediate Syncom study program. Advanced Syncom metamorphosed into NASA's Applications Technology Satellite program, while the Intermediate Syncom became the NASA-sponsored Intelsat II series. Between 1963 and 1968, Hughes launched thirteen communications satellites sponsored directly (Syncoms 1 through 3 and Applications Technology Satellites 1 through 5) or indirectly (Early Bird and four flights of the Intelsat II) by NASA.

If the government—namely, Congress and NASA—had not intervened, an AT&T-dominated medium-altitude system would almost certainly have been launched in the mid-1960s. The Hughes design eventually would have flown, but when, and with what success, is problematic. Government intervention had two main effects. First, the success of the Syncom series, and the NASA commitment to the geosynchronous orbit, made geosynchronous the logical choice for a commercial system. Second, the demonstration of the Hughes geosynchronous system gave Hughes an advantage over all of its potential competitors. This advantage grew as NASA persuaded Comsat to launch a version of Syncom using commercial frequencies (Early Bird) and offered to be an anchor tenant for the Intelsat II series. The NASA Applications Technology Satellite kept Hughes at the top of its technical form,[117] while Intelsat flew the TRW-built Intelsat III series.

Government intervention had very little effect on the development of new technology, however. Much of the electronics was developed originally at AT&T's Bell Telephone Laboratories and improved later by other companies using internal research and development funds. Hughes developed the basic "spinner" and "gyrostat" technologies using

---

117. It should be pointed out that future Hughes systems depended on the "gyrostat" principle developed at Hughes by Anthony Iorillo and demonstrated on the Army TACSAT.

its own funds. "Three-axis" technology derived not from NASA's sixth Applications Technology Satellite, but rather from the Television Infrared Operational Satellite and Defense Meteorological Satellite Program by RCA and internal studies carried out by Ford and others. Later government programs, notably the Applications Technology Satellite, had large effects on the application of geosynchronous technology. Some of these applications have never quite made it into practice (such as aeronautical communications and tracking), while others (geosynchronous weather satellites, for instance) have become ubiquitous.

There is a place for government involvement in commercial technologies, but myth and conventional wisdom must be examined carefully before taxpayer money is committed to the achievement of improbable goals. The government market has been a stronger force than direct sponsorship of research and development, and it has been stronger than government-funded "demonstrations" in the development of commercial technologies.

PART II:

Creating the Global, Regional, and National Systems

———⟫•⟪———

Europe

Chapter 10

Launching the European Telecommunications Satellite Program

by Arturo Russo

This chapter is based on research performed as part of the history project of the European Space Agency (ESA) in collaboration with John Krige and Lorenza Sebesta. For this history project, the author conducted a special study of the launching of the telecommunications satellite program of the European Space Research Organization (ESRO), a multinational space organization created in Europe in the early 1960s, whose activities ESA took over in 1975. The following is a discussion of the results of that study, which can be found in detail in two ESA History Study Reports.[1]

The United States dominated the history of satellite communications up to the end of the 1970s. Europe was a latecomer. Although the first plans for a joint European communications satellite program were elaborated in 1965, many years passed before the program was implemented and a European satellite finally was launched. Only at the end of 1971, in fact, did the ESRO member states approve a program aimed at developing an experimental satellite, and not until 1978 did that satellite, named the Orbiting Test Satellite (OTS), leave the launch pad. ESA approved the second phase of the telecommunications program in February 1978, with the aim of developing an operational satellite, named the European Communication Satellite (ECS). The first ECS was placed in orbit in 1983; three others followed in 1984, 1987, and 1988, respectively, thus completing the ECS system.

By the time the first ECS satellite began operating, however, six Intelsat V satellites already were orbiting over the Earth's oceans. Also, domestic communications satellites had been under development by private companies in the United States for ten years, and some 18 U.S. satellites were in orbit. Meanwhile, Intelsat itself was providing domestic services in more than 20 countries; Canada already had launched seven Anik satellites; Japan, India, and Indonesia also had acquired independent space communications capability; and in Europe, two Franco-German Symphonie satellites were approaching the end of their orbital life after several years of good performance. While not all these spacecraft were as complex and up to date as the ECS, the European system was certainly too late to play a major role in the competitive market of space communications.

The Big Picture

Why did it take such a long time to develop a European communications satellite program? To answer this question, one must first consider two main aspects. The first is the institutional framework in which the program was to be developed. Two multinational space organizations existed in Europe during the 1960s and early 1970s: one was devoted

1.   Arturo Russo, *The Early Development of the Telecommunications Satellite Program in ESRO (1965–1971)*, ESA HSR-9 (Noordwijk: ESA, May 1993); Arturo Russo, *ESRO's Telecommunications Program and the OTS Project (1970–1974)*, ESA HSR-13 (Noordwijk: ESA, February 1994). For the history of ESRO and ELDO, see John Krige and Arturo Russo, *Europe in Space, 1960–1973*, ESA SP-1172 (Noordwijk: ESA, September 1994).

to scientific research (ESRO), and a second, called the European Launcher Development Organization (ELDO), whose original mission was to build a rocket called Europa, became devoted to launching heavy satellites into near-Earth orbits. The eventual involvement of ESRO and ELDO in the new field of geostationary communications satellites implied a change in their charter and operational program. This change was not easy, owing to their different missions, structures, and memberships. Only six European countries, plus Australia, were members of ELDO, whose programs were defined essentially at the governmental level. Each member state was responsible for a specific task vis-à-vis the final goal of building the three-stage Europa vehicle and associated facilities, and each had a different level of financial and technical commitment. ESRO, on the other hand, consisted of ten member states, each of which contributed financially in proportion to their gross national product, and the European space science community defined its programs.

Figure 22

Approved in 1971, but not launched until 1978, the European Space Agency's experimental satellite, named the Orbiting Test Satellite (OTS), shown here, served as the forerunner of the space agency's subsequent communications satellites, known as the European Communication Satellite (ECS). (Courtesy of NASA)

ESRO member states were not equally interested in communications satellites; some of them were firmly against the organization's involvement in space applications. ELDO, for its part, was hampered by severe technical, financial, and managerial problems that resulted in a dramatic series of test launch failures of the Europa rocket. As a consequence, strong disagreements existed among ELDO member states about the viability of a European launcher for geostationary satellites. A collaborative European effort in satellite communications could only be developed within the framework of a comprehensive space policy in which national economic interests and political goals could be satisfied. The development of such a policy required many years of laborious negotiations.

The second reason why it took so much time to develop a European communications satellite program was the question of users. During the 1960s and 1970s, all European countries provided telecommunications services through state-owned operators. These state telecommunications administrations had the authority to act as the monopoly providers of all services considered at that time as suitable for a European satellite system—namely, international telephony, telex and data transmission, and international television distribution within the Eurovision network.

The state telecommunications administrations, while interested in supporting research and development studies of communications satellites, were hardly optimistic about the economic prospects of a European system. According to their estimates, in fact, the large investments required for procuring and launching satellites, as well as for building the necessary ground stations, would be greatly in excess of the savings achieved by transferring traffic from the terrestrial cable network to a satellite system. Over the short average distances between European centers, and without the problems posed by large oceans or vast undeveloped regions, it made no sense to put sophisticated transponders as high as 36,000 kilometers in the sky and to build huge antennas on the ground.

Two Package Deals

This, then, was the background to the launching of satellite telecommunications in Europe. Starting the program required a redefinition of the respective roles of ESRO and ELDO within the framework of European space policy. ESRO was the obvious candidate to develop communications satellites, based on its experience gained in the field of scientific research. Indeed, ESRO's executive and technical staff enthusiastically looked forward to working on satellite communications. A balanced program of both scientific and applications satellites implied, in fact, a more efficient use of capital resources, a more equitable distribution of industrial contracts among member states, and interesting jobs for recruiting the best engineers. ELDO, for its part, would be called on to develop appropriate launch vehicles.

Following the successful launch and operation of Early Bird (Intelsat I), Comsat's first satellite, in March 1965, ELDO member states agreed in July 1966 to fund a program to make the Europa rocket capable of launching small satellites into geostationary orbit. Later that year, ESRO was charged with studying an experimental satellite for telephony and television services in Europe and the Mediterranean. At the same time, the European Space Conference came into being as a political forum for discussing a comprehensive space policy for Europe. At its second meeting, held in Rome in June 1967, the conference set up a special committee to elaborate a long-term program that included scientific research, applications, and launchers.

The transition from the study phase to actual development, however, encountered three main difficulties. First, important elements of the European space science community expressed concern about the eventual extension of ESRO into applications satellites. Many scientists feared that economic and commercial interests rapidly would eclipse scientific objectives and that political factors would dictate program priorities, rather than the need to investigate scientifically interesting phenomena beyond the atmosphere.

The second difficulty derived from the diverging visions of the ESRO member states about the prospects of satellite communications in Europe. For some, notably France and Germany, there was no doubt that Europe should undertake a vigorous effort in this field, to reduce the perceived technological gap with the United States in this strategically important industrial sector. France went so far as to suggest that ESRO should concentrate on applications, leaving science to national space programs or multinational projects. Other ESRO member states held the opposite view. The British government, for example, following the advice of the powerful British Post Office, that country's state telecommunications administration, opposed any direct involvement in the space segment of satellite communications. They believed, in fact, that very few possibilities existed for autonomous European action in this field, because of both the strength of the American presence and the foreseeable small commercial demand for the kinds of satellites Europe could build and operate. For Britain, as well as for some of the smaller member states such as Switzerland and the Scandinavian countries, ESRO was doing well in science, and space research had to remain its main mission.

The third difficulty to the transition from the study phase to actual development of a European communications satellite system was the question of a launcher—namely, whether Europe should develop its own, more powerful launch vehicles or should rely on American rockets. Britain and France led the opposite camps. The British stressed the high cost and poor performance of the proposed ELDO launchers compared to American missiles; the French argued that Europe could not sustain a credible space policy without the availability of its own launchers.

The ambiguity of the U.S. position regarding the provision of their launchers fueled the controversy. NASA, in fact, had always been available to provide launch facilities for European scientific satellites, but it was by no means clear whether this availability would be extended to commercially oriented satellites. Eventually, the U.S. government stated that it would reserve the right to refuse launch facilities to European communications satellites potentially harmful to Intelsat's commercial interests. Britain claimed that the Americans should be trusted for their willingness to offer launch assistance to Europe whenever possible. In contrast, the American policy strengthened France's belief that Europe needed its own launchers to support a viable communications satellite program.

Political negotiations and technical studies dragged on for three years until the disagreement reached a breaking point at the fourth meeting of the European Space Conference, held in Brussels on 4 November 1970. Discussions were interrupted after the first day, as France, West Germany, and Belgium declared that they were prepared to go ahead on their own. There was no point, they said, in trying to define a European space policy when the partners' priorities differed so greatly. Britain announced its intention to leave ELDO, while France denounced the ESRO Convention.

After three more years of laborious negotiations, a solution emerged in the form of two package deals. These global agreements satisfied the national interests of all the countries by abandoning the ideal of a mandatory comprehensive space program. ESRO member states agreed on the first package deal in December 1971. It definitely transformed ESRO from an organization solely devoted to scientific research into one mainly involved in applications satellite programs, with only a minor fraction of its budget devoted to science. Only the scientific program remained mandatory, while all application programs were optional. The second package deal, agreed on in July 1973 during meetings of the European Space Conference, established the basic principles on which the new European Space Agency would be created two years later.

The main element of the second package deal was the à la carte program system: each ESA member state contributed to the space agency's various programs in proportion to its own political, economic, and industrial interests. Through this second package deal, France took prime responsibility for the Ariane rocket development program and agreed to contribute 63 percent of the cost. Similarly, West Germany took prime responsibility for

Spacelab, a manned scientific laboratory to be carried in the Space Shuttle cargo bay, and Britain took responsibility for a maritime communications satellite.

The ESA-EUTELSAT Agreement

It was within the framework established by the first package deal, however, that the satellite communications program finally came into being, with the support of eight of ESRO's member states (Belgium, Britain, Denmark, France, Italy, Sweden, Switzerland, and West Germany). It is important to note, however, that the program was divided into two different stages, each requiring approval by a qualified majority of the participating countries. Only the experimental phase of the program actually was approved; a decision on the operational phase was deferred until mid-1975.

The main reason for staggering the decision-making process was the persistent uncertainty about the commercial viability of the planned European communications satellite system. Unlike Intelsat, which was meeting a real need by improving communications between continents, a European satellite system seemed, on the evidence available, to be a luxury that Europe did not really need and that the state telecommunications administrations could not afford. According to a 1971 study prepared by the European Conference of Postal and Telecommunication Administrations (known by its French acronym, CEPT), the estimated operational cost of the system in the 1980s would be far in excess of the savings in the terrestrial cable network. Indeed, the CEPT study reported that terrestrial savings would not even cover the costs of the ground stations! The state telecommunications administrations made it clear that the development of the European space industry could not be financed by customers. The governments, in other words, had to pay the difference between the actual costs of the satellite system and those costs that users normally would have to pay.

The CEPT report raised great concern. It implied that potential satellite users would not use the system unless governments subsidized its operation. In the words of the British delegation to the ESRO Council: "Had the project under discussion been a U.K. national project it would certainly have been turned down in view of [this] report." In fact, the communications program was not a national project, but an international undertaking, as well as part of the much wider political context in which the first package deal was being discussed. The economic arguments in the CEPT report, however, could not be dismissed, so a program strategy was elaborated to allow the participating states to defer the decision on the program's operational phase.

Nonetheless, five years later the economics of European satellite communications were far from being settled. A 1976 CEPT study reported an estimated difference of about $200 million between the total operational cost of the satellite system and the terrestrial alternative. A solution seemed possible. The state telecommunications administrations would build and operate the ground stations, while management of the satellite segment would be entrusted to a new international organization, called EUTELSAT. ESA member states were expected to cover the shortfall in operating costs.

An agreement was not immediately at hand, however, and took another three years of harsh negotiations. The telecommunications administrations held a strong negotiating position. Indeed, they had never formally requested the introduction of a satellite system. Space policy, not the needs of the users, had motivated the ESA telecommunications program, and its member state governments could not help accepting the users' conditions to bring it to completion. So, in May 1979, ESA and EUTELSAT finally reached an agreement. ESA would contribute as much as 60 percent to EUTELSAT's operating costs.

Conclusion

From the users' point of view, it seems that none of the studies that examined the economics of the ECS program were able to demonstrate that its implementation would produce economic benefits in excess of the combined cost of the space segment and its associated ground station infrastructure. Wider economic arguments and political motivation provided the ultimate justification for program approval. Among those wider aspects are the following:

1. The assertion of Europe's political and technological independence from the two superpowers, which was a particularly key element of the French government's space policy

2. The recognition of the aerospace sector as strategically important for the development of advanced industrial technology

3. The need to qualify European industry for competitive participation in Intelsat procurement contracts

4. The understanding that the OTS and ECS programs were the heart of an evolutionary program leading to other applications fields, such as aeronautical and maritime telecommunications, direct television broadcasting, and Earth observation

5. The search for autonomy of political and cultural expression (as the influential French newspaper *Le Monde* declared in 1967: "The transmission of radio and television programs is one of the most supple and diversified means to assure a presence and influence abroad.")

6. The general drive toward European economic and technical integration

National governments had different views about the relative importance of these aspects, and these differences made launching the European communications satellite program such a long and complex process. In particular, the critical issue of launchers vis-à-vis the commercial prospects of application satellites was the real stumbling block that risked jeopardizing the entire European cooperative space effort. The 1973 package deal was a real landmark in the history of Europe in space, and it is fair to conclude by recalling that it was an Ariane rocket that put the ECS into orbit.

Chapter 11

U.S.-European Relations and the Decision to Build Ariane, the European Launch Vehicle[1]

by Lorenza Sebesta

Scholars generally recognize, although with different accents, that the U.S. policy on the availability of launchers for European communications satellites influenced the European decision to design and build its own launchers.[2] This decision, officially endorsed in July 1973, led to the construction of Ariane, which today, after more than a decade of technical reliability and good management, has assured itself the majority of the global commercial market.

What is still unclear, though, are the reasons for the U.S. position on launcher technology and facilities, and how those reasons evolved over time—from the first restrictive directive, National Security Action Memorandum (NSAM) 338 of September 1965, through the more flexible, and uncertain, position conceived in the second half of the 1960s, to the return to a more restrictive policy publicly announced by President Nixon in October 1972. Shaping this trajectory of changing U.S. policy were five different factors:

1.   The author would like to thank John Krige, head of the ESA Project at the European University Institute in Florence, Italy, and John Logsdon, director of the Space Policy Institute at George Washington University in Washington, D.C., for their insightful discussions of the topics addressed in this article. The author also thanks Richard Barnes, international space consultant in Washington, D.C., and André Lebeau, professor at the Conservatoire National des Arts et Métiers in Paris, for their invaluable criticism and comments. The form and expression of this disputed story are entirely the author's own responsibility. A more extensive analysis of the period under examination will be found in a book-length analytical history of U.S.-European space relations that the author is currently writing with John Logsdon. The author also would like to acknowledge the extremely valuable assistance of the NASA History and Security Offices in facilitating the declassification of documents essential to this work.

2.   For example, see John Logsdon, "International Involvement in the U.S. Space Station Program," *Space Policy,* February 1985, p. 18: "[T]he fact that there was resistance in providing that assistance reinforced the position of those in Europe (particularly in France) who were arguing for developing an independent European space capability." Peter Creola, referring to President Nixon's 9 October 1972 policy statement on the availability of American launchers in "European-U.S. Space Cooperation at the Crossroads," *Space Policy,* May 1990, p. 99, wrote: "The effect of this policy on Europe was decisive." See also Arturo Russo, "Launching Europe into Space: The Origin of the Ariane Rocket," paper read at the International Astronautical Federation Annual Meeting, 1995. In "La naissance d'Ariane," p. 85, in Emmanuel Chadeau, ed., *L'ambition technologique: Naissance d'Ariane* (Paris: Éditions Rive Droite, 1995), this point of view was expressed much more vigorously by André Lebeau (former President of CNES, the French space agency): "Il ne semble pas exagéré de dire que si les États-Unis avaient vendu sans conditions particulières les deux lancements de Symphonie, la décision d'engager le programme Ariane n'aurait jamais pu être obtenue. Une intransigeance maladroite, fondée sans doute sur l'idée que l'Europe serait de toutes façons incapable de ressusciter son programme de lanceurs, vints à point pour fournir un appui décisif aux promoteurs de L3S."

1. Rising concern about the "technological gap" between Europe and the United States

2. Technological breakthroughs in the field of communications satellites and launchers, their organizational consequences, and commercial concerns about these developments

3. The increasing importance of ballistic missiles as a central feature of NATO military strategy, as well as the U.S. nuclear nonproliferation policy

4. A thorough reassessment of European space policy

5. A worsening of U.S.-European relations coincident with the international economic crises of the early 1970s

This chapter examines the tremendous changes that these five factors wrought, as well as how NASA tried to cope with them.

The "Technological Gap"

During the late 1950s and early 1960s on both sides of the Atlantic Ocean, the notion of technology as a key to economic growth gained wider and wider acceptance at the highest levels of decision-making. As tariff barriers between the United States and Europe began to relax, factors other than tariffs took on added relevance to economic growth and international competition. In Europe, the chief body concerned with these questions was the Organization for Economic Cooperation and Development (OECD).[3] OECD decision-makers, inspired by the works of such economists as Robert Solow and Edward Denison, believed that the expansion of the labor force and capital, and their relative prices, did not explain economic growth by themselves, but that a "residual factor" accounted for a remarkable percentage of economic growth. This residual factor progressively came to be identified as knowledge, science, and technology.[4]

A 1965 OECD study pointed out that the United States and the Soviet Union controlled the bulk of the world's financial and human resources in the field of research and technology. In particular, a "technological gap" divided the United States from its western allies. Higher U.S. spending on research and development by the state (mostly the military) in "technology-intensive" sectors seemed to have a direct positive influence not only on U.S. economic growth, but on the position of U.S. firms in international markets and on the growing number of U.S. firms investing in Europe.[5] Europeans viewed a drive toward high-tech space applications, as well as electronics, computers, and atomic energy, as a possible tactic to solve the technology gap.[6]

---

3.   Jean-Jacques Salomon, *Science et Politique* (Paris: Seuil, 1970), pp. 51–54.
4.   Edward Denison, *Source of Economic Growth in the United States and the Alternatives before Us* (New York: Committee for Economic Development, 1962).
5.   Christopher Freeman and Anthony Young, *The Research and Development Effort in Western Europe, North America and the Soviet Union: An Experimental International Comparison of Research Expenditures and Manpower in 1962* (Paris: OECD, 1965), p. 70.
6.   Jean-Jacques Servan-Schreiber, *Le défi américain* (Paris: Denoel, 1967), pp. 119–125; see also David Beckler, Assistant to the Director, to Philip Hemily, Science Adviser, U.S. Mission to the OECD, letter, 3 June 1966, RG 359, National Archives, Washington, DC.

Europeans faced a dilemma. By allowing American capital into their countries, they were consigning their industry to a subsidiary role, at least in the technological sectors. The result would be technological dependence, uncertainty over the availability of supplies, and loss of freedom in formulating industrial policy. On the other hand, if Europeans refused to let Americans invest, they risked ending up double losers by denying themselves the capital needed to create jobs, as well as manufactured products.[7]

The United States recognized that the technology gap should be treated as "a problem with serious political overtones," as Secretary of State Dean Rusk reminded NASA Administrator James E. Webb in August 1966, because it was perceived as such by the Europeans.[8] The U.S. trend toward space cooperation with Europe during the second half of the 1960s had its roots in a willingness by the United States to reduce the political impact—and in the long term, the economic effects—of the technology gap.

As had happened during the U.S.-European "dollar gap," the State Department suggested that it was in the interest of the United States to have a strong Europe as a partner to increase the prospects for U.S. economic growth.[9] This farsighted political vision, however, came under attack during the late 1960s, as more and more American economic sectors began to face European competition. In Europe, the technology gap and the need to catch up with the United States in the space sector served one main political purpose: to convince Europeans to turn from science to technologically relevant, commercially viable endeavors and to cooperate on technologically advanced projects, such as Concorde, Airbus, communications satellites, and commercial space launchers. Authors, such as French journalist Jean Jacques Servan-Schreiber in his book *Le défi américain,*[10] not only suggested this policy, but European space organizations also endorsed it.[11]

Intelsat and NSAM 338

The International Telecommunication Satellite Organization (known as Intelsat) was set up in August 1964 as a single commercial global satellite system embracing voice, telegraph, high-speed data, facsimile, and television services. Early Bird, Intelsat's first successful geostationary communications satellite, confirmed the promising commercial potential of communications satellites in 1965.

Under the Intelsat interim agreements, its executive body was the American-based Communications Satellite Corporation (Comsat), which, as manager, proposed and implemented projects. Investment shares within Intelsat were determined by projections of long-distance traffic likely to be carried by satellites. Comsat received an initial 61-percent share against a 30-percent share for all European countries. Because the voting system was based on investment shares, Comsat enjoyed a de facto veto power and

7.   Alfred Grosser, *The Western Alliance: European-American Relations since 1945* (London: Macmillan, 1980), pp. 217–31.

8.   Dean Rusk to James Webb, letter, 29 August 1966, RG 255, 70-A-3458, box 7, NASA History Office, Washington, DC. On the need to reduce the political impact of the technological gap, see also the Interim Report of the Work of the Space Council's ad hoc Committee on Expanded International Cooperation, enclosure 1, statement concerning political objectives for expanded cooperation in space activities, presented for the chairman at the working group meeting on 20 October 1966, RG 255, 69-A-5089, box 5, NASA History Office.

9.   See, for example, the 20 October 1966 statement concerning political objectives for expanded cooperation in space activities, by the State Department, RG 255, 69-A-5089, box 5, NASA History Office.

10.   Servan-Schreiber, *Le défi américain,* pp. 119–25.

11.   Joachim Müller, "Historical Background and Start of the TELECOM Program," *Space Communications* 8 (1991): 105–40; John Krige and Arturo Russo, *Europe in Space, 1960–1973,* ESA SP-1172 (Noordwijk: ESA, September 1994), pp. 55–82.

maintained that power notwithstanding subsequent decreases in its controlling share as new countries joined Intelsat.[12]

Although Comsat's privileged role in Intelsat, according to one analyst, "assured efficiency and speed" in setting up a global satellite system, and its resources "proved critical to attracting interest on the part of developing countries in joining the enterprise,"[13] it also nourished U.S. hegemony in the field, which was rooted in an almost total monopoly of the industrial sector. The early entry into the market by such American firms as Bell Telephone Laboratories, RCA, and Hughes, combined with their ability to draw on studies performed by NASA, gave them an advantage in international competitive bidding on Intelsat contracts. One of the big controversial issues within Intelsat was Comsat's willingness to give priority to in-house research and development over international contracting. It was only under pressure from other Intelsat members that the percentage of outside contract expenditures progressively rose from 13 percent in 1968 to 50 percent by 1972.[14] By that time, however, with 52 percent of its capital from the United States, Intelsat spent 92 percent of its money in the United States.[15]

The White House was aware of the degree of European dissatisfaction with the U.S. monopoly of commercial satellite communications, as well as the danger that, through direct assistance from U.S. firms, foreign satellite activity might proliferate the development of competitive systems.[16] President Johnson, after lengthy negotiations with NASA and the Departments of State, Defense, and Commerce, approved NSAM 338, "Policy concerning U.S. assistance in the development of foreign communications satellite capabilities," in September 1965.[17]

The aim of NSAM 338 was, in its own words, "to guide government agencies in the dissemination of satellite technology and in the provision of assistance which is consistent with overall policies." It stipulated that the United States should not provide assistance to other countries that would significantly encourage the development of their communications satellite systems. The United States, moreover, would not entertain any requests by foreign nations involving technological assistance on satellites or launchers unless the assistance obtained would be used in accordance with Intelsat's rules, and provision of such services would be "conditioned upon express (written) assurances." Nonetheless, the United States *would* provide its allies satellite services for their security needs.[18] This was indeed a very tight political directive that did not leave much room for flexibility in future international negotiations.

---

12.   Richard Colino, "The INTELSAT System: An Overview," in Joel Alper and Joseph Pelton, eds., *The INTELSAT Global Satellite System* (New York: American Institute of Aeronautics and Astronautics, 1984), p. 62. See also Marcellus Snow, *The International Telecommunications Satellite Organization (INTELSAT)* (Baden-Baden: Nomos Verlagsgesellschaft, 1987), pp. 43–48.

13.   Colino, "The INTELSAT System," p. 62.

14.   Stephen Levy, "INTELSAT: Technology, Politics and the Transformation of a Regime," *International Organization* 29 (Summer 1975): 661–64.

15.   Address by Professor Hermann Bondi, former Director General of ESRO, on International Cooperation in Space, Goddard Dinner at Symposium AAS, 18 March 1971, RG 255, 74-734, box 15, NASA History Office; Michael E. Kinsley, *Outer Space and Inner Sanctums: Government, Business, and Satellite Communication* (New York: John Wiley & Sons, 1976).

16.   Jim D. O'Connell to Jack Valenti, Special Assistant to the President, letter, 7 May 1965, White House Confidential Files, TR 105, box 96, Lyndon B. Johnson Library, Austin, TX.

17.   RG 255, 69-A-5089, box 5, NASA History Office; also in WHCF (confidential files), TR 105, box 96, Lyndon B. Johnson Library.

18.   A bilateral U.S.-Britain military satellite agreement was signed in 1967, whereby Britain would build an all-British satellite for military communications with Australia and the Far East within the framework of a collaborative Skynet military space communications system. Krige and Russo, *Europe in Space*, p. 62.

Military Concerns and Nonproliferation Policy

The Atlantic Alliance, created in 1949, had two main objectives: social stability and military security. To fulfill both objectives, this alliance always had relied on America's arsenal of nuclear weapons. The United States postponed sharing the military burden with Europe by emphasizing the deterrent power of its nuclear arsenal.[19] This position implied that the United States maintained control over the ultimate decision to use nuclear arms.

The launch of the Soviet Union's Sputnik in October 1957 made the cost of this arrangement seem very high to Europeans. U.S. territory now was open to Soviet aggression by intercontinental ballistic missiles. Europeans wondered: Would the United States be willing to risk an attack against their own territory for the sake of Europe? Sputnik epitomized the double nature of missiles: the same launcher that had put a scientific satellite in orbit could become, with some modifications, the carrier of a nuclear warhead. The United States now was caught in an inescapable nuclear dilemma: antagonizing its allies was dangerous, because it fostered the development of independent nuclear forces; however, the U.S. Constitution made the president the head of the armed forces and, therefore, the individual with the last say on the use of nuclear weapons, so sharing that decision with other states did not seem to be authorized by the Constitution.

The dilemma was quite real, for the French had requested information on an intermediate nuclear ballistic missile system from the United States. Two weeks after Sputnik, the National Security Council directed the Eisenhower administration to discourage the production of nuclear weapons outside the nuclear "club" and, specifically, to "persuade France not to undertake independent production of such weapons."[20] The National Security Council was referring to General Charles de Gaulle's *force de frappe* and its strategy *tout azimut*. Upon his return to power in June 1958, de Gaulle had accelerated studies of both launchers and nuclear warheads. The first French atomic bomb (tested in 1960) was further proof of this determination.[21]

Between 1962 and 1965, the United States devised a hybrid formula, the Multilateral Force (MLF), to appease European requests for nuclear technology and to satisfy constitutional law and the Atlantic Alliance, but the MLF proved to be a slow, inexorable failure.[22] By the mid-1960s, nuclear issues were at the core of NATO difficulties. The United

19.   Thomas H. Etzold, "The End of the Beginning . . . NATO's Adoption of Nuclear Strategy," in Olau Riste, ed., *Western Security: The Formative Years* (New York: Columbia University Press, 1985), p. 291.

20.   National Security Council 5721/1, "U.S. Policy on France," 19 October 1957, in U.S. State Department, *Foreign Relations of the United States 1955–57* (Washington, DC: U.S. Government Printing Office, 1992), Vol. XXVII, *Western Europe and Canada*, pp. 189, 192. For the U.S.-French rift on nuclear issues, see McGeorge Bundy, *Danger and Survival: Choices about the Bomb in the First Fifty Years* (New York: Random House, 1988), pp. 472–87. More generally, see Pascaline Winand, *Eisenhower, Kennedy, and the United States of Europe* (Basingstoke: Macmillan, 1993), *passim*.

21.   For the *force de frappe* and the organizational changes it implied, see Roger Rhenter, "Implications de la politique de défense dans les domaines de l'industrie aéronautique et de l'espace," in Institut Charles de Gaulle, *De Gaulle et son siècle*, Vol. 4, *La sécurité et l'indépendance de la France* (Paris: La Documentation Française-Plon, 1992), pp. 160–63. For an updated review of French nuclear policy, Maurice Vaïsse, ed., *La France et l'atome* (Brussels: Bruylant, 1994).

22.   The MLF was intended to be a coordinated multinational deterrent nuclear force consisting of a fleet of submarines carrying Polaris missiles manned by crews of a minimum of three nationalities. Decision-making would be shared, but not the ultimate responsibility for the use of the warheads, which was in U.S. hands. See George Ball, *The Discipline of Power* (Boston: Little, Brown, 1968), especially the chapter titled "The Unfinished Business of Nuclear Management."

States strongly resented France's unwillingness to comply with Atlantic Alliance strategy and its stated wish to build up its own nuclear arsenal. The French nuclear strategy, however, was linked to its economic policy. "In politics as in strategy as in economics," de Gaulle asserted in a much publicized January 1963 speech that referred to U.S. policy, "monopoly quite naturally appears to him who holds it as the best possible system."[23] In fact, de Gaulle's attacks against the dollar gold standard, launched in February 1965, when France presented its dollars for conversion into gold, were followed in 1966 by France's withdrawal from NATO and, in 1967, by that country's first nuclear ballistic missile tests.

The meaning of French behavior acquired a much more disturbing twist in the context of the new global nonproliferation policy inaugurated by the United States during the 1960s. After the Cuban missile crisis of October 1962, the United States and the Soviet Union began to ease the international tensions that had reached a climax during that long week. The two superpowers agreed on, and formalized, common codes of conduct in the nuclear arena, including limitations on the testing and production of nuclear devices to prevent their proliferation. The Treaty Banning Nuclear Weapons Tests in Atmosphere, in Outer Space, and Under Water (better known as the Test Ban Treaty) and the hot-line agreements were signed in 1963, while the nonproliferation treaty was signed in 1968 only after protracted negotiations.

These international developments paralleled domestically the adoption of NSAM 294, "U.S. Nuclear and Strategic Delivery System Assistance to France," on the nonproliferation of "strategic delivery technology," including nuclear bombs and launchers.[24] Enforcement of this policy was assigned to Munitions Control at the State Department—the agency also responsible for controlling technological information to be sold abroad. In the context of this memorandum, then, the French independent course was interpreted not only as a refusal of American patronage, but, more dangerously, as an attempt to disrupt the architecture of American nonproliferation policy.[25]

Europe's Changing Space Policy

After Sputnik, and in parallel with the process that led to NASA's creation in 1958, European scientists began to call for the creation of a collaborative space organization. As a result, an agreement establishing the European Space Research Organization (ESRO) was signed in 1962, but it did not enter into force until 1964. ESRO was born out of many interest, including the scientists' desire to conduct specifically ambitious experiments that natural resources alone would not have permitted, as well as the wish of some scientists to be independent of national military authorities. The establishment of ESRO also benefited from the spirit of European unity following the Treaty of Rome that created the Common Market.[26] The ESRO agreement glossed over references to technological and

23.   "Excerpts from remarks by de Gaulle, news conference, 14 January 1963," *New York Times*, 15 January 1963, p. 2. See also Bundy, *Danger and Survival*, p. 492.

24.   NSAM 294, 20 April 1964, NSC Records, National Archives. Good indirect information on the content of the directive are to be found in James Webb, NASA Administrator, to Robert McNamara, Secretary of Defense, letter, 28 April 1966, James Webb, box 2, Lyndon B. Johnson Library.

25.   Frank Costigliola, *France and the U.S.: The Cold Alliance since World War II* (New York: Twayne's International History Series, 1992), p. 134.

26.   Krige and Russo, *Europe in Space*. Unless otherwise stated, all information relating to ESRO and ELDO is drawn from this work.

industrial concerns. The relevance of space technology to future economic development was not indisputable at the time, notwithstanding the initial propagandistic effort of some aerospace firms; moreover, industrial policy was still perceived as a prerogative of only national governments.

Starting in 1960, political negotiations on the building of a European satellite launcher were under way, leading to the 1964 ratification of an agreement to establish the European Launcher Development Organization (ELDO). ELDO was born out of Britain's willingness to "Europeanize" and convert to civilian use a military missile already in development, the Blue Streak. The mission of this launcher always had lurked in the background of ELDO, whose first concern was to acquire (or in the case of Britain to maintain) a technical expertise in a high-technology area at a bearable price.

The total spent by Europe on space in the mid-1960s was but a small fraction of the amount of money spent by the United States.[27] Total U.S. domination of the satellite field (communications, meteorology, and navigation) seemed to loom ahead in the 1970s. To compete with the United States, Europe needed to shift its emphasis from science to commercial and technological endeavors. The United States had provided launchers for national and European scientific satellites for free or at low costs. Why, then, should Europe follow the more costly and risky track of applications satellites and powerful launchers? That was the heart of the debate.

The British labor government (1964–1970) was the most skeptical of the European states. As early as 1966, Britain made it clear that, as far as launchers were concerned, it favored reliance on the United States. The British opined that Europa I, ELDO's first launcher, would be "obsolescent and uncompetitive in cost and performance with launchers produced by the U.S." by the end of the decade.[28] This debate disrupted ELDO negotiations in 1966 and again in 1968, and it formed the base for endless quarrels into the 1970s. British Minister of Technology Anthony Wedgwood-Benn summed up its essence when he stated that he was "very much alarmed at the thought that because a thing is European, and because a thing is international, this somehow excuses us from applying economic criteria."[29]

Britain eventually *did* stay in ELDO, which moved closer to the French position of independence from the United States. In July 1966, ELDO approved an upgrading of the European launcher; Europa II would be capable of putting into geosynchronous orbit a 170-kilogram satellite, compared to 140 kilograms with a Thor-Delta launcher.[30] In addition, on the insistence of the French, ELDO transferred its launch site from Woomera, Australia, to Kourou, French Guiana, where the French space agency was building its own launch complex.[31]

Still, no firm decision on the building of a European communications satellite was reached until December 1971. Among the hurdles to overcome in reaching that decision was the unclear legal framework for satellite operations—the Intelsat interim agreements were under renegotiation between 1969 and 1971—as well as the availability of launchers.[32]

27.   Müller, "TELECOM Program," pp. 379–81.
28.   Quoted in Krige and Russo, *Europe in Space,* pp. 74–75.
29.   Cited in John Krige, "Britain and European Space Policy in the 1960s and Early 1970s," *Science and Technology Policy* 5(2) (1992): 15. European cost estimations at that time made it clear that ELDO launchers were expected to be twice as expensive as their American counterparts. Müller, "TELECOM Program," p. 115.
30.   Jean-Pierre Causse, "Les lanceurs européens avant Ariane," in Chadeau, ed., *L'ambition technologique,* p. 24.
31.   Krige and Russo, *Europe in Space,* pp. 74–76.
32.   Arturo Russo, *ESRO's Telecommunications Programme and the OTS Project (1970–1974),* ESA HSR-13 (Noordwijk: ESA, February 1994), *passim.*

Both the Intelsat negotiations and the debate over the availability of launchers were inter-related, and Europe hoped to improve its bargaining position by building up a credible industrial and technical competence—and consequently a political presence—in the field of applications satellites.

That thinking was partly behind the Franco-German Symphonie program. This bina-tional project combined two national experimental communications satellites (called Saros 2 and Olympia) within a single spacecraft, whose launch was originally scheduled for 1970.[33] Symphonie, its supporters hoped, would put German and French industry in a preferred position when Europe built a communications satellite, but it also would be an asset during Intelsat renegotiations—not to mention a test of U.S. willingness to launch European commercial satellites.[34]

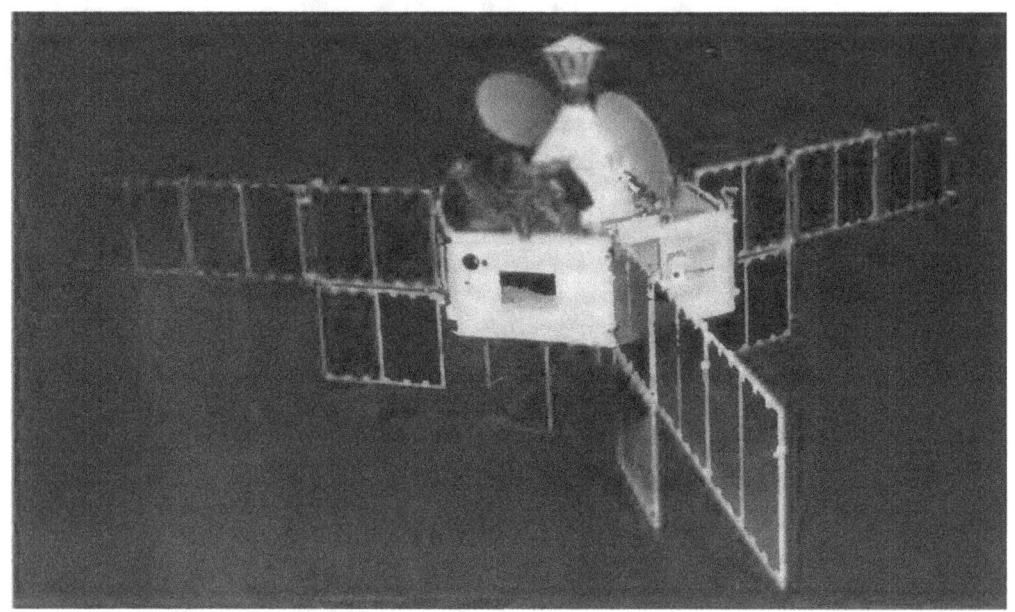

Figure 23
The Symphonie satellite was a binational satellite communications project sponsored by France and Germany. The project was a move to place the two countries in a favored position when Europe began to build communications satellites. NASA agreed to launch Symphonie with the strict understanding that it was to be only an experimental satellite. (Courtesy of NASA, photo no. 74-HC-635)

33.   Communiqué de presse, Symphonie, 28 April 1967, côte 81/244, article 188, liasse 517; confiden-tial note on the revision of French space policy on European launchers, no date (post-1966), no author (CNES or Minister of Foreign Affairs), côte 82/254, article 25, liasse 80, Archives Nationales, Fontainbleau.

34.   Nicholas G. Golovin to Donald Hornig, memorandum on trip to Europe, meeting with Bignier, 25 October 1967, RG 359, box 658, National Archives; White House to the President, memorandum, 8 February 1967, White House Confidential Files, TR 105, box 96, Lyndon B. Johnson Library.

Increased U.S.-European Cooperation?

During the mid-1960s, when space expenditures reached their historic peak, the United States sought to increase cooperation with Europe in space. The new interest in the use of space cooperation as a means toward achieving political objectives abroad put NASA in a very delicate situation. Namely, the danger was that cooperative projects would not always reinforce NASA's programmatic needs. In such instances, a NASA working group on increasing international cooperation in space reported that the State Department and the White House ought to justify such projects, "since NASA cannot itself justify a relaxation of its posture and programmatic needs."[35]

Cooperation in space between the United States and Europe had been developing since the late 1950s along the restrictive lines established by Arnold Frutkin, director of NASA international affairs starting in September 1959. The United States offered free launch services and space aboard spacecraft for European scientific experiments; joint undertakings were "purely" scientific. The partners considered this cooperative arrangement highly beneficial, and it gave rise to few occasions of rancor.[36]

Late in 1965, NASA proposed a new project to its European space partners. Known as the Advanced Cooperation Project, the proposal remained within the political directives set up by NSAM 338 (September 1965), yet raised the level of transatlantic cooperation in space. As part of the Advanced Cooperation Project, Europe would be responsible for developing a technologically advanced spacecraft for either a solar or Jupiter mission, and the United States would provide launch, tracking, and data collection services.

As a French official expressed to Charles Bohlen, the U.S. ambassador in France, the offer seemed to be but "a bone to nibble on."[37] Moreover, other critics perceived the Advanced Cooperation Project not as fostering space development, but as diverting Europe from the economic benefits to be derived from developing communications satellites.[38] The proposal failed to galvanize the European scientific community, and in the summer of 1966, ESRO officially declined it.

The United States, still wishing to increase international cooperation in space, transformed the Advanced Cooperation Project into a bilateral venture with West Germany known as Helios.[39] Cooperation with West Germany had become more critical to U.S. diplomatic policy, particularly following France's withdrawal from NATO,[40] and as West Germany came to be viewed as the "most faithful ally" of the United States in Europe.[41]

---

35.   Meeting of the Working Group on Expanded International Cooperation in Space Activities, "Summary Notes," 22 September 1966, RG 255, 69-A-5089, box 5, NASA History Office.

36.   Arnold Frutkin, *International Cooperation in Space* (Englewood Cliffs, NJ: Prentice-Hall, 1965); Homer Newell, *Beyond the Atmosphere: The Early Years of Space Science* (Washington, DC: NASA SP-4211, 1980); John Logsdon, "U.S.-European Cooperation in Space Science: A 25-Year Perspective," *Science* 223 (6 January 1984): 11–16; Sir Harrie S.W. Massey and Malcolm Owen Robins, *History of British Space Policy* (Cambridge: Cambridge University Press, 1986); Lorenza Sebesta, *U.S.-European Cooperation in Space during the Sixties*, ESA HSR-14 (Noordwijk: ESA, 1994).

37.   Charles Bohlen to Department of State, telegram, 8 March 1966, RG 255, 69-A-5089, box 5, NASA History Office.

38.   See, for instance, the report of the Western European Union cited in "Europe Accuses U.S. on Space Plans," *New York Times*, 8 June 1966, p. 4.

39.   Alexander Hocker to James Webb, letter, 3 August 1966, RG 255, 70-A-3458, box 7, NASA History Office.

40.   Dean Rusk to James Webb, letter, 29 August 1966, RG 255, 70-A-3458, box 7, folder 1; Daniel Margolis to Donald Hornig, memorandum, 13 December 1968, NAW, RG 359, box 755, NASA History Office.

41.   Costigliola, *France and the U.S.*, p. 148.

For its part, West Germany, through its participation in U.S. space efforts, was interested in acquiring a wide range of military-related technology that strictures dating from the end of World War II prohibited that country from producing. During the early 1960s, moreover, West Germany purchased military goods and services from the United States to offset the costs of stationing American occupational troops in that country. Now, the West German government hoped to substitute for at least part of those payments the cost of procuring and licensing high-tech American equipment to establish a pattern of technical cooperation and to reestablish German technical capabilities.[42]

Meanwhile, the United States continued to reconcile space cooperation with high-level deliberation. In March 1966, a special ad hoc committee of the National Aeronautics and Space Council, chaired by Deputy Under Secretary of State U. Alexis Johnson, was set up to advise President Johnson on the topic. The United States favored the work of ELDO, because its rocket programs tended to serve peaceful uses, were subject to international control, and absorbed personnel and financial resources that otherwise might be diverted into purely national programs that tended to concentrate on military systems. The dissolution of ELDO, the United States feared, might strengthen national military rocket programs and lead to a proliferation of launchers.[43] At the same time, argued Vice President Hubert H. Humphrey and Secretary of Defense Robert S. McNamara, any increased emphasis on the peaceful uses of space technology would go hand in hand with a decrease in European programs of independent military applications. The aim of U.S. policy, according to McNamara, should be to stimulate "foreign involvement in space technology as a means of diverting energies from the development of nuclear systems."[44] Moreover, in the case of France, encouragement to proceed with upper-stage liquid hydrogen-oxygen systems might divert resources from their *force de frappe* program, which used solid fuel propulsion technology.[45]

In addition to the dissolution of ELDO, the United States also was distressed over collaboration among ELDO, ESRO, and the European Telecommunications Satellite Committee (known by its French acronym CETS), because such collaboration could be a prelude to the creation of a combined competitive global space power. As Frutkin wrote in May 1966: "The greatest danger now is that the crises in space affairs in Europe will lead to a total redirection of European space effort in competition with the United States." If ELDO, ESRO, and CETS together established their own system of communications satellites, Europe would challenge and seriously disrupt Intelsat, which was dominated by the United States. "It seems very important," Frutkin declared, "in view of this possibility and in view of the difficult 1969 Intelsat renegotiation that everything be done to give the Europeans as little cause of concern as necessary regarding U.S. motivation. Certainly, no dog-in-manger attitude ought to be continued."[46] NASA Administrator James Webb added to Frutkin's concern. Once the Europeans completed construction of the Kourou launch complex, "the European nations could, if they wish," Webb opined, "be in a position to

42. Margolis to Hornig, memorandum, 13 December 1968.

43. Committee on Expanded International Cooperation in Space Activities, "Cooperation involving launchers and launching technology," first meeting, 17 May 1966, RG 255, 69-A-5089, box 5; Arnold Frutkin to James Webb, "Visit of Sir Solly Zuckerman," memorandum, 5 May 1966, RG 255, 69-A-5089, box 7, NASA History Office.

44. Vice President Hubert Humphrey to Donald Hornig, memorandum, 6 April 1966, NAW, RG 359, box 566; James Webb to Robert McNamara, letter, 28 April 1966, James Webb, box 2, Lyndon B. Johnson Library.

45. Webb to McNamara, letter, 28 April 1966; Michel Debré, *Gouverner: Mémoires, 1958–1962* (Paris: Albin Michel, 1988), p. 375.

46. Arnold Frutkin to James Webb, memorandum, 11 May 1966, RG 255, 69-A-5089, box 8, NASA History Office.

place in synchronous orbit an operable comsat spacecraft."[47] If the United States could not stop the creation of a European regional communications satellite system, the new Intelsat organization, then under renegotiation, might provide a framework in which to control the European system.[48]

The Limits of American Launcher Policy

American policy was caught in a dilemma. As NASA reported to the State Department, the United States was *"virtually at the limits of proposals for cooperation which [could] be made with any hope of success,* unless the U.S. should relax restrictions in the two areas of *prime* interest, vehicle technology and experimentation with comsat."[49] Those areas of interest fell under NSAMs 294 and 338, which NASA Administrator James Webb perceived were a "political irritant" to the Europeans and were "exacerbating existing political strains," especially with the French.[50] Revising NSAM 338 might improve U.S.-European relations, discourage Europeans from following a potentially competitive independent route, and even help the U.S. position in Intelsat negotiations scheduled to begin in 1969.[51]

The underlying idea was to liberalize U.S. policy on launching communications satellites. NSAM 338 had defined that policy in very strict terms and had left to the bodies and agencies of the newly formed Intelsat the responsibility to entangle the development of competitive international telecommunications satellites through a web of legal rules.[52] This idea was first embodied in a National Security Council directive approved as NSAM 354 by the president in July 1966 under the title "U.S. Cooperation with the European Launcher Organization (ELDO)."[53] The document called for positive support of ELDO and the provision of launch vehicles, components, and technology under precise conditions. For example, the technology could be used to improve the capability of communications satellites only in accordance with Intelsat agreements or to participate as an ally in the U.S. military satellite system. The technology was not to be used in nuclear missile systems nor to be passed on to other countries.[54]

In August 1966, in applying NSAM 354, the United States offered to support ELDO's development of a European launch vehicle. That support entailed the procurement of U.S. flight hardware, as well as assistance (such as technical information and personnel) in the design of subsequent ELDO projects using liquid-propellant upper stages. The

---

47.   James Webb to James O'Connell, Special Assistant to the President for Telecommunications, letter, 3 October 1966, RG 255, 70-A-3458, box 7, NASA History Office.

48.   Working Group of International Cooperation Subcommittee of the National Aeronautics and Space Council, "Summary Minutes," 19 May 1966, RG 255, 69-A-5089, box 5, NASA History Office.

49.   Emphasis in text. NASA to Department of State, "International Projects in Prospect," 19 May 1966, RG 255, 69-A-5089, box 5, NASA History Office.

50.   Webb to O'Connell, letter, 3 October 1966.

51.   No author, "Memorandum on Communications Satellite Technology," no date, 70-A-3458; Memorandum, Arnold Frutkin to Willis Shapley, 16 June 1966, 69-A-5089; both in box 7, RG 255, NASA History Office.

52.   Working Group on International Cooperation, "Summary Minutes," 19 May 1966.

53.   Arnold Frutkin to James Webb, "Space Council Task Group on Assistance to ELDO," supplementary notes on possible U.S. assistance to ELDO, memorandum, 14 June 1966, 70-A-3458, box 7; Third Meeting of the Working Group (to be held on 7 July 1966), 29 June 1966, and Fourth Meeting of the Working Group (to be held on 9 August 1966), 4 August 1966, 69-A-5089, box 5; all in RG 255, NASA History Office.

54.   Frutkin to Webb, "Space Council Task Group on Assistance to ELDO," memorandum, 14 June 1966; U.S. Cooperation with the European Launcher Development Organization (ELDO), 29 July 1966, RG 273, NSAM 354, National Archives.

United States also suggested the joint use of a high-energy upper stage to be developed in Europe, the sale of Scout, Thor, and Atlas rockets to the Europeans, and the provision for a fee for launch services for scientific and applications satellite projects.[55]

Formal discussions begun in September 1966 focused, at the request of ELDO, on general aspects of management (such as establishing adequate task definition, contractor selection, and contract supervision) and on certain specific technical problems relating to the injection of a satellite into geostationary orbit (namely, the ELDO-PAS program).[56] Subsequently, an ELDO team visited NASA Headquarters and the Goddard Space Flight Center, and various technical problems relating to ELDO-PAS were discussed.[57] Nonetheless, NASA eluded ELDO requests for technical comments on studies of the high-energy upper stages.[58]

In July 1967, the revision of NSAM 338, which had set strict limits on the supplying of launches for communications satellites to foreign entities, received President Johnson's endorsement.[59] Although substantial changes from the foregoing document were not readily noticeable, the change in perspective was evident from the start. Whereas the original text opened by declaring that "it is the policy of the United States to support development of a single global commercial communication satellite system to provide common carrier and public service communications," the revised policy declaration read: "The United States is committed to the encouragement of international cooperation in the exploration and use of outer space." In addition, the rules for transferring technology were slightly liberalized by the substitution of a more flexible expression, namely:

> [W]ithin the limits fixed by national security considerations and other pertinent regulations, the United States may decline to make available space technology to other nations when (a) such technology is critical to the development of a communication satellite capability and (b) it has been determined that this technology will be used in a manner inconsistent with the concept of and commitments to the continuing development of a single global commercial communications satellite system as embodied in the 1964 agreement.[60]

---

55. "Possibilities and Problems of Future U.S.-European Cooperation in the Space Field," remarks by Trevanion H.E. Nesbitt, Deputy Director, Office of Space and Environmental Science Affairs, Department of State, at the Meeting of EUROSPACE, Munich, Germany, 21 June 1968, Annex to ELDO/CM(July 68)WP/2, Historical Archives, European University Institute, Florence, Italy. Atlas, already phased out as a U.S. military vehicle, had a minimum of security difficulties (it used an old radio guidance system, for example), and it compared favorably against Blue Streak as a potential first stage of the European launcher. Frutkin to Webb, Space Council Task Group on Assistance to ELDO, memorandum, 14 June 1966.

56. Interim Response by ELDO to US Offer of Technical Assistance, by Clotaire Wood, NASA European Representative, 5 December 1966, RG 255, 69-A-5089, box 7; W.H. Stephens, Secretariat ELDO, to C. Wood, NASA Representative, U.S. Embassy, letter, 23 January 1967, RG 255, 70-A-3485, box 8, NASA History Office.

57. The ELDO team consisted of W.J. Mellors, ELDO, Assistant Technical Director, ELDO-PAS Project; T.W. Wood, ELDO, Head, PAS Vehicle Section; J.C. Poggi, Société pour l'Étude et la Realisation d'Engins Balistiques (SÉREB), Chief Engineer, PAS Project; and M. Lauroua, SÉREB, Head, PAS Vehicle Coordinated Department. Richard Barnes, Memorandum for the Record, 10 February 1967; Stephens to Arnold Frutkin, letter, with attached: "Annex on Questions on Injection of Spin Stabilized Satellites into Geostationary Orbits," 24 March 1967; W.J. Mellors to Gilbert Ousley, Technology Directorate, Goddard Space Flight Center, letter, 22 May 1967, RG 255, 70-A-3485, box 8, NASA History Office.

58. "ELDO-NASA relations, 1967, major events," attached to memorandum, Lloyd Jones to George Morris, 7 March 1968, RG 255, 72-A-3153, box 6, NASA History Office.

59. NSAM 338 revised, policy concerning U.S. assistance in the development of foreign communications satellite capabilities, 12 July 1967, National Security Archives, Washington, DC.

60. Foreign use of the national defense communications satellite system continued to be contemplated along the lines of the old text. A bilateral agreement with Britain along these lines was signed in 1967, whereby Britain would build an all-British satellite for military communications with Australia and the Far East within the framework of a collaborative Skynet military space communications system. Krige and Russo, *Europe in Space*, p. 62.

The principal new assumption of the document was the inevitable development of new regional communications systems. If the United States did not encourage those regional systems to join Intelsat, then, predicted Edward C. Welsh of the National Aeronautics and Space Council, "I would expect that the international system will be the one which breaks up and fails."[61] The goal of U.S. policy, therefore, was to attract regional systems into the Intelsat framework where they could be controlled,[62] but, as a NASA paper on the dissemination of technology overseas cautioned, "The health of Intelsat is assured in part by the feeling of the major Intelsat partners that they are indeed partners and not puppets in an organization dominated by the U.S."[63] If the United States were too stringent in imposing technology export controls, those nations might conclude that the United States did not intend to allow them to compete, and they might work together to create a competing satellite system, or even to defeat upcoming Intelsat negotiations.

The willingness of the United States to liberalize its space technology policy was put to a test in 1968, when the directors of the Franco-German Symphonie communications satellite program asked NASA to provide launch vehicles and service for two satellites. After consulting with the Department of State, NASA replied in October 1968 that it would launch the satellites for a fee, "if we could arrive at a mutual understanding of the experimental character of the project."[64] Also, regarding the eventual future use of the satellite system, the Europeans were asked to comply with Intelsat's rules. NASA's reply, carefully conceived within the logic of NSAM 338, was interpreted by Symphonie directors as a U.S. refusal to launch European communications satellites should they advance from an "experimental" to an operational phase.[65] Therefore, they endorsed the use of a European launcher, even if this prospect seemed less certain and more expensive.

The U.S. Space Program and the Permanent Intelsat Agreements

The year 1969 included the confluence of the U.S. space program and Intelsat negotiations. In the spring of 1969, Intelsat renegotiations opened, and in July, the Apollo 11 astronauts became the first humans on the Moon. In addition, U.S. offers to entice Europeans into participating in post-Apollo space programs were not entirely distinct from the Intelsat negotiations.

In October 1969, NASA Administrator Thomas O. Paine offered Europeans the opportunity to participate in the development and use of an ambitious set of space transportation and exploration projects—namely, a space station module, a reusable transportation system (the Space Shuttle), a tug to transfer payloads from the Shuttle to geosynchronous orbit, and a nuclear-powered rocket (NERVA) for long-distance interplanetary travel. If Europe could be convinced to abandon its "trouble-plagued and

61. Memorandum for the File, "Edward C. Welsh on Questions regarding Communications Satellite Policy," 25 November 1966, RG 255, 69-A-5089, box 5, NASA History Office.

62. Working Group of International Cooperation Subcommittee of the National Aeronautics and Space Council, "Summary Minutes," 19 May 1966.

63. NASA memorandum, "Control of Foreign Dissemination of Technology," 25 April 1966, RG 255, 69-A-5089, box 5, NASA History Office.

64. The citation comes from a retrospective summary of U.S. policy on launcher availability included in NASA Administrator Thomas O. Paine to Senator Clinton Anderson, letter, 9 September 1970, box 26, Paine Papers, Library of Congress, Manuscript Division, Washington, DC; Arnold Frutkin, interview with John Logsdon and Lorenza Sebesta, 8 November 1993, Washington, DC.

65. Robert Aubinière, interview with Lorenza Sebesta, 12 December 1991, Paris.

obsolescent" launcher program, Paine had argued in the summer, "European funds would be freed for more constructive cooperative purposes."[66]

However, as European negotiators explained during their first meeting to discuss Paine's offer held in September 1970, because Europe only had limited means, it would be unable to finance simultaneously the development of its own launchers (for communications and other applications satellites) and significant participation in the post-Apollo programs. Therefore, European entry into any such cooperative venture would have to be complemented by the granting of launchers on a commercial basis and without political conditions by the United States. The United States replied that, if Europe contributed substantially to post-Apollo programs, the Americans would provide launch services to Europe on a reimbursable basis "for any peaceful purpose consistent with existing international agreements." Furthermore, at the request of the European representatives, the American delegates clarified that the phrase "any peaceful purpose" could include commercial ventures capable of competing with American interests.[67]

The United States made it clear, moreover, that the Europeans would be required to contribute at least 10 percent of the overall development costs of the Space Shuttle. Those costs then were projected as being $10 billion over ten years; the European share would have been $1 billion spread out over the same period. Broadly speaking, Théodore Lefèvre (who was the president of the European Space Conference and the chief European negotiator) pointed out, that amount would correspond to expenditures on the development of the European launcher.[68]

At the same time as these NASA program talks were taking place, Intelsat negotiations were under way. The Europeans were striving to obtain a more equitable partnership within the system, and they succeeded in obtaining some good results.[69] Among the issues under discussion was the establishment of regional satellite systems outside the jurisdiction of the Intelsat network. The United States initially argued against regional satellites, but the Intelsat Definitive Agreements of 1971 opened the way for them to meet members' needs for international public communications services. In each case, though, members were to ensure the technical compatibility of the regional satellite with the Intelsat network and to avoid significant economic harm to the global system. Furthermore, Intelsat was not permitted to enforce sanctions against violators, nor were its recommendations considered binding, as the United States originally had demanded. This clause was all the more relevant, because the Definitive Agreements deprived Comsat, the American signatory, of the veto power it had enjoyed under the Interim Agreements.[70]

These significant American concessions, however, were offset by a modification of the interpretation of the voting formula in Article XIV (paragraph d). Drafted in ambiguous

66. Thomas O. Paine to the President, letter, 22 August 1969, box 23, Paine Papers, Manuscript Division, Library of Congress.

67. Statement by Piers van Eesbeek relating to the Washington talks (16–17 September 1970) between the European Space Conference (ESC) delegation and ESC authorities, 8 October 1970, CSE/CS (70) 23, Historical Archives, European University Institute.

68. Declaration by Théodore Lefèvre, 4 November 1970, CSE/CM (November 70) PV/1, Annex 1, Historical Archives, European University Institute. Some disagreement seemed to exist on this point. Some observers believed that the cost of European participation in NASA post-Apollo programs would be twice the cost of developing the European launcher.

69. Levy, "INTELSAT," pp. 655–80.

70. *Ibid.*, pp. 670–71. The text, with annexes, of the INTELSAT agreements, as well as the operating agreement and annex, all of which went into effect 12 February 1973, can be found in *Space Law and Related Documents: International Space Law Documents: U.S. Space Law Documents,* 101st Cong., 2d sess., June 1990, S. Print 101-98, pp. 211–318.

terms in order to reach a consensus on the text, the clause left Europeans wondering what kind of majority was needed for the approval of an international satellite outside the Intelsat network, a prerequisite at that time for its launch by the United States. In a letter dated 2 October 1970 to Théodore Lefèvre, Under Secretary of State Alexis Johnson stated that the United States was prepared to launch European satellites "in those cases where no negative finding is made by the appropriate Intelsat organ, regardless of the position taken by the U.S. in the vote."[71] The Europeans understood this somewhat "baroque" definition to mean that a two-thirds vote against the proposed satellite would be required to defeat it; if, on the other hand, less than two-thirds of the seventy-seven Intelsat members were opposed, the United States would be agreeable to launching it. In other words, Europe needed only a little more than one-third of the votes to obtain Intelsat permission to launch its satellite.

In a subsequent letter of 5 February 1971, Johnson clarified the U.S. offer, which was substantially limited. Instead of a two-thirds vote of the Intelsat assembly to defeat a proposed regional satellite, a two-thirds affirmative vote was needed to support the proposal.[72] According to NASA's Acting Administrator George M. Low, this reversal, if not accompanied by a specific commitment in advance by the United States to support the regional European communications satellite proposal within Intelsat, would "effectively kill the chances for post-Apollo participation by Europe."[73] The U.S. change of position was linked to pressures exerted by Comsat and the U.S. aerospace industry.[74]

The origin of this policy change cannot be understood without considering the role of the Office of Telecommunication Policy. Clay T. Whitehead, a young and resolute systems analyst from the Massachusetts Institute of Technology, had directed this office since its inception in September 1970. The aim of the office was to define American policy vis-à-vis satellite communications for overseas civilian operations, as well as to support the American aerospace industry against what was perceived as attempts by NASA and the State Department to endanger the U.S. monopoly in communications satellites. On 7 January 1971, in the much publicized "Statement of Government Policy on Satellite Telecommunications for International Civil Aviation Operations," the Office of Telecommunication Policy called for the international utilization (as opposed to international development and utilization, as NASA had proposed) of a specialized aeronautical communications satellite system (called Aerosat) for international civil aviation operations.[75] The office's statement, however, had wider implications. In February 1971, one

71. U. Alexis Johnson to Théodore Lefèvre, letter, 2 October 1970, p. 5, CSE/Comité ad hoc (71)9, 22 April 1971, Historical Archives, European University Institute.

72. U. Alexis Johnson to Théodore Lefèvre, letter, 5 February 1971, CSE/Comité ad hoc (71)10, Historical Archives, European University Institute.

73. Memorandum to the file (teleconference between Dr. George M. Low and Under Secretary Alexis Johnson), 13 January 1971, RG 255, 74-734, box 17, NASA History Office.

74. Michael Freudenheim, "Satellite Splits U.S., Europeans," *San Francisco Sunday Examiner and Chronicle,* 7 March 1971, cited in Burl Valentine, "Europe and the Post-Apollo Experience," *Research Policy* 1 (1971–1972): 117. For more detail on the pressure from Comsat, see Memorandum to the file (teleconference between Low and Johnson), 13 January 1971, RG 255, 74-734, box 17; Arnold Frutkin, Memorandum to the file (Lefèvre meeting preparation, Johnson/Charyk discussions), 25 January 1971, RG 255, 74-734, box 14, NASA History Office.

75. "Nixon Administration announces policy on aeronautical satellite communications," press release, 7 January 1971, Richard Nixon Project, White House Confidential Files, Subject Files, UT 1, box 14, Executive Office of the President, National Archives. This statement was supplemented by another one issued 19 March 1971, "The National Program on Satellite Telecommunications for International Civil Aviation Operations," attached to letter, Nilson to Hammarström, 2 April 1971, folder 50771, Historical Archives, European University Institute, Florence, Italy.

month later, Clay Whitehead heavily criticized U.S.-European negotiations on post-Apollo space programs, whose sole outcome, in his opinion, would be to give away "space launchers, space operations and related know how at 10 cents on the dollar" (a reference to the American proposal that Europe share 10 percent of the development costs).[76]

The Europeans reacted strongly to the new restrictive American policy. Lefèvre found it "confirmed neither by the joint preparatory work nor by the wording used in the text" (of the Intelsat agreements) and asked the United States for a further clarification of its position.[77] Not until September 1971, after the signing of the Definitive Intelsat Agreements, did Lefèvre receive the clarifications he had been asking for since March.[78]

According to those clarifications, the new U.S. position was that the availability of American launchers was not conditioned on European participation in post-Apollo space programs. Those two issues were now, for the first time, separate. The United States would offer launch services for satellites intended to provide international public communications services, including European regional satellites, provided that the Intelsat governing body approved this by a two-thirds majority vote. The proponents of a regional satellite, then, would bear the burden of persuading the General Assembly that the proposal would not cause significant economic harm to the global network and would be technologically compatible with Intelsat. The United States would consider a vote by Intelsat to be binding, contrary to the general interpretation of Article XIV.

The geographical area to be covered by any European regional satellite system, however, was another point of contention. The preliminary provisional satellite system outlined by ESRO Director General Hermann Bondi at the European Conference in Venice in September 1970 proposed voice, data, and television services within the member states of the European Conference of Postal and Telecommunication Administrations (CEPT), but only television services to countries of the European Broadcasting Area as defined by the International Telecommunications Union (ITU), which extended from Iceland to the North African coast and from Portugal to Lebanon and Israel. Representatives of the Office of Telecommunication Policy, the Federal Communications Commission, and the State Department's Bureau of Economic Affairs examined the proposal and judged that it "would appear to cause measurable, but not significant, economic harm to Intelsat." The United States, though, would not support the proposal if the services offered in the European Broadcasting Area embraced anything more than television. The United States felt that the satellite would cause significant economic harm to Intelsat and was, therefore, "clearly unacceptable" to the United States.[79]

Representatives of the Committee of Alternates of the European Space Conference (which included representatives of both ELDO and ESRO) warmly welcomed the "decoupling" of launcher availability and participation in NASA space programs.[80] The ESRO

---

76.   Clay Whitehead to Peter Flanigan, memorandum, 6 February 1971, Nixon Project, White House Confidential Files, Subject Files, box 2, National Archives.

77.   Théodore Lefèvre to U. Alexis Johnson, letter, 3 March 1971, CSE/Comité ad hoc (71)12, Historical Archives, European University Institute.

78.   On this and other aspects relating to the American decision-making process during the Intelsat negotiations, see Lorenza Sebesta, "The Politics of Technological Cooperation in Space: U.S.-European Negotiations on the Post-Apollo Programme," *History and Technology* 11 (1994): 317–41.

79.   Under Secretary of State Johnson to Minister Lefèvre, text of letter, 1 September 1971, Annex I, 8 November 1971, CSE/Comité ad hoc (71)18, Historical Archives, European University Institute. See also the Department of State telegram on European participation in the post-Apollo program, "Visit of Minister Lefèvre," 24 February 1971, RG 255, 74-734, box 16; Second Discussion with Representatives of the European Space Conference concerning European Participation in the post-Apollo Program, 8 February 1971, RG 255, 74-734, box 16, NASA History Office.

80.   Joint Meeting of the Committee of Alternates and the ad hoc Committee of Officials, minutes, 22 September 1971, CSE/CS(71)PV/27th October 1971, Historical Archives, European University Institute.

Council soon afterward adopted the so-called first package deal, which called for U.S. and European participation in Aerosat, the creation of a weather satellite program, and coverage of the communications satellite program to include the full European Broadcasting Area as defined by the ITU. The package deal also reaffirmed the priority of European launchers, although on the condition that their cost would not exceed 125 percent of comparable non-European ones. If the United States denied launch services, then the actual cost of production, as well as any development costs, would be permitted.[81]

Lefèvre now requested from Johnson further explanation of the U.S. stand on launcher availability, as well as American support for a European communications satellite within Intelsat, based on specific operational systems, missions, geographical coverage, frequency bands, and technical configurations.[82] In his reply of June 1972, Johnson pointed to three difficulties: (1) its economic impact (higher charges to users); (2) technical incompatibility (the satellite's orbit placed it too close to the United States coastline); and, most important of all, (3) the definition of the European region. Johnson explained once and for all that the United States would not support the concept of the European Broadcasting Area as defined by the ITU, which covered the former French colonies of North Africa, the western portion of the Soviet Union, and the lands bordering the eastern Mediterranean, including Iraq, but not Saudi Arabia.[83]

In October 1972, President Nixon laid out the U.S. position on the availability of launchers in the following terms: "United States launch assistance will be available to interested countries and international organizations for those satellite projects which are for peaceful purposes and are consistent with obligations under relevant international agreements and arrangements." With respect to communications satellites, Nixon declared:

1. The U.S. will provide appropriate launch assistance for those satellite systems on which Intelsat makes a favorable recommendation in accordance with Article XIV of its definitive arrangement. 2. If launch assistance is requested in the absence of a favorable recommendation by Intelsat, the United States will provide launch assistance for those systems which the United States had supported within Intelsat so long as the country or international entity requesting the assistance considers in good faith that it has met its relative obligations under Article XIV of the definite arrangement. 3. In those cases where requests for launch assistance are maintained in the absence of a favorable Intelsat recommendation and the United States had not supported the proposed system, the United States will reach a decision on such a request after taking into account the degree to which the proposed system would be modified in the light of the factors which were the basis for the lack of support within Intelsat.[84]

81.   Report by the Secretary General of the European Space Conference on the Status of European Space Programmes, 7 December 1972, CSE/CM (Dec. 72)5, Historical Archives, European University Institute. See also Arturo Russo, *The Early Development of the Telecommunications Satellite Programme in ESRO (1965–1971)*, ESA HSR-9 (Noordwijk: ESA, 1993).

82.   Théodore Lefèvre to U. Alexis Johnson, letter, 23 December 1971, Annex, to CSE/CS (72)1, 4 January 1972, Historical Archives, European University Institute. The entire exchange of correspondence between Lefèvre and Johnson up to this date is in CSE/Comité ad hoc (71) 22, 22 December 1971. On the European communications satellite program, see Russo, *The Early Development, passim.*

83.   The definition of the European Broadcasting Area is cited in Availability of Launchers for the European Communication Satellites Programme, 22 September 1972, ESRO/PB-TEL.(72)5, Historical Archives, European University Institute.

84.   Richard M. Nixon, "United States Policy Governing the Provision of Launch Assistance," 9 October 1972, Office of the Press Secretary, Washington, DC.

This declaration gave rise to dissimilar interpretations in Europe and the United States. Whereas the Europeans saw it as sanctioning the de facto binding character of any Intelsat recommendation, U.S. officials, in contrast, stressed the second and third points and emphasized U.S. resolution to offer the broadest guarantee of flexibility vis-à-vis recommendations made under Article XIV.[85]

After the launch failure of Europa II in November 1971 and then its cancellation in April 1973, the directors of Symphonie were without a launch vehicle. They then turned to both the United States and the Soviet Union. The Soviet Intercosmos space agency did not oppose the launch but stated it would not be technically feasible until 1976, which was too late for the Europeans. The United States, on the other hand, was able to promise a first launch window in 1975. After protracted negotiations, which remain an object of dispute today, an agreement was reached in June 1974. In this agreement, France and West Germany confirmed the experimental character of Symphonie, and if the satellite entered an operational phase, the two countries agreed to conform to any decisions reached within Intelsat.[86] Meanwhile, the European Space Conference had adopted the so-called second package deal, whose three main programs included the Ariane launcher.[87]

Conclusion

Between 1965 and 1973, NASA confronted two conflicting trends: the easing of controls on launch services and overseas technology transfers and the pressure from the telecommunications industry to tighten those controls. NASA had to shape those opposing trends into a coherent policy vis-à-vis European requests for communications satellite launch services. Some U.S. officials viewed sharing technology with Europe as achieving two policy goals: (1) blunting European criticism about the technology gap and (2) diverting resources from military rockets to civilian launchers as part of a global policy of nonproliferation. The United States abandoned this strategy when France acquired its own military launching capability.

Much of the tergiversation between the United States and Europe over communications satellites and launch services took place within the framework of Intelsat negotiations for the Definitive Agreements. The ambiguous wording of Article XIV provided the United States grounds on which alternately to soften or harden its position on the creation of a European satellite system outside the Intelsat network. Behind the hardening of the U.S. position was pressure from Comsat and the aerospace industry, which exerted pressure directly or through the Office of Telecommunication Policy.

85.   Creola, "European-U.S. Space Cooperation," p. 99; Richard Barnes to John Logsdon, letter, 28 February 1996.

86.   Note pour le Conseil d'Administration du CNES [Centre National d'Études Spatiales] par le Secrétaire executif français de Symphonie, "Situation des possibilités de lancement du satellite Symphonie," 17 September 1973, and CNES, Secrétariat executif Symphonie, Rapport de présentation, 25 Octobre 1973, côte 81/244, art. 188, liasse 517; DGRST, "Note sur le programme Symphonie," 18 June 1974, côte 81/244, art. 187, liasse 515, Archives Nationales, Fontainbleau. See also the exchange of notes in *United States Treaties and Other International Agreements*, Vol. 25, Part 3 (Washington, DC: U.S. Government Printing Office, 1975). The author is indebted to Richard Barnes for his reminder of this last publication. The two Symphonie satellites were placed in orbit in December 1974 and August 1975. Claude Carlier and Marcel Gilli, *Les trente premières années du CNES: L'Agence française de l'espace, 1962–1992* (Paris: La Documentation Française, 1994), pp. 227–30.

87.   Krige and Russo, *Europe in Space*, pp. 111–12.

Figure 24
The launch of Ariane 1 on 24 December 1979 from
the French space agency's complex at Kourou, French
Guiana. (Courtesy of NASA)

The decision to proceed with the Space Shuttle program also had an impact on U.S.-European cooperation. The Space Shuttle seemed to promise an extraordinary qualitative leap in launch systems and their cost effectiveness, and it also seemed to make any European rocket obsolescent. Thus, to a certain degree, the Space Shuttle decision reduced U.S. interest in preventing the Europeans from developing their own launch capabilities.

Equally important was a real shift in U.S. policy vis-à-vis Europe. Whereas the Kennedy and Johnson administrations had tried to appease the Europeans, as confrontations between the United States and the Soviet Union played in the background, Nixon shifted from confrontation to détente, thereby weakening the political weight of the Atlantic partnership. Then, in the summer of 1971, the United States experienced its first trade deficit since 1894, as well as a severe reduction of gold reserves, which led to the decision to stop selling gold to foreign banks and to abandon the so-called Bretton Woods system (the fixed gold-to-dollar exchange rate instituted following World War II). A 10-percent across-the-board import duty was just another troubling economic signal to Europe. As often occurs, economic crisis feeds isolationism, especially in the face of an expanding, competitive European Economic Community (Britain, Denmark, and Ireland soon became new members).[88]

By the first years of the 1970s, efforts to liberalize American policy on launch services and the sharing of technology had come full circle. They had failed in the face of prevailing internal economic interests, increasing competition from European industry, changing priorities in U.S. foreign policy, and European developments in both the military and space fields. From the European perspective, the unwillingness of the United States to provide firm assurances of launcher availability for communications satellites was but one factor that lead to the European decision to endorse France's L-IIIS launcher (later known as Ariane). That decision has to be understood within the context of strained U.S.-French relations—itself a legacy of the dissension

    88.   Costigliola, *France and the U.S.*, pp. 167–72. See also Pierre Melandri, *Une incertaine alliance: Les États-Unis et l'Europe, 1973–1983* (Paris: Publications de la Sorbonne, 1988), pp. 45–77, and the insightful account written by the American ambassador to the European Communities, Robert Schaetzel, *The Unhinged Alliance: America and the European Community* (New York: Harper, 1975), pp. 42–53.

between Charles de Gaulle and Lyndon Johnson—which endured well into the 1970s, as well as the confused nature of the European space field.

Institutional uncertainty regarding the future of the European space organization was particularly acute between 1966 and 1971. Financial commitments were weak in comparison to NASA; industrial experience with satellite technology was limited; international legislation on communications satellites was not yet defined; and the attitudes of potential satellite users were conservative because of the uncertainty of commercial revenues, the high costs of the system, and anticipated problems of technological reliability.[89] Not until December 1971 did the ESRO Council endorse the start of a communications satellite program, the Orbiting Test Satellite.

Furthermore, not all the European states agreed entirely on the objectives of regional space policy. Britain always criticized proposals to build a European launcher and preferred the less expensive route of relying on U.S. satellites, while Italy was interested only in projects that guaranteed contracts for its industry (this was apparently not the case of the European launcher). West Germany, after the failure of Europa II in November 1971, was eager to assume the prime financial burden and contract management for Spacelab, while withdrawing its earlier support for an independent European launcher. France and Belgium were the only countries that never varied from their support of a European launcher.

Even within France, though, not everybody was in favor of a European launcher. Nonetheless, those who supported it—de Gaulle and then his successor, Georges Pompidou—created a strong constituency, and they made good use of American launcher policy to improve their position. Of equal importance is the fact that the technicians who first conceived Ariane did not look for a technological breakthrough, which would have been politically and economically difficult to champion given historical circumstances. Instead, they sought to design a technically easy and reliable rocket, drawing partly on knowledge acquired through the development of Diamant.[90] A national launcher was neither financially possible nor strategically convenient: the project had to be European to distribute the financial burden and to secure future users.[91] Also, it had to be technologically uncomplicated to prevent the cost overruns that had haunted ELDO's past.

In the end, all of the reluctant European partners were induced to participate in the second package deal. The European decision to build Ariane had many roots and motives, among which was the unwillingness of the United States to guarantee availability of launchers for operational communications satellites. The decision to build Ariane, however, was not assured until the very end. The hectic bargaining that took place in July 1973 testified to the difficulty of the process up to the very last moment, and it dramatized the central role of international bargaining. If West Germany and Britain did not have their pet projects (Spacelab and maritime communications, respectively) to protect and to garner support for, the birth of Ariane probably would have been a far more traumatic delivery.

---

89. Müller, "TELECOM Program," pp. 105–40.

90. See the illuminating contribution of André Lebeau, "La naissance d'Ariane" in Chadeau, ed., *L'ambition technologique*, pp. 75–91, and the ensuing debate among eyewitnesses of the time, *ibid.*, pp. 95–108.

91. CNES, Rapport groupe sectorial 6, Programmes d'études et développement des lanceurs, 30 June 1970, côte 81/401, art. 70, liasse 179, Archives Nationales, Paris.

Chapter 12

The Formulation of British and European Policy Toward an International Satellite Telecommunications System: The Role of the British Foreign Office[1]

by Nigel Wright

Several accounts of the establishment of Intelsat have tended to present Britain, and the British Post Office specifically, as a villain, desiring to obstruct the early development of a satellite system to secure an extra lease on life for its large investments in transoceanic telephone cables.[2] By concentrating on the supposed intentions of the Post Office, such accounts have tended to ignore the role of other government departments in the formulation of British policy. Moreover, while identifying the foreign policy-oriented nature of the Kennedy administration's satellite communications goals, the literature on Intelsat has not looked in any detail at the parts played by overseas foreign ministries within the overall course of the intergovernmental negotiations that produced Intelsat.

This chapter provides a fuller picture of British government thinking by concentrating on the role of the British Foreign Office in the preliminary phase of the exploratory intergovernmental discussions that occurred during 1962. This phase preceded both the formation of the European Conference on Satellite Communications (CETS, the French acronym) in 1963 and the formal multilateral negotiations that began at the start of 1964 and led to the Intelsat Interim Agreements of mid-1964.[3]

Throughout 1962, although British officials were aware of the potential threat that an American satellite system posed for British and Commonwealth cable interests, their desire to protect those interests did not equate automatically with a desire to frustrate the early development of satellite communications. British officials realized from the outset that satellites and cables would play complementary roles within the overall system of international communications. In fact, rather than wanting to stall satellite communications, many in Britain, both in and out of government, believed that Britain should construct its own satellite system in collaboration with the Commonwealth and Europe. This

---

1.   The research on which this chapter is based was supported by a Science and Engineering Research Council/Economic and Social Research Council studentship. The author would like to thank Philip Gummett for his suggestions and comments on an earlier draft.
2.   For examples of this tendency, see Delbert D. Smith, *Communication by Satellite: A Vision in Retrospect* (Leyden, MA: A.W. Sijthoff, 1976), p. 136; Walter McDougall, . . . *The Heavens and the Earth: A Political History of the Space Age* (New York: Basic Books, 1985), pp. 356–57.
3.   The formal period of negotiation is covered in Judith Tegger Kildow, *Intelsat: Policy Maker's Dilemma* (Lexington, MA: D.C. Heath, 1973); Jonathan F. Galloway, *The Politics and Technology of Satellite Communications* (Lexington, MA: Lexington Books, 1972). These works focus mainly on developments within the United States, however.

system would have been both separate from and competitive with a U.S.-initiated system. Foreign Office officials, however, considered that British interests would be better served if Britain were to cooperate fully with the United States in constructing a single world satellite system in which as many countries as possible were represented. It was inevitable that a single system would be led by the United States in its early years. Nonetheless, Britain's cable interests could still be safeguarded from within that system. That safeguard could be achieved if Britain insisted on taking an active role in the system's design and operation, thereby ensuring that it fully met the requirements of Britain and the Commonwealth, as well as Western Europe and the rest of the world.

Britain and the United States

By the early 1960s, Britain had long held a prominent and leading position in international telecommunications.[4] After World War II, Britain maintained this position within the framework of the Commonwealth Telecommunications Partnership, established under the terms of the Commonwealth Telegraphs Agreement of 1948.[5] During the early 1960s, the Commonwealth accounted for about 20 percent of the world's total intercontinental telecommunications traffic, with the United Kingdom alone accounting for 10 percent. This strong position was based on an extensive network of cable and radio facilities. These facilities were set to be extensively modernized through the 1960s with the installation, in successive stages, of a comprehensive system of high-capacity repeatered submarine telephone cables, which were a recent innovation first introduced in the mid-1950s. Spanning both the Atlantic and Pacific Oceans, and connecting together each of the major Commonwealth countries, the new network was expected to meet Commonwealth traffic growth into the 1970s.[6] The overall Commonwealth network, comprising the international communications facilities of each member country, was coordinated by the Commonwealth Telecommunications Board, through which each party to the 1948 Agreement was required to consult with its partners before extending or adding to its part of the network.

By 1960, several British government departments, including the Post Office and the armed forces, had expressed an interest in the practical and commercial potential of communication by satellite. As yet, however, the whole subject was ringed with various technical and commercial uncertainties. Very little experimentation had been done with actual satellites, even in the United States, and the commercial feasibility of satellite communications had yet to be demonstrated practically. Nevertheless, the Post Office, responsible

---

4.   The total net outstanding investment represented by the assets making up the Commonwealth system was approximately £110 million (more than $300 million). "Means of Associating Commonwealth Countries with a European Regional Organisation," undated draft, circa late August 1963, FO 371/171060, GPI173/289, Public Record Office, London (hereafter "London PRO"). For the origins and subsequent development of the Commonwealth network, see Daniel R. Headrick, *The Invisible Weapon: Telecommunications and International Politics 1851–1945* (Oxford, Eng.: Oxford University Press, 1991); Hugh Barty-King, *Girdle Round the Earth: The Story of Cable and Wireless and its Predecessors to Mark the Group's Jubilee 1929–1979* (London: Heinemann, 1979).

5.   The original parties to the 1948 Agreement were Australia, Canada, India, New Zealand, South Africa, Southern Rhodesia, and the United Kingdom. By the end of 1962, Ceylon, Cyprus, Ghana, Malaysia, and Nigeria had joined, while South Africa had withdrawn from the Commonwealth on becoming a republic in 1961.

6.   The Commonwealth "round-the-world" telephone cable scheme had been accepted by the Commonwealth governments during 1958. It had been estimated to cost a total of £88 million—a figure that included the Anglo-Canadian share of the first transatlantic telephone cable, TAT-1, laid in partnership with AT&T. Britain had agreed to contribute 50 percent of this total. "Proposed Commonwealth Pacific Telephone Cable System," *Post Office Electrical Engineers' Journal* 53 (April 1960): 42–44. This scheme included cables to South Africa that subsequently were constructed outside of the Commonwealth network.

for Britain's commercial telecommunications operations, realized that the U.S. lead in space and satellite technology, if translated into a U.S. monopoly over commercial satellite communications operations, could become a threat to the commercial viability of the Commonwealth network.[7] Despite this concern, the Post Office did not expect that satellites would ever wholly replace cables. Rather, it expected that the two media would have "complementary" roles within the overall system of international telecommunications.[8]

Mindful of the various uncertainties, the British authorities were keen to stay abreast of U.S. technical developments and, wherever possible, to become associated with U.S. experimental programs. British officials solicited technical discussions with their counterparts in the U.S. government. As a result, a joint civil-military mission headed by Major General Leslie de Malapert Thuillier of the Cabinet Office traveled to the United States during October and November 1960. During this visit, NASA officials invited the Post Office to participate in NASA's program of satellite communications experiments.[9] A memorandum of understanding was signed the following February, thereby laying the groundwork for Post Office involvement in NASA's experimental active satellite program, Relay. Later, following an agreement between NASA and AT&T, the Post Office also agreed to participate in the trials of AT&T's own experimental satellite project, Telstar.[10]

As its contribution to the Relay and Telstar tests, the Post Office undertook to design and construct an experimental ground station in Britain. Following extensive surveys of the southwest of England, a site was chosen at Goonhilly Downs on the Lizard peninsula in Cornwall.[11] The United States had built a large aperture horn antenna at Andover, Maine, which was protected from heavy local winter snows by a radome. The Goonhilly antenna, by contrast, took the form of an uncovered steerable eighty-five-foot-diameter (twenty-six-meter-diameter) paraboloid dish constructed to withstand gales rather than snow. Designed by Husband and Co. of Sheffield, Goonhilly benefited greatly from that company's earlier experience with the 250-foot (seventy-six-meter) Jodrell Bank radio telescope. Although initially constructed for experimental tests with Telstar and Relay, the Goonhilly antenna was intended to be compatible with a range of satellite designs and orbits and so could serve a future operational system. After its completion in mid-1962,

---

7.   Herbert Schiller has strongly suggested that a satellite communications monopoly was, in fact, a conscious goal of many U.S. policy makers, who exhibited "a compulsive drive to transfer permanently to American hands the former British communications superiority." Herbert I. Schiller, *Mass Communications and American Empire* (New York: Augustus M. Kelley, 1970), p. 136.

8.   For expressions of this view, see Parliamentary Debates, House of Commons, V.632, written answers, 21 December 1960, col. 189-90, and V.659, Oral, 8 May 1962, col. 204-5. The conclusion that both satellites and cables would be accommodated in a complementary relationship has the obvious precedent of the accommodation between shortwave radio and telegraph cables in the late 1920s. See Headrick, *Invisible Weapon*, chapter 11.

9.   "Discussions with the National Aeronautics and Space Administration," Post Office memorandum, 28 August 1961, CAB 134/2144, JTS(61)6, London PRO. Initially, the Post Office was less enthusiastic about this offer than we might now expect. Captain Booth of the Post Office had received NASA's invitation very coolly in the eyes of others among the Thuillier delegation. Booth apparently was unaware that NASA's authority extended to the sphere of satellite communications and regarded AT&T as the more authoritative organization within the United States. See D. Gibson, minute, 9 February 1961, FO 371/157381, IAS 131/8, London PRO. See also note 10 below. It should be recalled that in December 1960, the Eisenhower administration had issued a statement of its preferred satellite communications policy, in which it emphasized the traditional role of private industry in U.S. communications.

10.   AT&T and the Post Office were long-standing business partners, having introduced the first transatlantic radiotelephone services in the late 1920s and the first transatlantic telephone cable in the 1950s. In fact, AT&T already had asked the Post Office to participate in experimental satellite trials before NASA's decision to become involved in similar trials. Postmaster General to Prime Minister, 28 July and 6 September 1960, PREM 11/3098, London PRO.

11.   Technical details of the Goonhilly station can be found in F.J.D. Taylor, ed., *The Goonhilly Project: Communication-Satellite Ground Station* (London: The Institution of Electrical Engineers, 1964).

Goonhilly's construction was believed to have cost only a fifth of the price of the U.S. station at Andover.[12] British officials were hopeful that British industry would be able to sell and build similar stations around the world for use with an eventual commercial system.

On 24 July 1961, President Kennedy issued an invitation to the nations of the world "to participate in a communication satellite system, in the interest of world peace and a close brotherhood among peoples throughout the world."[13] He had adopted the goal of an international satellite communications system, in part at least, as a means by which to restore and maintain U.S. technological prestige in the wake of the recent space successes of the Soviet Union.[14] Soon afterwards, U.S. officials explained their view that there should be a single global satellite system offering nondiscriminatory access to all countries. Only one such system could be financially viable, they claimed, because rival competing systems almost certainly would cause each and every system to suffer a financial loss. However, they did not make clear the form that they intended international participation to take. There was thus some suspicion that the United States intended to build, launch, and control a system entirely by itself with other nations' participation being limited solely to the role of users of the system, thereby cementing a U.S. monopoly.

Kennedy's invitation focused British thinking. Britain would have to act quickly to forestall the perceived U.S. threat to the Commonwealth network. Ideally, Britain would need to act before the Unites States had taken its own planning to the point where that country was ready to initiate concrete action. Before it could initiate any action of its own, however, the British government was bound to consider the interests of its Commonwealth telecommunications partners and to consult with them accordingly. For this purpose, a Commonwealth Satellite Communications Conference was arranged to be held in London in the spring of 1962.[15]

The Post Office had the primary responsibility for civil telecommunications, and it had the task of coordinating the details and agenda for the Commonwealth Conference. The views of other interested government departments had to be taken into account when determining British goals and policy. To achieve this, an interdepartmental working party was established at the beginning of 1962 under the chairmanship of Sir Robert Harvey, the Post Office Deputy Director General.

In addition to the Post Office, two other departments took a special interest in satellite communications—namely, the Ministry of Aviation and the Foreign Office. The individual positions that these three took into the working party stemmed from a number of different considerations and criteria, although there was considerable overlap among at least some of their specific goals.

---

12.   Ronald C. Hope-Jones, "Satellite Communications: the Salient Facts," 15 March 1963, FO 371/171048, GP1173/48, London PRO.

13.   Galloway, *The Politics and Technology*, p. 26.

14.   Vernon Van Dyke, *Pride and Power: The Rationale of the Space Program* (Urbana: University of Illinois Press, 1964), p. 25.

15.   The Post Office originally had intended to seek its Commonwealth partners' views during 1961, but the conference arranged for the end of that year was postponed until spring 1962, to allow for the conclusion of the negotiations toward the establishment of ELDO, thereby allowing the conference to meet with the knowledge that an independent European launcher would become available in due course.

The Ministry of Aviation

The Ministry of Aviation's interest in satellite communications arose from its role as the department responsible for Britain's contribution to the European Launcher Development Organization (ELDO). In this capacity, it intended to advance the interests of both ELDO and the British aerospace industry.

ELDO itself was in large part a product of the British government's desire to foster closer political ties to the nations of Western Europe, especially those of the Common Market.[16] Its origins lay with the cancellation of Britain's Blue Streak missile project in 1960, when the possibility was raised of continuing the development of the basic rocket, stripped of its specifically military aspects, to produce a satellite launcher.[17] The only justification that British officials could find for proceeding with this expensive project, rather than relying on U.S.-built launchers, was that an independent British launcher could be used to launch commercial communications satellites. This reasoning reflected a belief, widespread in both government and industry, that while the United States would be willing to make launchers available to other countries for scientific satellites, and to Britain for military satellites, they would be unwilling to launch foreign communications satellites and thereby prevent the emergence of commercial competition with a U.S.-owned system.[18]

Subsequently, Blue Streak was enlisted in Britain's attempts to improve its relations with Western Europe. In fact, by the end of 1960, the Cabinet had determined that the development of Blue Streak should be continued only within the context of European collaboration, and this option was pursued.[19]

The commercial potential of satellite communications was still regarded as the most significant role for the actual launcher. The Minister of Aviation, Peter Thorneycroft, stressed this potential as he toured Commonwealth and European capitals in search of support for a collaborative project. Commonwealth countries expressed little interest, however, leading Thorneycroft to concentrate his efforts within Europe.[20] After a protracted period of negotiation, those efforts led to the establishment of ELDO, whose Convention was opened for signature early in 1962.[21]

ELDO's initial program did not include any plans for a complementary program of satellite communications research. This absence reflected British and French Treasury opposition, along with the objections of the British Post Office, which was concerned by the likely complications that a European satellite program would introduce into the existing Commonwealth partnership. Nevertheless, the Ministry of Aviation retained its interest in satellite communications and was intent on using a European launcher to launch satellites in the later stages of whatever international system was established. The Ministry of

---

16. Initially, the British government's intention had been to foster closer links between the European Common Market and the countries of the European Free Trade Association. A formal decision to apply for actual membership in the Common Market was not taken until mid-1961. John Campbell, *Edward Heath: A Biography* (London: Pimlico, 1994), pp. 113, 116.

17. For a fuller account of ELDO's origins and the part played by the Macmillan government's European aspirations, see John Krige, *The Launch of ELDO*, ESA HSR-7 (Noordwijk: ESA, March 1993).

18. Ad Hoc Ministerial Committee on Blue Streak (GEN 716), minutes of first meeting, 6 July 1960, CAB 130/173, London PRO.

19. Ad Hoc Ministerial Committee on Blue Streak (GEN 716), minutes of third meeting, 5 December 1960, CAB 130/173, London PRO.

20. The Commonwealth and European governments shared a mutual antipathy toward Commonwealth-European collaboration. However, Australia was included in the membership of ELDO by virtue of the extensive Woomera rocket range from where the European launcher was to be fired.

21. For detailed accounts of these negotiations, see Krige, *The Launch of ELDO;* Michelangelo De Maria, *The History of ELDO Part 1: 1961–1964*, ESA HSR-10 (Noordwijk: ESA, September 1993).

Aviation's officials expressed doubt that ELDO would be allowed a role in a U.S.-led system. For this reason, they expressed a strong preference for a separate Commonwealth-European satellite system within which ELDO could be given a guaranteed place. If desired, the operation of this separate system could then be coordinated with its U.S. counterpart such that both would effectively become subsystems within an overall world system.

The Post Office

The Post Office position derived from an interaction between its membership in the Commonwealth Telecommunications Partnership and its statutory obligation to conduct its operations according to strictly commercial criteria. Initially, the Post Office favored the establishment of an independent Commonwealth-European satellite system, because it believed that the design of a U.S. system would not adequately meet Commonwealth requirements. The Post Office was aware that U.S. organizations, including AT&T, wanted to establish a satellite system at the earliest possible opportunity and that proposals had been made for low-altitude systems using as many as 50 satellites in random polar or inclined orbits.[22] The Post Office saw such a system as best serving the needs of the Northern Hemisphere and the profitable transatlantic telecommunications routes. It did not believe that such systems would provide the broad global coverage that was necessary to meet fully the needs of traffic among the Commonwealth countries, which were widely dispersed over both hemispheres.

Both the Post Office and the Ministry of Aviation had made studies of possible satellite systems. Each had concluded that the broadest coverage would be best provided by a system of position-controlled, as distinct from randomly orbiting, satellites in medium-altitude equatorial orbits.[23] Launching such a system would be well within the expected capabilities of the ELDO launcher. However, such orbits could be best reached using launch sites on the equator, and for this reason, the studies had raised considerable doubt about the suitability of the Australian Woomera launch site, from where the ELDO rocket would be fired initially.

The Post Office realized that the U.S. lead in developing both launchers and satellites meant that the first commercial satellite communications system to be orbited would almost inevitably be U.S. in origin. Despite this, its studies had suggested that traffic growth would be such as to allow a second system to be established profitably in the mid-1970s. The Post Office believed, therefore, that its participation in a U.S. initial system should be minimal and take the form of a lease of circuits rather than a share of ownership, while Britain and Europe built up their capability to construct an independent second system.[24]

The obligation to fulfill its purposes according to commercial criteria had specific consequences for the Post Office's views concerning the procurement of equipment for

---

22.   AT&T proposed such a system to the Federal Communications Commission in July 1960. Smith, *Communication by Satellite,* p. 58. The company's experimental Telstar satellites were intended to be the forerunners of a system of this type.

23.   Long Distance Communications by Satellite, subcommittee of the Joint Civil and Services Telecommunications Committee, Memorandum JTS(61)11, 31 October 1961, CAB 134/2144, London PRO. The current thinking of the Post Office with regard to the design of a satellite system is outlined, albeit without specific reference to Commonwealth requirements, in W.J. Bray, "Satellite Communication Systems," *Post Office Electrical Engineers' Journal* 55 (July 1962): 97–104.

24.   Ronald C. Hope-Jones, "Commonwealth Conference on Satellite Communications," 23 March 1962, FO 371/165275, GT2/54, London PRO.

the second system. The assessment that a second system could be profitable depended on a number of assumptions, one of which was that its costs should include only the cost of purchasing launchers and not the full cost of developing the European launcher.[25] Moreover, the Post Office wanted to insist that the European launcher would be used only so long as its purchase cost was no more than that for an equivalent U.S.-built launcher—a condition that the Post Office extended to all forms of relevant equipment.[26]

The Foreign Office

Officials within the Foreign Office took a wholly different approach to that of their colleagues in the Post Office and the Ministry of Aviation. While the Post Office had approached the matter from the perspective of a commercial telecommunications organization, and the Ministry of Aviation from the standpoint of the British and European aerospace industries, Foreign Office officials adopted a much broader view. Alongside the technical and commercial issues, they recognized many political problems with potentially conflicting implications for Britain's relations with the Commonwealth, Europe, and the United States. For the Foreign Office, it was as important to resolve and reconcile these problems as it was to secure a guaranteed and prominent role for British and European industry in the construction of a satellite system.

Foreign Office officials understood the importance with which their American counterparts viewed the foreign policy aspects of satellite communications and their desire for a prestigious technological coup within the context of the Cold War. They realized that if Britain was to build a separate and competing satellite system of its own, it would effectively be rejecting U.S. foreign policy goals. As the leading ally of the United States, Britain should look to supporting those goals and should cooperate fully with the United States from the outset. This did not mean that Britain should follow the United States blindly and forego its own industrial and communications interests. Rather, Britain should seek to temper any threat to its interests from within the single world system, by influencing the system's design and operation in such a way as to accommodate those interests. It could also further both its own and U.S. goals by persuading the European and Commonwealth countries that they, too, should join the single world system. If the latter nations were to accept similar policy goals to Britain, then all would have a stronger bargaining position if they were to enter negotiations with the United States as a bloc. It was important, therefore, that Britain should be prepared to show decisive leadership to both the Commonwealth and Western Europe. For the Foreign Office, in the words of an internal memorandum, the establishment of a global satellite system was a "major exercise in international cooperation" in which:

---

25. Maurice Dean to the Secretary of State for Air, memorandum, 12 May 1961, AIR 8/2255, London PRO. The Post Office was quite adamant that it would not contribute to the cost of the European launcher or even to a proposed program of basic satellite research into techniques common to a range of different satellite applications.

26. During 1963, the Post Office's commercial bias was compared with that of Comsat in the United States; the only difference between them was judged to be that whereas Comsat's commercial bias favored U.S. industry, that of the Post Office did not favor British industry. See Edward Heath to Julian Amery, enclosure, 5 July 1963, FO 371/171055, GP1173/182, London PRO. It should be noted, however, that the Post Office's commercial bias meant only that it was not prepared to subsidize European industry by paying noncompetitive prices; this did not preclude the possibility of subsidy from elsewhere in government.

> *The part that the United Kingdom plays . . . should be commensurate with our standing as a Great Power and will have an important bearing on the development of our relations with the United States, Europe, and the Commonwealth. It will be of great importance for us to ensure that we are not pulled one way by our desire to cooperate with the United States, another by our desire to act as good Europeans, and yet another by our desire to protect Commonwealth interests. It will not be easy to reconcile these divergent interests, and we shall greatly improve our chances of doing so if we devote as much constructive thought as possible to the whole problem and then give a decisive lead, both to Europe and to the Commonwealth.*[27]

For the immediate future, however, the Foreign Office needed first to get its views accepted by the British government itself. Only then would it have the authority to begin to influence Commonwealth and European opinion. The first step in this process was to persuade the officials of other British departments—a task pursued within the interdepartmental working party established to determine British policy in advance of the forthcoming Commonwealth Conference.

The Interdepartmental Working Party

The working party first met in January 1962. At that time, the thinking of officials in Whitehall was dominated largely by the concept of an independent Commonwealth-European satellite system. This concept also was reflected in the working party. Consistent with the positions outlined above, the Post Office and the Ministry of Aviation favored an independent system.[28] Only the Foreign Office representative, Ronald C. Hope-Jones, offered any significant support for a single system.

Hope-Jones offered a number of arguments against an independent system.[29] He argued that a second satellite system would be unlikely to succeed because countries that built ground stations for the U.S. initial system would be unlikely to duplicate their investment with a second station to use with British satellites. Moreover, the Post Office's favorable assessment of a second system's commercial viability had depended heavily on the assumption that it would take 50 percent of the profitable transatlantic traffic.[30] In its turn, this assumption depended heavily on U.S. goodwill toward the second system—a questionable assumption when the second system inevitably would undermine the profitability of the first. With regard to a U.S. system meeting Commonwealth requirements, Hope-Jones saw no reason why the United States would not want to configure it to meet the needs of the Commonwealth and others. There was some indication that U.S. plan-

---

27.   Cabinet Combined Communications-Electronics Committee, CCC(63)4, Memorandum by the Foreign Office, undated, circa February 1963, CAB 134/1451, London PRO. Although dating from 1963, this quote nevertheless is of a piece with Foreign Office comments through 1962.

28.   Ronald Hope-Jones to M.D. Butler, letter, 26 November 1962, FO 371/165249, London PRO.

29.   Hope-Jones summarized his arguments in an internal Foreign Office report during March. See "Commonwealth Conference on Satellite Communications," 13 March 1962, FO 371/165275, GT2/54, London PRO. More generally, Hope-Jones reported his belief that "American thinking . . . [with regard to the practicalities of international cooperation in the field of satellite communications] . . . is not nearly as far advanced [in progress] as is sometimes supposed, and I believe that if we could get a mandate from the Commonwealth to discuss our ideas with European countries and could then obtain their support in principle for a joint approach to the United States, we should regain a great deal of the initiative now held by the Americans."

30.   At that time, transatlantic routes accounted for almost 80 percent of total intercontinental telecommunications traffic. Thus, the key to profitability for any satellite system would be the extent of its access to transatlantic traffic and revenues. Parliamentary Debates, House of Commons, V.674, 29 March 1963, col. 1719.

ners already had appreciated British views on the desirability of the equatorial orbit. If this was so, an American system also might use that orbit, in which case a second system would be largely a duplication of the first.

Hope-Jones successfully persuaded the Post Office to accept that a single system would be preferable to having two or more competing satellite systems in existence. With the Post Office having shifted its ground, the overall balance of opinion within the working party shifted similarly. Consequently, the British delegation to the Commonwealth Conference was briefed to encourage Commonwealth acceptance for the desirability of broad international cooperation with the United States, involving both the Commonwealth and Europe. At the same time, the conference should be encouraged to endorse further exploratory talks with Europe and with the United States. British ministers were not yet ready to rule out the possibility of a separate British satellite system, however. The Minister of Aviation, among others, was concerned that whatever satellite system emerged should provide a guaranteed role for the ELDO launcher. Thus, while ministers endorsed the official brief for the British delegation, they did not want to settle final policy until after the full round of exploratory talks. Only then, with a clearer idea of foreign intentions, would they feel ready to decide where the balance of Britain's best interests lay.

The Commonwealth Conference on Satellite Communications

The Commonwealth Conference met in London from 28 March to 13 April 1962. It concluded that satellite communications were technically feasible, but that there was still much uncertainty regarding the bases for detailed system design and economic estimates. Nevertheless, the conference felt that a commercial satellite system should serve as many countries as was possible and that "the ideal arrangements would be for the Commonwealth to play a full part in a broadly based system of satellite communications, rather than become wholly dependent on the United States, or alternately seek to go it alone in competition with the United States."[31] Thus the Commonwealth should seek to cooperate fully with the United States and with European countries.[32]

The conference's endorsement of a Commonwealth approach to Europe, alongside that to the United States, had been secured only through strong Foreign Office pressure. Commonwealth representatives had shown an extreme reluctance to endorse cooperation with Western Europe, fearing that telecommunications would become one more economic field in which the balance of British attention would shift away from the Commonwealth and toward Europe.[33] Australia and Canada would have preferred a more limited form of Commonwealth cooperation with the United States. For the smaller, less developed countries, meanwhile, satellites held no immediate interest, because the Commonwealth telephone cable scheme was expected to meet their needs well into the 1970s.

For political reasons connected to their general policy of seeking closer links with Europe, British ministers and officials, including those in the Foreign Office, had intended that, following the Commonwealth Conference, exploratory talks should be held with Europe in advance of any similar talks with the United States. This plan proved to be

---

31.   Cabinet Economic Policy Committee, E.A.(62)80, memorandum by the Postmaster General, 13 June 1962, CAB 134/1696, London PRO.

32.   Registry number 22325/62, File 2A, SWP 21, British Telecom Archives, London.

33.   Foreign Office brief, undated, circa March 1963, FO 371/171030, GP1023/3(A), London PRO.

impractical. Discussion within the Commonwealth Conference had been constrained by a lack of detailed knowledge of U.S. plans and intentions. This uncertainty made it difficult to discuss anything concrete, and the same limitations equally would affect talks with Europe.[34] For practical reasons, therefore, it was accepted that talks with the United States should be initiated first, and preparations went forward on that basis.

Alongside the purely practical considerations influencing the decision to make the initial approach to the United States, the Foreign Office also believed that an initial approach to Europe would restrict Britain's opportunities for making the U.S. authorities aware of Commonwealth requirements. It was concerned that the other Europeans might insist that the approach to the United States be made by Europe as a whole, speaking with one voice. If the rest of Europe did not accept Britain's Commonwealth concerns, that single voice might well be antithetical to British interests. Such a move would produce a clear conflict between Britain's Commonwealth interests and its desire for a close association with Europe. The risk of conflict would be lessened by stating Commonwealth needs to the United States first, thereby avoiding a need to seek European agreement on this point.

Discussions Between Britain, Canada, and the United States

In accordance with the Commonwealth Conference recommendations, Britain and Canada jointly initiated talks with the United States; the U.S. State Department agreed to a meeting in Washington on 29–31 October 1962. An earlier meeting had not been possible because of the lengthy passage through Congress of the Communications Satellite Act. The U.S. Communications Satellite Corporation (Comsat) envisaged by the act had yet to be established, while the law itself had left many aspects of future U.S. policy and intentions quite vague. Amidst these domestic uncertainties, the State Department did not want to be seen to have preempted future U.S. policy. Its officials stressed that the discussions would be purely exploratory and that in no way should American comments be taken as expressions of future American policy.[35] For the British, too, the talks were not intended to be anything more than exploratory, given that the British ministers themselves had not yet decided British policy.[36]

The British came away from Washington with the feeling that the discussions had been "friendly and completely frank," with the United States officials having given satisfactory answers to nearly all of the questions asked of them.[37] The Americans assured the British that they did not intend that other countries should be limited solely to the role of users of a U.S.-built satellite system.

There was general agreement with a British statement that the development of a full global satellite system would be evolutionary in character and that it would be constructed in successive stages, the first of which would be provided by the United States to cover the busiest traffic routes. This "evolutionary concept" represented Britain's wish to reconcile its desire to support a single global system with the desire that British and European industry should be able to provide equipment for the system. Thus, although the United States would provide the initial stage, the system's coverage and capacity would

---

34.   Hope-Jones to Butler, 26 November 1962.
35.   Minute to the Postmaster General, 13 November 1962, Registry no. 22325/62, File 2A, SWP 26(Revise), British Telecom Archives.
36.   R.J.P. Harvey, minutes, 22 October 1962, FO 371/165249, GP173/39, London PRO.
37.   Satellite working party, minutes of the eighth meeting, 8 November 1962, Registry no. 22325/62, File 4, British Telecom Archives.

be expanded in subsequent phases, and the system might use equipment from other countries in a number of different, but complementary orbits. The British stressed their belief that as time went by, and as Britain and other nations developed the appropriate technological capabilities, they should be allowed to supply the necessary equipment for the later stages alongside U.S. industry.[38]

Although U.S. officials talked freely about full international participation, they would not agree to references in a British-prepared summary of the talks that mentioned both "international cooperation in the provision and development of a satellite system" and "the need for decisions about design and construction to be taken jointly by the countries participating in the system" lest these be taken as an expression of U.S. policy. In their place, the United States suggested weaker phrasing: "broad international participation" and "the need for consultation in design and construction matters."[39]

Coming away from these talks, Sir Robert Harvey, the leader of the British delegation, had felt that matters now were much more uncertain regarding future U.S. policy.[40] Ronald Hope-Jones of the Foreign Office was far less pessimistic, however. He recognized that U.S. officials had been faced with domestic constraints that prevented them from being more forthcoming in their acceptance of Anglo-Canadian views.[41]

British-European Discussions

The Washington talks were followed by exploratory talks with Europe. The British Post Office arranged for these to be under the auspices of the Telecommunications Commission of the European Conference of Postal and Telecommunications Administrations (CEPT, French acronym). The meeting was held in Cologne on 12–14 December 1962. Gilbert Carter of the U.S. State Department informally addressed the Telecommunications Commission. Carter outlined U.S. thinking on a "single global system," constructed through "full international cooperation." His presentation strongly influenced the subsequent course of CEPT discussions.

The Europeans expressed support for the U.S. vision of the future development of satellite communications. None showed any desire to establish a separate European satellite system. At the same time, they did not intend to become wholly dependent on the United States, and they were intent on playing as full a part as possible in the design and management of the single global system, as well as in the construction of its later stages. CEPT members rejected as premature French and Swedish proposals for the establishment of a formal European satellite organization, limiting themselves to forming an ad hoc committee of CEPT itself. This committee would study the technical aspects of the problem in greater detail and would draft the basis of a common CEPT position ahead of talks with the United States.[42]

Although the CEPT meeting had effectively endorsed the Foreign Office's views, the latter nevertheless was unhappy with the choice of CEPT as the focal point for the exploratory talks with Europe. The Foreign Office felt that CEPT members, who were drawn from the telecommunications authorities of their respective governments, would

---

38.   Ronald C. Hope-Jones to J.F. Hosie, 5 November 1962, FO 371/165249, GP173/27, London PRO.
39.   Foreign Office brief, undated, circa March 1963.
40.   *Ibid.*
41.   Ronald C. Hope-Jones, handwritten sleeve note, 5 November 1962, FO 371/165249, GP173/26, London PRO. Hope-Jones was sure that "the Americans are now in no doubt about our aspirations and know that they will have to accommodate them in working out their own plans for a single global system to be achieved by full international participation."
42.   Registry No. 22325/62, File 2A, SWP 31(Final), British Telecom Archives, London.

be insufficiently aware of the political dimensions of the problem.[43] Underlying this dissatisfaction was a much deeper difference of approach between the Foreign Office and the Post Office. The Post Office objected to the former's desire that the making of satellite communications policy should be removed from the technical level of the telecommunications authorities,[44] wishing instead to conduct matters according to traditional international telecommunications practices in which foreign policy issues had played very little part.

The Foreign Office knew that traditional practices would be inappropriate to the specific issues raised by satellite communications. It realized that the Post Office could not be relied on to insist on a role for European industry. The Foreign Office also feared that French President Charles de Gaulle might want to insist on an independent European satellite system and thereby prevent a broader U.S.-European-Commonwealth system from taking shape.[45] The Post Office would be unable to handle that situation. By the end of 1962, while it had not as yet expressed any preference for an independent system, the French government had indicated that it would be treating the whole subject as primarily a political issue. The French foreign ministry, the Quai d'Orsay, would soon be given a prominent role in determining French policy.[46] Thus, while the CEPT meeting had indicated an apparent European readiness to endorse a single global system, with even the French representatives signaling their approval, its position had been derived largely from technical considerations. The CEPT meeting's views might yet be overruled once European governments became more fully aware of the wider political, commercial, and industrial issues involved in establishing a satellite organization.[47]

Conclusion

The round of intergovernmental discussions outlined above was wholly exploratory, and no firm decisions were made during its course. Throughout, however, Foreign Office officials worked to secure broad acceptance at home and abroad for their own view of how international satellite communications should be developed—namely, through a fully

---

43.   Ronald C. Hope-Jones, minute, 21 December 1962, FO 371/171046, GP1173/3, London PRO. A U.S. State Department official had commented that the Post Office seemed to regard satellites as little more than "a telegraph pole in the sky." Hope-Jones stressed that satellite communications had to be thought of as a space activity just as much as a communications activity. As a space activity, it was the most likely field in which to secure a fruitful international cooperation in the peaceful uses of outer space, which was a major interest of the United Nations.

44.   This difference was not unique to Britain. Gilbert Carter of the U.S. State Department had noted similar conflicts between the telecommunications authorities and foreign ministries within several European countries. The same differences also existed in the United States between the State Department and Comsat, with the latter having adopted the working attitudes of the traditional communications carriers. While the State Department wanted satellite communications to be handled through fully multilateral negotiations, Comsat was pressing for matters to be settled through bilateral agreements. Galloway, *The Politics and Technology*, p. 103.

45.   Hope-Jones to Butler, 26 November 1962.

46.   When the French learned of the Anglo-Canadian visit to Washington, they assumed that Britain had proposed a formal U.S.-Commonwealth collaboration in preference to a European collaboration centered on ELDO, resulting in "sadness in high quarters" within the French government (that is, General de Gaulle). Paris Embassy Telegram No. 362, 7 November 1962, FO 371/165249, GP173/30, London PRO.

47.   M.D. Butler to Hope-Jones, letter, 7 December 1962, FO 371/165251, GP173/63, London PRO. An official of the Quai d'Orsay had told Butler, a member of the British Embassy staff in Paris, that as yet the French government had not decided its satellite communications policy, which in any event would have to be approved by General de Gaulle. The Quai d'Orsay had no knowledge of a paper tabled at the CEPT meeting by the French telecommunications authority. This paper had not been authorized and, as such, could only be regarded as giving the personal views of the French telecommunications authority representatives who had written it.

participatory international cooperative effort in which all members would have a right to participate in the design, construction, management, and operation of a single world satellite system.

By the end of 1962, support for the Foreign Office view of how satellite communications should develop had been expressed by the Commonwealth and by the European telecommunications authorities represented in CEPT. Although a number of British and Continental European politicians and industrialists still wanted an independent European satellite system with which to compete with a U.S. system, no European government thus far had expressed a desire to establish a separate system. At the same time, the Post Office, despite its support for the goal of a single system, and despite the obvious foreign policy implications of the subject, was anxious to deny foreign ministries any role in the settling of policy. Instead, it wished to proceed by the same commercial practices that had governed the introduction of telephone cables a few years previously. This attitude was shared by other European telecommunications authorities and by Comsat. Nevertheless, during the course of 1963, despite such internal opposition, European governments would opt to pursue negotiations with the United States on something similar to the full multilateral basis that had been recommended by the Foreign Office, thereby rejecting an independent and wholly competitive response to the United States.

Chapter 13

Originating Communications Satellite Systems: The Interactions of Technological Change, Domestic Politics, and Foreign Policy

by Jonathan F. Galloway

Technological change can have both evolutionary and revolutionary consequences. Some participants in the policy process in the 1960s viewed satellite communications as ushering in a new era of global peace and understanding. Others viewed the new technology as threatening vested interests and creating new entrepreneurial opportunities.

The policy-making process that led to the establishment of the original communications satellite institutions—for example, Comsat, the Defense Satellite Communications System, Intelsat, and Intersputnik—was an arena of cooperation, competition, and conflict. The actors in that policy process were national and international, and thus the process was, in many respects, transnational.

The rationality of decision making cannot be understood holistically or comprehensively, but as pragmatic, incremental, and muddled. Dominating decision making was the breaking down of national borders and barriers to entry into emerging global communications markets. Yet, paradoxically, the creation of advanced defense communications satellite systems during the Cold War era reflected the drive to defend the territorial integrity of states and military alliances.

Looking back on the years between Sputnik and the establishment of the Intelsat Definite Agreements in 1971, the three themes of the 1972 book by this author[1] can be seen in a slightly different light. Those themes were: (1) revolutionary or evolutionary technological change, (2) the breakdown of barriers between making and understanding foreign and domestic policy, and (3) models of rationality in policy making appropriate to changing contexts. Those themes remain very relevant, despite the more colorful and even dramatic new vocabulary introduced by the likes of Newt Gingrich, the Tofflers,[2] and Kenichi Ohmae.[3] The economic and technological aspects of globalization often are thwarted by, or exist in paradoxical and ambivalent contiguity with, the forces of resurgent localism and the economic doctrine of neomercantilism. The world is not a tidy place. There is chaos, and at the edge of chaos one new world order is not emerging. So it was in the formative period of satellite communications.

---

1.   Jonathan F. Galloway, *The Politics and Technology of Satellite Communications* (Lexington, MA: Lexington Books, 1972).

2.   Alvin and Heidi Toffler, in *Creating a New Civilization: The Politics of the Third Wave* (Atlanta: Turner Publishing, 1994), p. 33, talk about "the Third Wave," the end of nationalism, and the coming of globalization and borderless states.

3.   Kenichi Ohmae, *The Borderless World* (New York: Harper Collins, 1990).

Foreign Policy and Domestic Bargaining[4]

The Communications Satellite Act of 1962 provided for foreign participation "in the establishment and use" of a communications satellite system and authorized the corporation formed by that act (Comsat) to operate the system "itself or in conjunction with foreign governments or business entities."[5] What form would participation take? The statutory language was open ended. Partnership could comprise system use, ownership, or research and development. Which particular forms of cooperation would evolve was not mandated by the law, yet it was clear that operational relationships would have serious consequences for domestic and foreign policy. Thus the substance and the form of United States foreign policy goals to promote leadership, international peace and understanding, and the rapid establishment of a global system were entangled in satellite communications policy decision making and negotiations.

To cooperate with other countries, it was first necessary for the United States to develop a cohesive approach toward outstanding issues. Developing that approach entailed a gradual working out of the relationship between the U.S. government and Comsat. However, early consultations and briefings at the international level occurred at the same time as the crystallization of domestic policy. While it is sometimes fashionable for analytic purposes to separate domestic and international discussions, in this case the separation is not legitimate: foreign policy consultations affected the domestic bargaining position of the United States from the spring of 1963 to the summer of 1964. The following is a discussion of domestic developments whose consequences affected the formation of U.S. policy toward international arrangements for communications satellites. Attention is focused on the establishment of Comsat and its relations with the State Department and other agencies responsible for foreign policy considerations.

After the incorporation of Comsat in the District of Columbia, President Kennedy appointed thirteen Incorporators on 15 October 1962. Although the Incorporators were political appointees, Kennedy combined politics with expertise. The Incorporators chose Philip L. Graham, the publisher of *The Washington Post*, as their first chairman on 22 October 1962. Graham headed Comsat for only three months, resigning in January 1963 for health reasons. During those three months, however, he strenuously objected to what he saw as the interference of the State Department in the international aspects of Comsat's plans. The State Department had been briefing European telecommunications administrations on American policy, and Graham thought that was Comsat's prerogative.

This attitude persisted with Graham's successor, Leo D. Welch, who became chairman and chief executive officer in February 1963.[6] Welch formerly had been chairman of the board of Standard Oil of New Jersey. He was dedicated to the proposition that the private sector should determine policy. Welch viewed the State Department not as an ally, but as an obstacle. This was also true of his attitude toward the Federal Communications Commission (FCC). He charged the FCC with the "invasion of the managerial functions

4.   This section is based on Galloway, *The Politics and Technology*, pp. 80–104.

5.   The Communications Satellite Act of 1962, sections 201(a)(5) and 305(a)(1).

6.   At the same time, Kennedy appointed Dr. Joseph V. Charyk as president and principal operating officer of Comsat. Charyk was formerly under secretary of the Air Force, a businessman, and a university scientist. Also, in February 1963, Comsat's articles of incorporation were approved. The articles and bylaws may be found in Senate Committee on Commerce, *Communications Satellite Incorporation Hearings*, 88th Cong., 1st sess., pp. 35–39.

of the corporation" in connection with the FCC's directions on how the corporation should disburse its funds.[7]

On the other hand, Comsat's relationship with NASA was always cordial. During its incorporation period, when its financial resources consisted only of borrowed funds,[8] Comsat depended largely on NASA programs for its research and development. Close and constant cooperation thus was a prime necessity for economic survival, but NASA also had the responsibility, under the Communications Satellite Act, of cooperating with Comsat in research and development.

On the whole, though, relations between the government and Comsat were somewhat less than ideal, as a result not of personality conflicts but of opposing views based on differences of opinion. While the government looked forward to close cooperation and regulation of Comsat, Welch saw Comsat as a traditional private enterprise. Government officials saw Comsat as an entirely different entity, more like a public utility, or even a creature of the government. Under the Communications Satellite Act, Comsat was to serve public as well as private purposes. Moreover, Comsat was not subject to regulation by the FCC, like all other carriers, but by the director of telecommunications management and the President, in the exercise of his foreign affairs powers, as well.

The initial estrangement between the State Department and Comsat was especially crucial for foreign policy and foreign relations. Comsat was under no firm statutory responsibility to consult with State. Thus, Comsat leaders could choose to follow their own individual preferences. The ability of State to persuade Welch of its prerogatives also was hampered by the State Department's weak power position in domestic politics. To overcome this disability, State tried to convince Comsat that its views were also the Europeans' views and that Comsat was not only a domestic corporation, but the chosen instrument of American foreign policy.

While positions between Comsat and the government were not polarized, there were four issues whose resolution involved differences of opinion: (1) ownership; (2) the form of an international agreement (bilateral or multilateral); (3) Comsat's management role (complete or partial); and (4) the role of government in the negotiation of international agreements to establish the system. This last issue, in particular, pitted Comsat, some of whose officials thought it should undertake negotiations, against the State Department.

To resolve these issues, Comsat and the State Department agreed in June 1963 to work out principles for international participation. During the same month, President Kennedy established the Ad Hoc Communications Satellite Group under the joint chairmanship of Deputy Attorney General Nicholas Katzenbach and the President's Special Assistant for Science and Technology Dr. Jerome B. Wiesner. This group exercised the responsibilities of the director of telecommunications management because of a vacancy in that office.[9] A subcommittee, composed of representatives from the Departments of State and Justice, the FCC, the Space Council, and the office of the director of telecommunications management, kept abreast of Comsat's draft principles for international cooperation.

Initially, the ideas of the State Department on the problem of foreign participation did not coincide with those of Comsat. Welch wished to establish a Comsat-owned system through a series of bilateral intergovernmental agreements in which Comsat was the pri-

---

7.   Leo D. Welch to E. William Henry, Chairman of FCC, letter, 7 August 1963, reprinted as Attachment C in U.S. Congress, House, Committee on Interstate and Foreign Commerce, *Communications Satellite Act of 1962—The First Year*, 88th Cong., 1st sess., 1963, H. Rept. 809, p. 19 (hereinafter cited as H. Rept. 809).

8.   In 1963, the FCC permitted Comsat "to borrow $1,900,000 pursuant to a line of credit agreement for $5 million entered into by the corporation with 10 commercial banks." H. Rept. 809, p. 11.

9.   House Committee on Interstate and Foreign Commerce, H. Rept. 809, p. 29.

mary negotiator. This conception of international cooperation clashed with what State viewed as politically feasible and desirable. To State, it seemed that some form of joint ownership would be necessary. If other nations were to have no control over the system, why should they cooperate with it?

The State Department also thought that it would be necessary to conclude both inter-governmental and commercial agreements. As early as the fall of 1962, William Carter of State's Bureau of Economic Affairs considered that negotiations on two levels would have to be undertaken to expedite the establishment of an international system. On one level, discussions between governments would conclude an agreement on the principles of international cooperation in space communications; on the other level, technical and business accords between the telecommunication administrations of the various countries would be arranged. These accords would be analogous to traditional agreements for the management of cables and radio. The necessity for the two-level approach was based on State's assumption that other nations would be unwilling to join a venture whose political and economic consequences were completely removed from their influence. That assumption was based on Carter's experience in discussing satellite communications with foreign countries in 1962 and 1963.

The chairman and president of Comsat were not advocates of the same line of reasoning, even though they also had engaged in international consultations. In May and June of 1963, they had held meetings with representatives of Western European nations, Canada, and Japan.[10] While they had not reached any definite conclusions, they favored a Comsat-owned, Comsat-negotiated venture.

Nonetheless, within Comsat, Edwin J. Istvan, who came to Comsat from the Office of Space Systems in the Department of Defense (DoD), drafted a set of principles for international participation that was nearer to the views of State. Istvan based his set of principles on three propositions:

1. Partnership in the international system could take one of three forms: (a) space segment ownership, (b) communication terminal ownership, or (c) mutual aid to assist developing nations.

2. System ownership should be established by an agreement between Comsat and the participating foreign entity; the U.S. government would not play a prime role.

3. Comsat would manage the system to the extent of controlling its own percentage of ownership and controlling contract awards.

Welch rejected Istvan's proposal; he maintained that the best system would be technically integrated and centrally owned and managed.

The Ad Hoc Communications Satellite Group thought differently, however. In the course of Comsat's discussions with the group, it became evident that the government favored an approach closer to that outlined by Istvan. Istvan played a crucial role by pointing to the possibility of separating space segment ownership from station ownership. The government accepted this idea, but not the second or third. The government contended that it had developed the capability at its own expense and therefore should have a central role in determining how it was to be organized on an international, as well as on a domestic, basis. The State Department felt that other nations would not be willing to

---

10.   Comsat, "Statement Relating to Anticipated Participation by Foreign Governments or Business Entities in Such a System," reprinted in H. Rept. 809, p. 6.

establish an international agreement for communications satellites along the same lines as the cable arrangements. Negotiations between governments would be required to lessen fears that communications satellite technology would disrupt existing investments before they had been amortized.

The Ad Hoc Communications Satellite Group objected to Istvan's third idea (Comsat management) not so much for its substance as its form. Other nations would be more likely to join a system, the group pointed out, if Comsat's management role were played down. The State Department thought that Comsat's management role should be worked out in negotiations rather than proclaimed beforehand. As the United States was the only country with developed potential, it would be natural for other nations to agree to Comsat's primary position.

The State Department was especially concerned about presenting a flexible and attractive proposal to other nations to promote the establishment of a single global system rather than a number of competing systems. Writing in May 1963, the State Department's Deputy Assistant Secretary for International Organization Affairs stated that "economic, technical, and political considerations all point to the desirability of a single system." A single system would avoid wasteful duplication of expensive satellite and ground facilities, would enhance the possibility of "fruitful exchange of communication between all countries," would avoid "destructive competition," would facilitate technical compatibility between satellites and ground terminals, would assure the best use of scarce frequency spectra, and would promote operational efficiency and flexibility in routing messages.[11]

In light of these advantages, and considering its opinions on foreign attitudes, it is no surprise that the State Department objected to Welch's ideas and wished to modify those of the preliminary draft drawn up by Istvan. It was of the utmost importance to project an image of partnership rather than of paternalism, according to the State Department; otherwise, U.S. policy would lend weight to Communist charges of exploitation by capitalist monopolies and, furthermore, could encourage the establishment of competing systems.

An integral part of the government's stand for a more flexible negotiating position was its contention that Comsat should undertake its discussions with foreign telecommunications administrations in light of consultations with interested government agencies. The implications of various decisions were too far reaching to be considered merely business responsibilities. For instance, the State Department thought that system choice "will have important consequences on coverage and burden-sharing, the availability of satellites for communication between and within countries, and the cost of ground installations which nations will have to build."[12] In addition, the State Department asserted:

> The establishment of global satellite communications will involve international discussion of other questions: the handling of research, development, manufacture, and launch; participation in ownership of ground terminals and satellites; allocation of satellite channels between use and users; means of determining the number and location of ground terminals; technical standardization; rate-making; assistance to less developed countries; possible public uses of the system by government information agencies of the U.N.; and means of facilitating the exchange of programs between nations.[13]

---

11. Richard N. Gardner, "Space Meteorology and Communications: A Challenge to Science and Diplomacy," *Department of State Bulletin* 48 (13 May 1963): 774. It may be asked why competition is beneficial within the United States and disastrous outside.

12. *Ibid.*

13. *Ibid.*

Decisions on all these problems would require the closest possible cooperation between Comsat and the government, according to the State Department. In many cases, the government's role would be regulatory in addition to being promotional. However, the difficulties in evolving a working statement of principles continued into the fall of 1963.

During the summer, Comsat discussed its draft principles of foreign participation with representatives of the Departments of State, Justice, and Defense, NASA, and the FCC, as well as with the President. As a result of these discussions, Comsat slightly revised its statement of principles, but not their substance. Welch still insisted that Comsat be the owner of the entire system. Comsat also discussed its principles with the common carriers and interested committees in Congress, but in the course of these discussions, no significant abridgment or change emerged.

The chair of the Senate Foreign Relations Committee, Senator J. William Fulbright (D-AR), expressed dismay that government participation in the preparation of the principles was not more apparent. In reply, Welch expressed the opinion that it was. Other committees, however, were not opposed to Comsat's position. Thus, as the time to start actual bargaining with other countries approached, the U.S. negotiating position lacked cohesion.

It is somewhat surprising that Welch's views had not moved more toward center in light of European attitudes. After a June 1963 meeting of potential partners in London, the British government sent the State Department an aide memoire in which the British related American policy as they understood it. This brief stated that the United States favored shared ownership and pluralistic decision making. The statement was approved by State and cleared through Joseph Charyk, Comsat's president, before being sent back to the British. Yet Welch was still arguing for 100-percent Comsat control. Furthermore, as late as September 1963, he (and the carriers) objected to the Istvan proposal of two agreements—one governmental, the other between communications entities. Although the government, Congress, and the carriers cleared a Comsat draft on negotiating principles on 4 October 1963, government clearance did not mean approval.

At the Extraordinary Administrative Radio Conference in Geneva, an informal U.S. policy group was hard at work trying to convince member countries of the International Telecommunications Union that the proposed system was not an American monopoly in disguise. Later in October, a meeting between the United States and European nations produced the harmony in the U.S. position that was lacking as a result of purely domestic discussions. The European group, known as the European Conference on Satellite Communications (CETS, the French acronym), had formed to negotiate with the United States. The CETS nations made it known that they were unwilling to negotiate with Comsat on a bilateral basis. This blunt fact of necessity changed Welch's outlook on the character of an international system.

Comsat now agreed to principles that pointed to joint ownership and to negotiations conducted by a joint American team representing equally the government and Comsat. Henceforth, Comsat worked more closely with the State Department at the international level. It was agreed, moreover, that a joint negotiating team headed by Welch and Abram Chayes, the State Department's legal advisor, would represent the United States in forthcoming discussions with European and other nations interested in establishing a global communications satellite system. Comsat and the government also agreed on an explicit formulation of principles to serve as the U.S. negotiating position. Thus, the stage was set for a harmonious approach to establishing a global system.

Possibly reinforcing close relations between Comsat and the government was the appointment of several notable government officials to Comsat positions. In January 1964, NASA General Counsel John A. Johnson became vice president in charge of Comsat international affairs. Lewis Meyer, formerly an Air Force deputy for financial analysis, became Comsat's finance coordinator. Richard Colino, previously with the State Department and the FCC, became an assistant to John Johnson. In addition, Louis B. Early and Siegfried H. Reiger, both of whom had been with the RAND Corporation, joined Comsat to serve as chief of economic analysis and manager of systems analysis, respectively. While one might have expected the government to play an active role in promoting staffing of this kind, this was not the case. Former government personnel came to Comsat because the firm was anxious to find a nucleus of people with satellite communications experience. Such a nucleus could be found only in the government or among certain selected carriers and manufacturers, such as AT&T and RCA.

Foreign Negotiations for International Arrangements

The space age put the smaller industrialized nations to a great disadvantage compared to the two superpowers. The allies of the United States in Western Europe and elsewhere could develop their own capabilities in space only through a symbiotic relationship with NASA. Cooperative experiments most often were arranged on a bilateral basis. Early during the space age, though, Western Europeans saw the advantage of building multilateral organizations to pool their resources and thus develop a technological base independent of the United States.

At the intergovernmental level, discussions leading to the establishment of the European Space Research Organization (ESRO) began in 1960 and entered into force on 20 March 1964. The stated purpose of ESRO was "to provide for, and to promote, collaboration among European states in space research and technology exclusively for peaceful purposes." The composition of the organization included neutral nations, so the proviso "exclusively for peaceful purposes" had more than rhetorical importance. ESRO proposed no military projects and restricted its activities to those of scientific importance, rather than extending itself into the race for prestige, security, and wealth. A second European multilateral space organization was the European Launcher Development Organization (ELDO), conceived as a result of Anglo-French discussions early in 1961. The convention establishing it entered into force on 29 February 1964. ELDO planned to develop programs in meteorology, navigation, telecommunications, and scientific research.

A nongovernmental, multilateral, nonprofit organization of considerable importance was Eurospace. Its membership included 146 aerospace companies, including eight U.S. companies as associate members. Formed in 1961 on the initiative of certain British and French industrial groups, Eurospace's principal objective was to create a Western European industrial complex capable of providing expert assistance and advice on space programs to governments, supranational bodies, and private interests.[14]

These three multilateral organizations represented Western Europe's organizational response to the space age. They had much wider interests than space communication per se. In contrast to the United States, where communications carriers were owned by private companies and regulated by the government, the principal communications agencies in foreign countries at that time were government administrations.

The prime source of European experience in international communications, and the first organization to cooperate with the United States in such space communications

---

14. The following information on ESRO, ELDO, and Eurospace is drawn from British Information Services, *Britain and Space Research* (London: Central Office of Information, 1965). See also U.S. Congress, Senate, Committee on Aeronautical and Space Sciences, *International Cooperation and Organization for Outer Space*, 89th Cong., 1st sess., 1965, Staff Report, Document 56, pp. 105–17, 123–28.

experiments as Echo and Telstar, was the British Post Office. As early as 1960, the Commonwealth Telecommunications Board's technical and traffic meeting discussed radio communications via satellite. The British government subsequently invited its Commonwealth partners to the Commonwealth Satellite Communications Conference held in London in April 1962. According to the British Information Services:

> The Conference recognized that it might be some years before satellite systems could become technically feasible, and, having regard to the research and development work being undertaken in the United States and elsewhere, that any commercial satellite system should serve as large a number of countries as possible and should have maximum flexibility. The Conference also acknowledged that satellite communications and submarine telephone cable systems would be complementary one to another, and had regard, throughout its discussion, to the projected submarine cable developments both by the Commonwealth and by other countries.[15]

While the British viewed satellite communications as a possibility for the 1970s, that view did not mean that they were oblivious to the potential economic threat to cables. Shortly after the enactment of the Communications Satellite Act of 1962, Britain and Canada suggested that exploratory talks be held with the United States on the development of an operational system.[16] The Department of State concluded that talks would be beneficial, and from 27 to 29 October 1962, representatives of the British Post Office and Foreign Office, the Canadian Ministry of Transport and Department of External Affairs, and an interagency group headed by the U.S. State Department met. The British and Canadians "emphasized their desire to participate fully in the technical development, ownership and management of the system."[17] They thought, however, that another generation of cables could be laid before a space communications system would be operational. The United States took this opportunity to explain the purposes of the Communications Satellite Act, emphasizing the desirability of a single global system as opposed to competing national systems.

The British reported that they were expecting to submit a summary of these discussions to the European Conference of Postal and Telecommunications Administrations (CEPT, the French acronym) to be held in Cologne in December 1962. The State Department thought this move would only relate American ideas secondhand. Therefore, the State Department sent an American team to Europe to brief CEPT members prior to the December conference. During these meetings, and in the course of the conference, it became evident that the European countries were quite excited about participating in a joint system. The form of participation was not set out precisely at this time, however. More detailed consideration at the international level would have to await the formation of specific proposals by Comsat. The role of the State Department was to lay the groundwork for these discussions.

During these preliminary meetings, the principal State Department representative, Gilbert Carter, perceived a pattern of internal competition within several European countries.[18] Past international telecommunications agreements had been negotiated on what

---

15. British Information Services, *Britain and Commonwealth Telecommunications* (London: Central Office of Information, 1963), p. 26.
16. Department of State, *Summary of Activities of Department of State Relating to the Communications Satellite Act of 1962* (September 20, 1963). Reprinted in H. Rept. 809, pp. 25–27.
17. *Ibid.*
18. U.S. Congress, House, Committee on Government Operations, *Military Operations Subcommittee, Hearings, Satellite Communications—1964*, Part 2, 88th Cong., 2d sess., 1964, p. 661, hereinafter cited as *Satellite Communications Hearings*.

might be called a nonpolitical basis. European telecommunications administrations had bargained with AT&T and other carriers regarding the laying of new cables or the starting up of radio services. Foreign offices were not involved in these negotiations. However, the economic and political consequences of communications satellites were far from predictable, so it was natural for foreign offices to become involved. The resolution of questions affecting broad national interests could not be left to traditional means. Telecommunications experts, on the other hand, wanted to preserve the old way of doing business; they believed that diplomats only created difficulties.

This domestic alignment observed by Carter in certain European countries corresponded to the pattern within the United States. To one trained in international relations, this pattern pointed to the existence of a subculture in the international environment whose representatives had more in common with each other on a functional level than they did with nationals of their own countries who performed different tasks. The transnational cohesiveness of communications administrations, though, did not serve as a barrier to participation by others. The introduction of satellite communications into the international environment could not be isolated, as was the introduction of telephone cables in 1956. Broader national interests were at stake.

In Great Britain, the Air Ministry, the Foreign Office, and aerospace companies had a great interest in space communications. This broad range of interest also was the pattern in France, West Germany, and other countries. The rationale that had led to the creation of Eurospace, ELDO, and ESRO worked in terms of space communications, too. Europeans felt that in view of the U.S. technological leadership, their interests would be served best if they could speak with one voice rather than many.

The first concrete manifestation of a European regional approach came at the CEPT conference held in Cologne in December 1962. The Telecommunications Committee, one of the permanent organs of the conference, set up a committee to: (1) study all problems relating to the organized participation of all interested European countries and the operation of a single world network of communications satellites; (2) establish the basis of discussions to be held between the CEPT countries and the United States for the establishment and operation of a single world network of communications satellites; and (3) determine the basis for a world organization to manage such a network. The creation of this CEPT committee signified that Europe would take a regional approach to the U.S. proposal rather than negotiate a series of bilateral arrangements and that Western Europe accepted, in broad outline, the idea of a single world system.

After the 1962 conference in Cologne, Western Europe proceeded in high gear to organize a European say in space communications. At a July 1963 intergovernmental meeting in London, CETS set up a structure consisting of steering, organizational, and space technology committees. The committee set up by CEPT in December 1962 became the advisor to CETS on technical matters. Subsequently, in November 1963, CETS agreed that it "should be set up to provide to the extent possible, a counterpart to the U.S. Communications Satellite Corp. [Comsat]."[19]

The existence of a European regional approach set the stage for definitive negotiations to establish an operational global system. As a prelude to formal meetings, a team of American officials from the State Department, the FCC, and Comsat met with technical experts from CETS to discuss the outlook for satellite communications in the near future.[20] The first formal meeting took place in Rome in February 1964. No drafts were

---

19. Department of State, "Summary of European Regional Organization in the Communications Satellite Field," reprinted in *U.S. Congressional Record*, 88th Cong., 2d sess., p. 175.
20. *Satellite Communications Hearings, Part 1*, p. 316.

tabled at this meeting, but a thorough discussion of the general principles for the proposed system's framework resulted.

One of the basic system characteristics agreed on at the Rome meeting was the idea of an interim system. The participants felt that establishing a permanent international plan would be premature in light of the many economic, political, and technical unknowns. Another area of agreement was that Comsat should act as system manager on behalf of all the other participants.[21] While ownership would be joint, Comsat would assure its commanding place by managing the system. On the other hand, Comsat was not to have a free hand. The idea of an international steering committee to oversee developments was discussed.

As the Rome conference ended, considerable agreement on all but three major areas emerged: (1) the duration of the interim agreements; (2) the allocation of ownership quotas; and (3) the voting procedure. The alignment of opinion on these issues pitted the United States against CETS. The Europeans wanted a short interim arrangement, while the United States wanted one of considerable duration. The rationale of the American position was that a longer interim period would provide more experience on which to base the permanent arrangement. The United States also wanted a longer interim period to retain its initial leadership position. U.S. negotiators feared that the time needed to negotiate a permanent arrangement would be so great that it would run counter to the mandate of the Communications Satellite Act of 1962.

The Europeans, on the other hand, hoped to increase their satellite ability at the earliest possible time and therefore pushed for a short agreement. The idea that short interim arrangements would lead to an increase in Europe's stature may have been wrong, however. As the system was to be global, the increasing number of non-European nations in the venture might align to work against the increase in European influence. Nonetheless, it was the position of CETS at the Rome conference that the interim arrangements should last for only about three years.

The Rome conference also included the ownership issue, which revolved around the amount of capital the Europeans wished to invest in the system. The greater the amount of a country's capital investment, the greater its percentage of ownership and thus, potentially, control and rate of return. Expressing their optimism on the prospects for satellite communications, the Europeans wished to invest more money than the United States thought they should.[22] Moreover, the U.S. negotiating team argued for substantial majority ownership by Comsat.

The voting issue concerned the formula by which the system's steering committee (which came to be known as the Interim Communications Satellite Committee) would decide on various matters regarding system management. Because the United States was to have majority ownership, it naturally would be able to control the majority of votes under the system of weighted voting that everyone envisaged as legitimate. The problem lay in the amount of control that the Europeans wanted. Important decisions would involve weighted voting encompassing a greater number of votes than the United States alone possessed. Any other decision-making formula would destroy the concept of partnership. The issue, therefore, was not weighted voting, but how much of a majority, in addition to the votes of the United States, would be required to pass a resolution. The United States feared that CETS might vote as a bloc, and that would constitute a veto even under a weighted voting arrangement. The Europeans feared that, if they were not given

21. *Satellite Communications Hearings, Part 2*, p. 663.
22. See the testimony of Abram Chayes in *Satellite Communications Hearings, Part 1*, p. 363.

some assurance on the criteria for making decisions on procurement, on ground stations, or on other matters, their standing as partners would be meaningless.

These three issues—the duration of the interim arrangements, ownership, and voting—dominated the discussions and negotiations that followed the Rome conference in February 1964. Additional problems, such as system use by DoD and possible membership by the Soviet Union, were deliberated, but they did not impinge directly on the progress of events leading to the establishment of the interim system. Those two problems were handled at a future date, as discussed later in this chapter.

The next meeting following the Rome conference occurred in London from 6 to 8 April 1964. At that time, the U.S. negotiating team advanced two agreements—one governmental and the other commercial. The United States felt that Europe, Australia, Canada, and Japan would insist on an intergovernmental agreement rather than a purely commercial arrangement as had been characteristic in cable consortia,[23] and therefore the U.S. team came prepared with a draft of a simple intergovernmental "umbrella" agreement.[24] It received tentative approval from the Europeans.

A period of intense drafting activity then ensued. In May 1964, two meetings took place to compare drafts (in London) and to study traffic statistics (in Montreal). The London meeting was not a negotiating session, but an attempt to develop language for those points on which no substantial differences had arisen. The Montreal meeting on traffic projections was important, because ownership in the interim system was to reflect the actual distribution of international telecommunications traffic. To measure that traffic, the 1963 International Telecommunications Union's projections for the year 1968 served as a base. Those statistics were reexamined to determine what share of total international traffic could be handled legitimately by communications satellites. The Montreal meeting agreed that the United States had more than 50 percent of that total traffic, but did not determine how much more.

The next meeting to discuss substantive differences occurred in London on 15 and 16 June 1964. The issues of interim agreement duration and system ownership reached resolution. Those attending decided that the planned Interim Communications Satellite Committee would submit a report no later than 1 January 1969 to all parties recommending what changes, if any, should be made in the "Interim Arrangements." An international conference would consider the report, but if no changes were agreed to, the Interim Arrangements would remain in effect. This compromise gave the United States the assurance of at least five years of experience before laying out the definite arrangements. The Europeans were satisfied because five years was not so long as to prejudice their position in the coming decade, when they hoped to have developed a greater technological base in satellite communications technology.[25]

The issue of ownership allocations was settled by giving Comsat a 61-percent undivided interest of the space system. One should bear in mind that the ground stations were not included, as they were to remain in national hands. The total European participation, as well as that of the Canadians, Japanese, and Australians, was 39 percent.[26] A difficult facet of the allocation decision was how to arrange for the accession of new members to the Interim Arrangements. The Europeans pushed to have the United States agree that, up to a certain point, all new signatories' shares should come out of the U.S. quota. The United States successfully resisted this move; instead, it was decided that everyone's share

---

23.   *Ibid.,* p. 661.
24.   *Ibid.,* p. 664.
25.   See Article IX of the Interim Arrangements, 2 U.S.T. 1705, T.I.A.S. 5646.
26.   See Annex of Special Agreement, in Interim Arrangements.

would be cut on a pro-rata basis as new participants joined.[27] Comsat's quota, however, could not fall below 50.6 percent, thus assuring the firm a majority voice during the period of the Interim Arrangements.

At the end of June 1964, only one important issue remained: voting. A conference in Washington, D.C., held from 21 to 24 July 1964, resolved that issue. On important decisions (for example, system choice, rates, ground station standards, or contract letting), the Interim Communications Satellite Committee voted by a weighted majority of 12.5 votes above those controlled by Comsat. All other decisions were by simple majority vote, but all parties committed themselves to try to arrive at unanimous agreements. In addition, if a decision concerning the budget, the placing of a contract, or the launching of a satellite were delayed more than 60 days, a majority vote of only 8.5 above those votes controlled by Comsat was required to pass.[28] For this system of weighted voting to be acceptable to the United States, the CETS had to convince the Americans that it would not vote as a bloc; otherwise, the formula gave the Europeans a veto.

The voting formula was related to the crucial matter of contract awards. Article X of the agreement specified that "the Committee and the Corporation [Comsat] as manager shall . . . seek to ensure that contracts are so distributed that equipment is designed, developed and procured in the States whose Governments are Parties to this Agreement in approximate proportion to the respective quotas of their corresponding signatories to the Special Agreement." This particular provision was very difficult to negotiate. European industry had lobbied through organizations such as Eurospace to pressure their governments into negotiating for a plan that would assure them substantial access to the procurement process.[29] The Europeans wanted an allocation for procurement on a national basis according to quotas of capital contribution.[30] The compromise provision allowed for geographical distribution only when competitive bids were comparable in terms of the best equipment for the best price.

It was obvious to most participants, though, that Europe would be able to contribute substantially only in the 1970s. Almost the entire initial system would be built by the United States. Already, contracts had been signed with Hughes for the Early Bird satellite in March 1965, and Comsat had requested design contracts for an initial system from Hughes, AT&T, RCA, Space Technology Laboratories (TRW), and ITT. Comsat was to make a choice from among these alternatives in the fall of 1965—long before the Europeans would possess the level of technology required.[31] Nonetheless, the voting procedures and the procurement provision assured the Europeans that they could be partners in research and development as well as in system use.

Thirteen governments and Vatican City initialed the "Agreement Establishing Interim Arrangements for a Global Commercial Communications Satellite System" on 24 July 1964. The agreement then was opened for signature by the governments or their designated communications entities on 20 August. Thus, a joint venture for space communications, known as Intelsat (the International Telecommunications Satellite Consortium), came into existence.

---

27. *Satellite Communications Hearings, Part 2*, p. 667.
28. Article V (e) of the Interim Arrangements.
29. *Satellite Communications Hearings, Part 2*, p. 681.
30. *Ibid.*, p. 100.
31. *Ibid.*, p. 739.

Foreign Relations and National Security[32]

The prospect of a shared system between the Interim Arrangements and DoD was laid to rest in July 1964, but the issues involved arose again in connection with DoD use of Intelsat for administrative traffic. Similarly, these issues surfaced in terms of the NASA-Comsat agreement concerning communications for the Apollo lunar program and the FCC's "authorized user" decision in the summer of 1966. The use of the Intelsat system by DoD involved three questions: (1) government relations with private enterprise (the subsidy issue); (2) requirements and capabilities of the military communications system; and (3) foreign relations matters, such as the promotion of U.S. leadership in communications satellite developments.

In January 1965, a controversy arose as to whether DoD should use its own system for both bulk traffic and urgent national security needs. Comsat was afraid that if DoD used its own system for all its traffic, the firm's financial future would be dim. The military share of all U.S. overseas traffic amounted to 30 percent. Comsat made an offer to DoD after the demise of the shared system plan. The new proposal would cost $50 million, but DoD turned it down because it cost half as much as the Philco system.[33] However, the Pentagon did assure Comsat that it would not put all DoD traffic through its own system. Dr. Eugene Fubini, Deputy Director of Defense Research and Engineering, estimated that "95 or 90 percent of our traffic will not go through the military system."[34] As it turned out, though, two years later DoD was planning to have a third of its needs met internally and two thirds by lease from commercial carriers.[35] By January 1971, in fact, the Defense Satellite Communications System (DSCS) was using 44.2 million two-way channel miles, 62 percent of which were leased from private carriers and the other 38 percent government owned. Private carriers thus did not receive the share of business they had anticipated.

For communications satellite service in particular, the DoD spent $22,746,360 through commercial carriers in fiscal 1970, while spending $111 million on its own DSCS I satellite. In addition, the DSCS II satellite was to have 1,300 full duplex voice channels, compared to DSCS I's five to twelve channels. Furthermore, DoD funding for space communications rose from $62.9 million in fiscal 1972 to $192 million in fiscal 1973. This dramatic increase arose from the start of a new mobile communications satellite system called FLEETSAT or the Fleet Satellite communications system.[36]

The interest of this issue lies in evaluating the importance of government assistance to Comsat as a means of promoting the foreign policy goals of the Communications Satellite Act of 1962. The different roles of the corporation created a paradox. Leaving aside the pros and cons of the financial controversy, one can make the following policy-oriented observations. If Comsat is a private enterprise, it would be contrary to free enterprise philosophy for the government to assist it, especially if the government could provide its own services more economically. On the other hand, if Comsat is viewed as the chosen instrument of American foreign policy in space communications, it would be contrary to the national interest to refuse assistance to Comsat, without which its financial and

32. This section is based on Galloway, *The Politics and Technology*, pp. 114–18.

33. U.S. Congress, Senate, Committee on Aeronautical and Space Sciences, *Hearings*, *NASA Authorization for Fiscal Year 1966*, 89th Cong., 1st sess., 1965, p. 612.

34. *Ibid.*, p. 610.

35. U.S. Congress, House, Committee on Government Operations, Military Operations Subcommittee, *Hearings*, *Military Communications—1968*, 90th Cong., 2d sess., 1968, p. 25.

36. Katherine Johnsen, "DoD Details New Satcom System," *Aviation Week & Space Technology*, 24 June 1968, p. 26–27; Philip J. Klass, "New DoD Satcoms Nearing Launch," *Aviation Week & Space Technology*, 11 January 1971, p. 40–45; "Industry Observer," *Aviation Week & Space Technology*, 10 April 1972, p. 9.

political image and stature abroad might be damaged. As Comsat was a little of both, one can readily understand why the subsidy issue was a matter of both domestic and foreign policy.

The same general analysis applies to the criticisms voiced against the 1965 contract that NASA, through the National Communications System (NCS), made with Comsat to provide communications for the Apollo program. President Kennedy established the NCS in 1963 to unify and integrate all government communications capabilities. The House Military Operations Subcommittee was a forceful political actor on this issue.[37] The Senate Committee on Aeronautical and Space Sciences also addressed the program.[38] Concerning the subsidy issue, Representative Fernand J. St. Germain (D-RI) observed: "This is supposed to be a public corporation, and as I recall, the logic behind this and the arguments behind this were we want this to be a public corporation because the Government shouldn't be in private business, and here, lo and behold, the Government, one agency in itself, is going to buy 45 percent."[39]

Representative William S. Moorhead (D-PA) asked NASA Associate Administrator Robert C. Seamans, Jr., whether, when the opportunity to get a regular customer from the DoD had disappeared, Comsat had not "turned to NASA to bail them out of a hole?"[40] But Comsat had not requested the Apollo communications program. NASA, within the framework of the NCS, had examined all the alternatives and had, on its own, requested the NCS manager to contact Comsat.[41]

The U.S. government had documents to prove that its choice was made after all due deliberation—as distinguished from the Comsat-DoD case. The House Military Operations Subcommittee was especially concerned about the real economies involved, however.[42] Herbert Roback, the staff administrator, observed that the Communications Satellite Act provided that a separate government system could be created in the national interest and that one of these interests was economy.[43] He was concerned in particular that a large part of the decision-making process was classified. The feeling of the Military Operations Subcommittee seemed to be that too much secrecy might cover up hidden purposes. The subcommittee was willing to accept government subsidy of Comsat, if it served to strengthen Comsat's position in the international bargaining to establish definitive global satellite communications arrangements after 1969.

The Military Operations Subcommittee also felt that the secrecy surrounding the negotiations between Comsat and the NCS should be removed because the use of taxpayers' money ought to be the subject of public debate. On the other hand, public debate might undermine the purpose of the contract, if in reality its purpose were to subsidize Comsat. Foreign governments would recognize that government subsidies to the firm were a reality. Perhaps this was the reason for the government's secrecy. Nonetheless, the government provided evidence that the contract with Comsat was the most efficient and economical alternative available.[44] The outside observer is left with a partial picture.

---

37.   U.S. Congress, House, Committee on Government Operations, Military Operations, Subcommittee, *Hearings, Missile and Space Ground Support Operations*, 89th Cong., 2d sess., 1965, hereinafter cited as *Missile and Space Ground Support Hearings*.

38.   U.S. Congress, Senate, Committee on Aeronautical and Space Sciences, *Hearings, National Communications Satellite Programs*, 89th Cong., 2d sess., 1966, *passim*.

39.   *Missile and Space Ground Support Hearings*, p. 29.

40.   *Ibid.*, p. 35.

41.   *Ibid.*, p. 150.

42.   *Ibid.*, pp. 26–27, *et passim*.

43.   *Ibid.*, p. 77.

44.   *Ibid.*, pp. 150–51, *et passim*. The contract as finally negotiated in early 1966 called for a three-year contract totaling more than $38.9 million.

The third and most important problem that arose following the demise of the joint Comsat-DoD proposal concerned the "authorized user" issue. Section 305(b)(4) of the Communications Satellite Act authorized Comsat "to contract with authorized users, including the United States Government, for the services of the communications satellite system." One interpretation of this section would allow direct Comsat-DoD contracting with no necessity for FCC approval. This interpretation was that of the drafters of the legislation, according to Edward Welsh, Executive Secretary of the National Aeronautics and Space Council. Another interpretation required the FCC to determine who were "authorized users" and under what circumstances the government would fit this category. DoD proceeded on the basis of the first interpretation when, in January 1966, it notified Comsat of its need for thirty voice circuits in the Pacific.[45] The FCC operated on the basis of the second assumption when it instituted a public inquiry, "In the Matter of Authorized Entities and Authorized Users Under the Communications Satellite Act of 1962" (Docket No. 16058), in the spring of 1965.[46]

The Defense Communications Agency (DCA), which coordinated separate communications systems of the armed services, considered Comsat to be the best carrier. It notified the other international carriers of its traffic needs in the Pacific a full three months after it had notified Comsat.[47] It was not surprising that, after the submission of proposals from five carriers, Comsat's proved to be the best. DCA signed a contract with Comsat on 26 July 1966. It also was not surprising that the FCC, acting as the protector of the status quo, decided that Comsat could deal only with the government directly "in unique and exceptional circumstances." On 21 July 1966, the FCC issued a "Memorandum, Opinion and Statement of Policy" to this effect.[48] The FCC believed that Comsat was "primarily a carrier's carrier" and judged "that if the Government or others were to obtain (service) directly from Comsat, there would be serious adverse affects upon the well-being of the commercial telecommunications industry and the general public it serves."

There was a clear difference of opinion, if not a conflict of interest, between the FCC and DoD. One way to resolve the difference would have been for DCA to notify the FCC that the circumstances were "so unique and exceptional as to require service directly from Comsat." Although DCA felt that way, the statement was not included in the contract; otherwise, the FCC would not have approved it.[49]

This issue received congressional attention when the House Military Operations Subcommittee stepped in to recommend "that the DCA assign the Comsat contract to one or more American international carriers, based upon an across-the-board substantial reduction in charges for satellite and cable circuits in the Pacific area."[50] All agreed to this solution. The cost saving to the government by reducing composite rates on all 128 cable circuits leased in the Pacific by the DoD was greater than $6 million per year, in contrast to the $1.6 million in savings that would have been realized by dealing directly with Comsat for thirty circuits.[51]

---

45.   U.S. Congress, House, Committee on Government Operations, Military Operations Subcommittee, *Hearings, Government Use of Satellite Communications*, 89th Cong., 2d sess., 1966, pp. 44–45.

46.   *Ibid.*, pp. 703–05.

47.   *Ibid.*, pp. 45, 81.

48.   Reprinted in *ibid.*, pp. 706–18.

49.   Reprinted in *ibid.*, p. 717.

50.   U.S. Congress, House, Committee on Government Operations, *Forty-third Report, Government Use of Satellite Communications*, 89th Cong., 2d sess., 1966, H. Rept. No. 1836, p. 7.

51.   U.S. Congress, House, Committee on Government Operations, *Seventh Report, Government Use of Satellite Communications, 1967*, 90th Cong., 1st sess., 1967, H. Rept. No. 613, p. 10.

One can conclude from this episode that a new technology (communications satellites) was slowed down to accommodate established interests (cables). Nonetheless, the results of incrementalism and the consensus approach produced economic benefits in the short run and avoided a sudden, potentially destabilizing decision. The FCC's "authorized user" decision showed that the issue of cables versus satellites may never be unequivocally resolved. A mix of the two, although slowly modified over time, will persist. By 1972, in fact, the new office responsible for coordinating governmental communications, the Office of Telecommunication Policy, recognized that the heyday of decision making through systems analysis had passed. Rational decisions in areas where one does not have all the facts, and where the facts changed from day to day, required day-to-day decision making. This conclusion seemed particularly apropos a generation later, as the Joint Chiefs of Staff consider substantial reductions in DoD's reliance on communications satellites because of advances in fiber optic cables.[52]

Intelsat and Intersputnik[53]

Satellite communications and foreign policy problems also entailed strained relations with the Soviet Union. During the 1962 meetings of the Legal Subcommittee of the United Nations Committee on the Peaceful Uses of Outer Space, the Soviets objected to the Americans' Project West Ford and the operations of Comsat. Their objections seemed to render unlikely the conclusion of an agreement on legal principles governing human activities in space. The Soviet draft of legal principles relating to satellite communications asserted that space cooperation was incumbent on all states and that the implementation of any measure that in any way might hinder the peaceful exploration and use of outer space by other countries should be permitted only after prior discussion of and agreement on such measures among the countries concerned. The Soviets also declared that all space activity should be carried out "solely and exclusively by States."[54]

The Soviets directed the draft of principles against Project West Ford. Needless to say, the United States found it unacceptable, not because it wished to carry out this project or others without taking a proper regard for the opinions of humanity, but because it did not wish to subject national programs to a Communist veto. The United States objected to the declaration that all space activity should be carried out "solely and exclusively by States" because it undermined Comsat as a somewhat private entity. By stressing the exclusiveness of state responsibility for and control of space endeavors, the Soviets meant to downgrade the role of private enterprise or, as the Soviets would call it, exploitative monopoly. Clearly, the Soviet draft of principles witnessed the intrusion of the Cold War into what had seemed a promising move toward cooperation in the form of the Kennedy-Khrushchev letters. Nevertheless, the end of 1963 saw a spirit of agreement. The Soviet Union agreed to omit the consideration of the principles they had so strongly urged in the United Nations Legal Subcommittee. The General Assembly was able to unanimously pass Resolution 1962 on 13 December 1963, but the following year ushered in another period of renewed Cold War competition.

The Soviet Union evolved a position extremely critical of Comsat and Intelsat. The Soviets were slow in voicing this antagonism, however, and most likely came to it after a

---

52. Pat Cooper and Robert Holzer, "DOD Eyes Satellite Alternative," *Space News* 6 (7–13 August 1995): 1, 28.

53. This section is based on Galloway, *The Politics and Technology*, pp. 127–32.

54. United Nations Document A/AC.105/6, 9 July 1962. The Soviet draft also included principles on the rescue of astronauts, an agreed point.

period of protracted deliberations. In February 1963, the American Embassy in Moscow delivered a note to the Soviet government suggesting that it might be useful to discuss a commercial communications satellite system.[55] The Soviets, though, considered such a meeting premature. They also had this attitude during the Extraordinary Administrative Radio Conference later in the year.

In 1964, however, the United States initiated concrete negotiations with various Western European and other nations. In February 1964, perhaps for reasons of intelligence, the Soviets expressed an interest in consultations with the United States. They suggested that a meeting be held in Geneva, during the May and June 1964 meeting of the Technical Subcommittee on the Peaceful Uses of Outer Space.

The United States agreed to this proposal; after all, the Communications Satellite Act mandated agreements with other nations. Therefore, in June 1964, a U.S. team of government and Comsat officials flew to Geneva and briefed the Russian delegation to the concurrent London negotiations on the Interim Arrangements in terms of the negotiations for establishing a global commercial system. The Russians asked a few questions concerning whether the allocation decisions of the Extraordinary Administrative Radio Conference were definitive enough to warrant the engineering and construction of an operational system. This line of inquiry reflected the Soviet view that operational arrangements were premature. Consequently, at the end of the meeting, the Soviet delegation pronounced that the United States was still in "an experimental phase"; the negotiations were "an American or U.S. inspired experimental program," and the Soviet Union was "not really very interested in it at this time." They concluded: "We will continue to do some experimental work on our own. And perhaps at some time in the future we will get together and talk about the whole project again."[56]

This rather opaque, lukewarm statement did not reflect the ideological opposition to private enterprise that characterized Soviet attitudes later in the year. Perhaps an explanation for this statement lies in the agreement concluded by Dr. Anatoli Blagonravov of the Soviet Academy of Sciences and NASA Deputy Administrator Hugh L. Dryden concerning U.S.-Soviet cooperation in space. By leaving the door open, Blagonravov may have expressed an attitude contrary to that of his political superiors. The Soviets might have had much to gain from cooperating with the Interim Arrangements, such as the pooling of resources and the cutting of costs. For political reasons, though, the arguments of economy often have gone unheeded.

In the fall of 1964, at meetings of the International Institute of Space Law and of the United Nations Committee on the Peaceful Uses of Outer Space, the Soviet Union and other Communist countries expressed strong opposition to Intelsat as a device for transferring space communications into the hands of U.S. private capital. In Warsaw, Dr. I.I. Cheprov read a paper in which he contended that it was "the duty of lawyers imposed upon them . . . by resolutions of the U.N. General Assembly to see to it that communications by means of satellites become 'available to the nations of the world as soon as practicable on a global and non-discriminatory basis. . .' (Res. 1721/XVI)."[57] Cheprov contended that the Interim Arrangements were a means of perpetuating the Comsat

55. Testimony of William Gilbert Carter in *Satellite Communications Hearings, Part 2*, p. 665. Much of the following information on U.S. discussions with the Soviet Union comes from this source, pp. 665–66, and from William Gilbert Carter, interview with Jonathan F. Galloway, Department of State, Washington, DC, 18 March 1965.

56. Testimony of William Gilbert Carter, *ibid.*, p. 666.

57. I.I. Cheprov, "Some Legal Problems of International Space Communications," *Proceedings of the Seventh Colloquium on the Law of Outer Space* (Vienna: Springer-Verlag, 1965).

Figure 25

The Soviet Union created a satellite communications system, called Intersputnik, that rivaled Intelsat. Its technological basis was the Molniya satellite, shown above. Critical of Comsat, the Soviet Union argued at the United Nations and in other forums against the role of private enterprise in space communications. (Courtesy of NASA, photo from unnumbered folder, "USSR Molniya general," NASA History Office)

monopoly. Not only would the partners in the Interim Arrangements be victims of monopoly domination, but so would the Americans themselves: "It is well known that private enterprise in the United States has a long experience of discriminating against Americans themselves, so how can we expect this enterprise to be unbiased and just on the international scene?" Cheprov quoted the opponents of the Communications Satellite Act as evidence against existing American policy.

These charges echoed at the meetings of the United Nations Committee on the Peaceful Uses of Outer Space held between 26 October and 6 November 1964. The delegates of Communist bloc countries criticized Intelsat as a profit system for the benefit of the undeveloped countries at the expense of the developing nations.[58] Second, they criticized the Interim Arrangements as a violation of the principle of sovereign equality because of the procedures for weighted voting. Third, they reproached them as an attempt to create an organization outside the more appropriate organizational and political framework of the United Nations and the International Telecommunications Union. These

58.   See United Nations Document A/AC.105/PV.26-35 for this and the following ideas.

charges, couched in traditional Cold War semantics, were placed in a different perspective by the dispassionate analyses of the representatives of the United States and Britain.

The U.S. representative pointed out that a system based on profit was not necessarily exploitative: "Perhaps what we refer to as a 'profit motive' in this country is not as far removed as it might first appear to be from what sometimes is translated from the Soviet press as being 'profit incentive.' " Thatcher went on to suggest that the Communist delegates might be mistakenly confusing participation in Intelsat with access: "It is clear from the agreements themselves . . . that participation in these arrangements is open to all state members of the ITU [International Telecommunications Union] and that whether or not a State is a participating member, access to the system is on a completely free and non-discriminatory basis to all states."

Western delegates did not address the Soviet charges at that time. However, several considerations in this regard may be mentioned. First, the principle of sovereign equality in some arrangements is not incompatible with the principle of weighted voting in others. Political treaties are usually based on sovereign equality, although even in that case the United Nations Security Council can be viewed as an exception. Functional arrangements may involve special formulas reflecting unique situations and responsibilities, as with the case of the World Bank, the International Monetary Fund, and Intelsat. Thus, weighted voting does not per se lead to exploitation.

Second, Intelsat was not divorced from the framework of the United Nations or the International Telecommunications Union. The Extraordinary Administrative Radio Conference of the International Telecommunications Union established the technical framework for the formation of a system of communications satellites. The United Nations does not discourage but rather—in Chapter VIII of its Charter—encourages the formation of regional arrangements whose principles harmonize with those of the Charter. It was somewhat strange for the Soviet Union to criticize the United States for carrying on operations beyond the fringe, when the Russians themselves often appeared as the consummate devotees of self-insulation. However, the tactic of calling for increased international cooperation for others, while reneging on international collaboration itself, was a familiar Soviet practice. In fact, what one saw in this instance was another form of competition—competition to see who cooperated the most. Because both the United States and the Soviet Union continually pronounced their cooperative intentions, it can be interpreted as a mark of prestige to have a better record of international collaboration. As NASA Administrator James E. Webb stated: "Despite its protestations of peaceful interests, the Soviet space program can show no comparable (to our own) engagement in cooperative relationships."[59]

To overcome this cooperative weakness, the Soviet Union proposed its own international satellite communications system, Intersputnik. On 5 August 1968, the Soviet Union, Bulgaria, Cuba, Czechoslovakia, Hungary, Mongolia, Poland, and Romania submitted a draft agreement for Intersputnik to the United Nations.[60] The proposal took advantage of several apparent weaknesses in the Intelsat approach. In the first place, the only members of Intersputnik were states, and the voting procedure was one vote per state. This approach obviously had appeal to many countries, for it contrasted favorably with the image of Intelsat as being dominated by a private U.S. corporation. Furthermore, the Preamble to the draft contemplated the establishment of a system that would provide direct broadcasting from satellites. Such a system could bypass the need to establish expensive ground stations to relay signals from satellites in space. Thus, if such a direct broadcast system were implied, Intersputnik could greatly aid the less developed countries

59.   *NASA News*, 25 January 1965, pp. 13–14.
60.   United Nations Document A/AC.105/46, 9 August 1968.

by enabling them to circumvent the need to establish a costly communications infrastructure. Finally, the Intersputnik draft agreement allowed for more than one international communications satellite system, whereas one of the main criticisms of Intelsat was that the signers of the Interim Arrangements committed themselves to a single system.

Unfortunately for the Soviet Union, the Intersputnik draft had several drawbacks. First, during the same month the eight Communist states proposed their own system, the Soviets launched their invasion of Czechoslovakia. The callousness of Soviet behavior toward their own ally overshadowed any possible propaganda benefits from Intersputnik.

Another drawback to Intersputnik involved a number of ambiguities in the draft proposal. One concerned system ownership. Article 3(2) stated that the satellites could be the property of Intersputnik or leased to Intersputnik by its member states. As Stephen E. Doyle wrote, this provision "could be construed as a Soviet effort to ensure that the USSR would be able to provide at least some of the space segment requirements of the international system to the organization on a lease basis with the USSR collecting the rent."[61] It follows that, in contrast to Intelsat, Intersputnik would be a coordinating umbrella organization rather than one that is the sole owner of the space segment. However, similar to Intelsat, Article 3(4) of the draft stated that "the ground complex shall be the property of the States which have constructed it in their territory." A further ambiguity in the Intersputnik proposal concerned the lack of any requirement that there be nondiscriminatory access for all users—a requirement clearly stated in the Preamble to the Intelsat agreement. On the other hand, Article 10 of the Communist draft stated that "the distribution of communications channels among states members of the organization shall be made on the basis of their need for communications channels." Paragraph 2 of the same article stipulated, however, that states had to pay for its channels at fixed rates, which was certainly a strange, albeit understandable, requirement for a Communist organization ideologically committed to the proposition "From each according to his ability, to each according to his needs."

Yet another ambiguity in the Intersputnik draft concerned the organization's executive and governing structure. Each state had one vote, and actions required a majority of two-thirds; consequently, a minority of member states could stall an initiative backed by a simple majority.[62] In addition, unlike Intelsat, Intersputnik lacked an international executive body, but instead had a secretary general whose responsibilities seemed to include much executive as well as administrative authority.

In addition to these ambiguities, a third major drawback to the Soviet Intersputnik draft proposal was that no states took the Soviets up on their offer four years after its submission. It was true that the draft treaty of 1968 was superseded by a treaty in final form, which was deposited at the United Nations on 15 November 1971, but that treaty had not come into force by mid-1972. Furthermore, four years later, Intersputnik still had not begun actual operations. The Soviet Union had established a domestic system, Orbita, which was quite successful and extensive, but their interest in international space communications lacked the vigor of their other programs of space exploration and exploitation.[63]

Nonetheless, the Soviets had plans to launch a synchronous communications satellite, to be called Statsionar 1, over the Indian Ocean. This position would allow it to cover pop-

---

61.   Stephen E. Doyle, "An Analysis of the Socialist States' Proposal for INTERSPUTNIK: An International Communication Satellite System," *Villanova Law Review* 15 (Fall 1969): 83–105, quote on p. 88.

62.   In Intelsat, this problem is overcome by lowering the majority required for passage on important matters after sixty days. See Agreement, Article V(d).

63.   For a detailed analysis of the entire Soviet space program through 1970, see U.S. Congress, Senate, Committee on Aeronautical and Space Sciences, *Soviet Space Programs, 1966–70*, 92nd Cong., 1st sess., 1971.

ulated areas between Britain and Japan. Also, the Soviets helped construct ground stations abroad in Mongolia, Cuba, the United Arab Republic, and Mali,[64] but it was unlikely that the system would have been able to compete on a point-to-point basis with Intelsat. The only potential threat to U.S. leadership in Intelsat could have come, but never did, through the establishment of a direct-broadcasting system.

In summarizing the competitive features of U.S.-Soviet relationships regarding satellite communications, one must emphasize that the United States had a commanding lead in exploiting the early technology for purposes of international communications. The character of the U.S. leadership was such that it damaged the prestige but not the security of the Soviet Union. The Soviets were not impervious to prestige competition, but through 1972, they confined their responses to ideological thrusts and limited international experiments.[65]

If the Soviet Union had been able to establish a direct-broadcast television service to the newer nations, they might have had at their disposal a powerful instrument of propaganda and mass communication.[66] In the early years, U.S. policy makers were not blind to direct-broadcast possibilities. In fact, considerable discussion of such a system had taken place in the United Nations and in Congress.[67] These systems, though, did not come into their own until the late 1980s. By then, the Cold War was coming to an end, and competition in communications satellites was mainly a market phenomena, not one between enemy states. In fact, by the 1990s, Intelsat was facing the prospect of privatization, and Intersputnik was thinking of modeling itself on Intelsat's early life history!

Conclusion

The origins of communications satellite policies, programs, and institutions in the 1960s and early 1970s set off an avalanche of new communications satellite techniques that continue to this day. In some cases, the original players have fallen victim to their own successes, or they are making their way with the same names in increasingly deregulated, privatized, and interconnected global networks and webs. Comsat still exists, but events have overtaken its foreign policy goals. The Defense Satellite Communications System is ongoing, but its role after the Cold War has changed. Intelsat exists as an organization set up by treaty, but it may split up into two or more parts. Meanwhile, the Soviet Union has disintegrated, but Intersputnik continues to operate while fashioning itself on the model of the old Intelsat.

What lessons does the political scientist draw from the history of communications satellites over the last three decades? Do we really have twenty-twenty hindsight now, or do our glasses filter out certain perspectives that are crucial for understanding technological

---

64.   Thomas L. Shillinglaw, "The Soviet Union and International Satellite Telecommunications," *Stanford Journal of International Studies* 5 (June 1970): 199–226.

65.   On 29 November 1965, color television experiments between the Soviet Union and France began using the French color television system and the Molniya satellite. *Spaceflight* 8 (January 1966): 1.

66.   Just how powerful is a matter of conjecture. Mass media without the use of person-to-person communications may be a poor tool for manipulating attitudes and behavior. For a spectrum of opinion, see Ithiel de Sola Pool, "The Mass Media and Their Interpersonal Social Functions in the Process of Modernization," in Lewis Anthony Dexter and David M. White, eds., *People, Society, and Mass Communications* (New York: Free Press, 1964), pp. 429–43.

67.   In 1968, the General Assembly established a Working Group on Direct Broadcast Satellites as part of the Committee on the Peaceful Uses of Outer Space. Also, see U.S. Congress, House of Representatives, Committee on Foreign Affairs, Subcommittee on National Security Policy and Scientific Developments, *Hearings, Satellite Broadcasting: Implications for Foreign Policy*, 91st Cong., 1st sess., 1969.

change, domestic policy, foreign relations, and national sovereignty at the dawn of the twenty-first century?

From the vantage point of today, the themes stressed in the author's 1972 work on communications satellites remain salient and even more pronounced. The importance of technological change is manifest, and the distinction between evolutionary and revolutionary change pointed out in 1972 is still probably correct. A technological change can be evolutionary—for instance, the move from geostationary satellites to low-Earth-orbit satellites. Yet there can be revolutionary economic impacts on industry structure and no political effects—say, in terms of the balance of power or the national interest. In the main, the consequences of technological innovation in communications satellites have been evolutionary, and it is certainly ironic that cables, not coaxial but fiber optic, are making a comeback.

From another perspective, policy making in a fast-changing environment may be shocked and compromised by the system's changing structure. Such an environment will affect rationality in decision making, as well as the boundaries of political systems—how wide they are and how porous and tenuous they are. Another effect will be on democracy. If globalization processes become encoded in the technological logic of global communications satellite systems, then locally based checks and balances and systems of accountability may be undermined, or put under great stress. However, checks and balances and accountability may reappear at the global level in terms of market-based incentives and the requirements of international regimes.

In today's world, we see great forces pushing in both directions—toward global economic logic and toward local political control. Nationalism seems to be both resurgent and atavistic. We live in a time of paradoxes and ambivalences. Communications satellite technology is one small part of this total technological, economic, and political reality. It is a microcosm, and its history and its lessons point to a better understanding of our age, which may be called "an age of ages"—the space age, the nuclear age, the computer age, and the information age. One must not forget, too, that this is the age of democracy. Perhaps that is the most pertinent lesson of the history of communications satellites: if the technology has globalizing effects, eventually felt at the level of the individual, then communications satellites, as part of the worldwide communications revolution, will be part of the world movement toward democracy.

PART III:

The Unfolding of the World System

Geography, Politics, and Culture

Chapter 14

The Pursuit of Equality: The Role of the Ionosphere and Satellite Communications in Canadian Development

by Bert C. Blevis

Canada is a vast country. It spans six time zones, and much of its area is dominated by rugged terrain and inhospitable climates. As a result, large stretches of the country are sparsely populated. Most of the population is concentrated in a thin strip along the U.S. border, and even that is clustered heavily in the most southerly portion of the country, extending from Windsor, Ontario, in the west to Montreal, Quebec, in the east. Until the advent of satellite communications, the great rural and remote expanses of the country lacked access to telecommunications and other services taken for granted in the southern urban areas. Well before satellites arrived on the scene, however, Canadian researchers vigorously pursued other avenues of communication to bring the nation together through the enhancement of long-distance communications.

Early Ionospheric Studies

Canada's early interest in the ionosphere stemmed from two sources (Table 1). One was to understand the phenomenon that gives rise to the beautiful displays of auroral borealis (the northern lights), so often visible over Canada. The auroral borealis is especially prevalent during certain parts of the solar cycle and at particular times of the year. The other source was the attempt to understand the vagaries of the ionosphere—particularly the causes of disturbances and blackouts experienced in shortwave radio propagation.

Indeed, the early lunar radar work carried out in Canada in the late 1950s was not so much to look at the Moon as a means of communication, or even to explore its surface. Rather, it was to use the Moon as a reflecting (or, more correctly, a scattering) object beyond the ionosphere to study the effects of the Earth's ionospheric layers on electromagnetic radiation passing through them, including the Faraday rotation of the plane of polarization. The historic message from U.S. President Dwight D. Eisenhower to Canadian Prime Minister John Diefenbaker to commemorate the opening of the Prince Albert Radar Laboratory on 6 June 1959 was transmitted via the Moon. Canadian ionospheric concerns during the 1950s also extended to the problems of using radar to detect any missiles that might pass over the country's polar region.

This early attention to the Moon, as well as contemporary studies of reflections from meteor trails, led to a predilection to consider passive satellites, particularly for secure communications. Other passive communications were investigated, including large reflecting balloons such as Echo, clouds of dipoles placed into the Earth's orbit (Lincoln Laboratory's Project West Ford), and various proposals for multifaceted satellites.

Table 1

Canada's Early Ionospheric Programs

Date	Program
1839	First magnetic observatory established at the University of Toronto
1882–83	First International Polar Year; measurements of magnetic and auroral phenomena
12 December 1901	Marconi's transatlantic transmission to Signal Hill, Newfoundland
1932–33	Second International Polar Year; field stations established in north; eclipse measurements
1941	Ionosonde installed north of Ottawa by the National Research Council
1947	Formation of the Canadian Defence Research Board; continuing ionospheric studies at the Defence Research Board's Telecommunications Establishment
4 October 1957	Sputnik I, the first artificial Earth satellite (Soviet Union), International Geophysical Year
January 1958	Explorer I, the first American satellite; discovery of the Van Allen belts
July 1958	Satellite proposals invited by the Space Science Board of U.S. National Academy of Sciences
31 December 1958	Canadian proposal submitted for topside sounding satellite
11 March 1959	NASA approval in principle for topside sounder
6 June 1959	Lunar communications demonstration at the opening of the Prince Albert Radar Laboratory
16 December 1959	Letter of Agreement between NASA and the Defence Research Board for Alouette
1960	Echo I
29 September 1962	Alouette 1, Canada's top-side sounder and the first satellite by other than the Soviet Union or the United States
1962	Telstar I, the first transatlantic television via satellite
June 1962	First Black Brant IIIA atmospheric research rocket launched

Table 1 (continued)

Date	Program
23 May 1963	International Satellites for Ionospheric Studies (ISIS) program, through a Canada-U.S. memorandum of understanding for a joint program to launch four additional satellites
August 1963	Agreement with NASA for Canadian participation in the testing of experimental communications satellites, including a commitment to build a ground station
29 November 1965	Alouette 2
February 1967	Recommendation that the prime Canadian objective in space technology be its applications to domestic telecommunications and resource survey
30 January 1969	ISIS 1
August 1969	Formation of the Department of Communications
1 September 1969	Establishment of Telesat Canada
16 May 1970	First aircraft-to-aircraft voice transmission using the sixth Lincoln Experimental Satellite (LES-6)
31 March 1971	ISIS 2
20 April 1971	NASA and the Canadian Department of Communications sign an agreement to begin the Communications Technology Satellite (CTS) program.
November 1972	Anik A1, the first domestic geostationary communications satellite (C-band)
April 1973	Anik A2
May 1975	Anik A
17 January 1976	CTS/Hermes, the first high-powered, Ku-band satellite and the world's most powerful to date
15 December 1978	Anik B, the first commercial hybrid satellite, operating in both the C-band and the Ku-band
September 1987	Emmy Award to NASA and the Department of Communications for the Hermes contribution

Alouette/ISIS

When Sputnik I was launched on 4 October 1957, scientists at Canada's Defence Research Telecommunications Establishment were among the first to monitor its transmissions and determine its orbit.[1] In July 1958, the Space Science Board of the U.S. National Academy of Sciences invited proposals for scientific experiments to be conducted with satellites. Canadian scientists were eager to participate. On the last day of 1958, the Defence Research Telecommunications Establishment submitted a formal proposal to the newly formed National Aeronautics and Space Administration (NASA) for a top-side sounding satellite. NASA accepted the proposal in principle on 11 March 1959, and the project (known as Alouette 1) became a joint undertaking between Canada and the United States through a letter of agreement between NASA and the Defence Research Board on 16 December 1959.

With the launch of Alouette 1 on 29 September 1962 (just before midnight Vandenberg Air Force Base local time on 28 September), Canada became the third country in space after the Soviet Union and the United States. The stage was now set for the unfolding of Canada's space program.[2] However, the government's decision not to develop satellite launch facilities in Canada restricted the space program to projects achievable only through international arrangements with foreign space agencies, such as NASA and the European Space Agency.

The principal experiment on the Alouette spacecraft was the ionospheric topside sounder, whose two rigid dipole antennas extended twenty-three meters and forty-five meters tip to tip, respectively. Other experiments included the measurement of cosmic radio noise, very-low-frequency radio emissions, and energetic charged particles.

Within a short time after the successful launch of Alouette 1, Canada initiated negotiations with NASA for additional scientific satellites. On 23 May 1963, those negotiations led to the creation of the International Satellites for Ionospheric Studies (ISIS) program, consisting of Alouette 2, ISIS 1, and ISIS 2. Alouette 2 (launched 29 November 1965) was a modified version of Alouette 1 and included a probe experiment and an expanded sounder frequency range. The two ISIS spacecraft (launched 30 January 1969 and 31 March 1971, respectively) incorporated additional equipment furnished by the United States.

An international working group, called the Topside Sounder Working Group (but later renamed the ISIS Working Group), was set up in 1960 to provide guidance to the program. In addition to the United States and Canada, Australia, Britain, Finland, France, India, Japan, New Zealand, and Norway became involved in the program.

Before Alouette 1, scientists had virtually no direct knowledge of the ionosphere above approximately 300 kilometers. The Alouette satellite instruments provided information on electron distributions, their temporal and spatial variations, their irregularities and resonances, the influence of incoming charged particles, cosmic and solar noise, polar cap absorption, solar wind penetration, and ion species in the Earth's atmosphere.

The two Alouette and two ISIS satellites were extremely complex spacecraft for their time. They set records for longevity and established a precedent for a long history of inter-

---

1.   For an interesting history of the Canadian space program, see Doris H. Jelly, *Canada: 25 Years in Space* (Montreal: Polyscience Publications, in cooperation with the National Museum of Science and Technology, 1988).
2.   For details of the development of the Alouette, ISIS, and Hermes satellites, as well as a number of interesting anecdotes, see Theodore R. Hartz and Irvine Paghis, *Spacebound* (Ottawa: Government of Canada, Department of Communications, 1982).

national cooperation in space. Alouette 1 was designed to have a lifetime of one year. In private, project scientists hoped to gain at least a month of data from Alouette 1, but in reality, the satellite provided data for more than ten years. The ISIS satellites remained operational for almost twenty years. In 1987, Alouette 1 was selected as one of the ten greatest engineering achievements in Canada in the past century; in 1993, the Institute of Electrical and Electronic Engineers designated the Alouette/ISIS program an International Milestone of Electrical Engineering. Scientists from ten nations published approximately 700 scholarly papers that described the results of the Alouette/ISIS program. This was perhaps the most prolific of any such program.

A fourth satellite in the ISIS program (ISIS-C) was to have been launched under the agreement with the United States, but it was abandoned in 1969. Canadian space policy had been undergoing some fundamental changes in the previous years. In 1963, for example, the Canadian government decided that the technology that had been resident in the Defence Research Telecommunications Establishment until then was to be transferred to industry during the ISIS program as a means of augmenting the Canadian industrial space capability. Then came a February 1967 government report (the so-called Chapman report), which recommended redirecting space technology research to specific applications—in particular, communications and remote sensing. The objective was to place elements of space technology vital to Canada under Canadian control, as well as to foster a Canadian space industrial capability to meet Canadian needs and to address export markets.

A Global First

Even armed with all the new knowledge about the ionosphere, the emergent communications satellite technology offered the only practical solution for providing reliable telecommunications—particularly radio and television—to the 20 percent of Canadians who had no possibility of sharing in the communications and information revolution. To achieve the Canadian government's highest objective of providing basic telecommunications services to all, Telesat Canada—initially half owned by government and half by the private sector—was created on 1 September 1969.[3]

The launch of Anik A in November 1972, and the inauguration of service in 1973, placed Canada in the forefront—the first country in the world to implement a domestic commercial geostationary satellite system. To complete the initial system, two other Anik A satellites were launched, one in April 1973 and the other in May 1975. Similar to other communications satellite systems of the time, all three Anik A satellites operated at C-band. The Anik B satellite, launched 15 December 1978, was intended to provide Telesat Canada with capacity as a backup for the Anik A series.

Hermes

After a long series of discussions, the Canadian Department of Communications and NASA signed an agreement on 20 April 1971 to undertake a joint program called the Communications Technology Satellite (CTS). The CTS was to replace the fourth spacecraft originally planned as part of the ISIS program. Because of the number of technological challenges that had to be overcome, however, the CTS program was not

---

3.   Hon. C.M. Drury, *A Domestic Satellite Communications System for Canada* (Ottawa: Queen's Printer, 1968), which was a Canadian government white paper, argued that "a domestic satellite communications system is of vital importance for the growth, prosperity and unity of Canada, and should be established as a matter of priority."

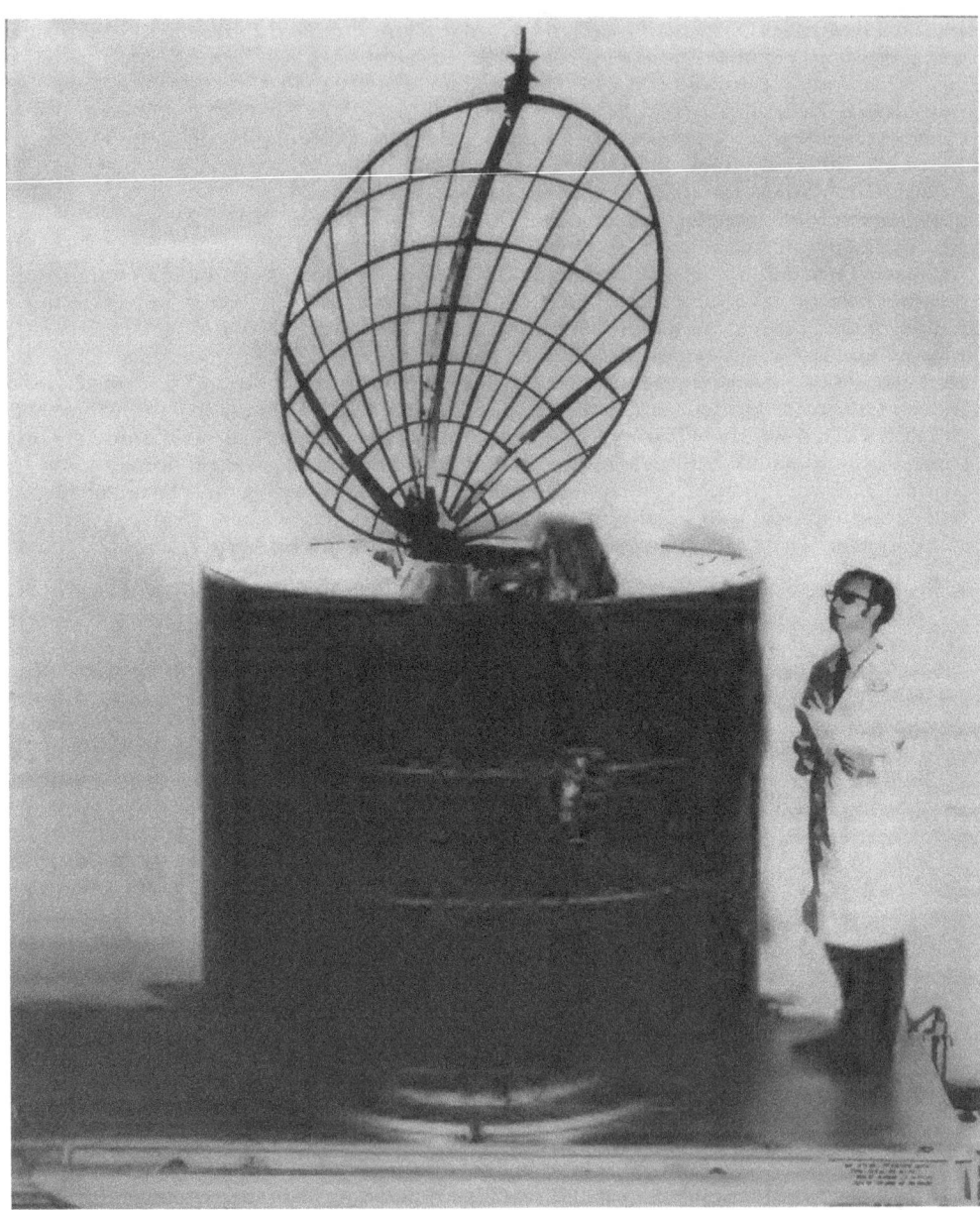

Figure 26
Launched in 1972, Anik A (above) made Canada the first country ever to inaugurate a domestic commercial geostationary satellite system. (Courtesy of NASA)

without its detractors. Nonetheless, the Canadian government agreed to take the lead and undertake the development work at its Communications Research Centre. This center originally had been the Defence Research Telecommunications Establishment; it was transferred to the Department of Communications in August 1969 and renamed.

The objective of the CTS program was to advance the state of the art of communications satellites by developing a system capable of operating at higher powers and higher frequencies than existing systems. Such a system thereby would make direct communications possible with low-cost (at the time) ground stations in individual homes and communities. The program also aspired to develop and flight-test a three-axis stabilization system to maintain accurate antenna pointing and to conduct communications and technological experiments with the system. An additional objective of the CTS program was to improve Canadian industrial capability in the design and manufacture of spacecraft and satellite subsystems. The prevailing political environment influenced the communications experiments, focusing on an evaluation of the economic, social, and political impact of the introduction of new services in the future. Those services included the provision of medical and educational two-way television services to remote areas, community interaction, the delivery of government services, and direct-television broadcasting.

Canada designed, built, and operated the CTS spacecraft. The United States provided the high-power traveling-wave tube for the satellite transponder, as well as test and launch services for the spacecraft and rocket. It is interesting to note also that the original Canadian proposal included a supplementary L-band mobile satellite communications payload. That payload was dropped in favor of the higher frequency (twelve- and fourteen-gigahertz) communications package when it was learned that the launch vehicle provided by the United States precluded the inclusion of both systems.

Canada and the United States shared the use of the satellite equally. Subsequently, in May 1972, the European Space Research Organization (now the European Space Agency) also participated in the CTS program. The Europeans agreed to provide several components for use in future European communications satellites, including a twenty-watt traveling-wave transmitting tube, in return for developing the solar arrays at no cost to Canada.

The CTS, launched 17 January 1976, was renamed Hermes in Canada when service began on 21 May 1976. The inauguration of service included a one-hour color television teleconference between NASA's Lewis Research Center in Cleveland, Ohio, and the Communications Research Centre in Ottawa, Ontario. Canada chose the name from classical Greek mythology. Hermes, the son of Zeus, was considered to be the god of science and invention, as well as eloquence and dreams. The Hermes spacecraft was the most powerful civilian spacecraft (as measured by effective isotropically radiated power) yet launched. Designed to endure two years in space, Hermes operated for almost four years before it was lost.

During those four years, the satellite fulfilled all of its objectives. Hermes performed various communications experiments that proved its usefulness in providing medical and educational services to remote areas ("telemedicine" and "tele-education"), in promoting community interaction, in delivering government services, and in demonstrating direct-to-home television using very small reflector antennas. It is believed that the first ever direct-to-home satellite television broadcast was that of a Canadian hockey game in May 1978. The broadcast was transmitted via the Hermes satellite to a sixty-centimeter dish antenna set up at the home of a Canadian embassy official in Lima, Peru, during a reception for an international meeting. In August 1979, Hermes, repositioned over the Pacific Ocean, served in a joint satellite communications workshop with Australia and in a demonstration of direct-television broadcasting in Papua, New Guinea. Subsequently, control of the satellite was lost, and all communications ceased.

Nonetheless, for Hermes's accomplishments in the field of television broadcasting and its applications, the Communications Research Centre and NASA jointly received an Emmy Award from the National Academy of Television Arts and Sciences in 1987. Because of the need to have some back-up and follow-on capability for Hermes in the event of failure of any one of its many innovative subsystems, the Canadian government arranged with Telesat Canada to include fourteen-gigahertz uplink/twelve-gigahertz downlink transponders on its Anik B satellite. As a result, Anik B was the first satellite in the world to operate in both the C-band and the Ku-band.

Mobile Satellite Communications

As early as 1967, Canadian researchers were involved in trials with the United States on the use of ultrahigh-frequency (UHF) satellites for mobile services primarily for defense operations. The use of UHF frequencies on U.S. experimental military satellites, such as the LES-5, LES-6, and TACSAT, offered the possibility of small, lightweight mobile or transportable terminals. Several demonstrations were carried out in land, maritime, and aeronautical environments. The world's first direct aircraft-to-aircraft voice communications via satellite took place on 16 May 1970 between two Canadian Department of National Defence aircraft.

The Canadian Department of National Defence continued to use U.S. satellites; Canada never implemented a military UHF mobile satellite system. Nonetheless, the early Canadian efforts led to a proposal, although never implemented, to include a UHF transponder on the Hermes spacecraft. The idea of a Department of National Defence MUSAT (Mobile UHF Satellite) eventually resulted in a proposal for a civilian mobile communications satellite called MSAT. Ultimately, with the encouragement of the Canadian government, and in cooperation with NASA, Telesat Canada and several American companies (which later formed the American Mobile Satellite Corporation) undertook the development of two satellites to provide mobile satellite services on a commercial basis in North America. Later, the responsibility for Canada's involvement in the project was transferred to a private firm, TMI Communications.

Conclusion

The rest of Canada's communications satellite program is more recent history. Canada was one of the founding nations of the satellite-aided search-and-rescue system known as COSPAS/SARSAT for Cosmicheskaya Sistemya Poiska Avariynych Sudov (Space System for Search of Distressed Vessels in Russian) and Search and Rescue Satellite-Aided Tracking, which became an almost instant success after the launch of the first spacecraft in 1982. Researchers at the Communications Research Centre participated in the European Space Agency's large, high-powered, multipurpose Olympus communications satellite. These researchers continue to make major contributions to the development of new communications technologies, to carry out studies on the next generation of satellite communications systems, and to explore new ways of providing access for all Canadians to the new multimedia information superhighway.

Telesat has gone on to launch its Ku-band Anik C, C-band Anik D, and the hybrid Anik E series of satellites. Anik E made history when both of the two other satellites failed during an unusual solar event. Telesat subsequently has brought them back to full operation. Mobile communications satellite studies at the Communications Research Centre have led to the manufacture of American Mobile Satellite Corporation's AMSC-1 satellite, which was launched in 1995, and of MSAT-1, launched on 20 April 1996 and placed in geostationary orbit at 1075° West.

Teleglobe, Canada's international telecommunications carrier, is a signatory to and a major player in Intelsat and Inmarsat. It also has become a partner in the Orbcomm Little LEO (low-Earth orbit) satellite and now is participating with TRW in the Big LEO satellite program called Odyssey.

Canada, spurred on by the necessity of providing for the social, economic, and political needs of a population widely dispersed over a vast, and sometimes inhospitable, terrain, and subject to a harsh climate, played a major role in the exploration of the ionosphere and in the early international development of satellite communications. It has maintained its leading role through succeeding generations of new satellites and technological progress. Canada can be expected to remain at the forefront of satellite communications technology well into the future.

Chapter 15

The Long March to Space: Satellite Communications in China

by Zhu Yilin

China entered the satellite field in 1965 and has been developing applied satellite systems for economic, scientific, technological, educational, and cultural uses since the launch, on 24 April 1970, of the first Chinese satellite, Dong Fang Hong-1 (DFH-1), named for the Maoist victory song. By the end of 1996, China had launched thirty-seven of its own satellites, including sixteen returnable remote-sensing satellites, seven communications satellites, two meteorological satellites, and twelve scientific and technical experimental satellites, as part of three applied satellite series: low-Earth-orbit returnable remote-sensing satellites, geostationary Earth orbit communications satellites, and Sun-synchronous orbit meteorological satellites (see Table 2). This chapter describes the development of, and the benefits derived from, China's communications satellites.

China entered the field of satellite communications to improve the relatively under-developed state of its communications infrastructure. In June 1970, the Chinese Academy of Space Technology proposed a preliminary technical plan for a communications satellite. Not until February 1975, however, did the State Council of China approve the "Report on the Development of Chinese Satellite Communications," which had been outlined by the State Planning Commission and the National Defense Science and Technology Commission. Approval from the State Council of China meant that the task of developing the country's satellite communications was incorporated in state planning.

Experimental Communications Satellites

China launched its first experimental communications satellite (the so-called Test Satellite-1) on a Long March 3 rocket on 29 January 1984, placing it in a nonsynchronous orbit. Subsequently, the country launched its first geosynchronous experimental communications satellite on 8 April 1984 and parked it in an orbit at 125° E above the equator (see Table 2). The satellite began its trial operation in May 1984 and subsequently has operated normally. It provides China with a variety of services, such as communications in remote districts; the management of water conservation and electric power; telephone, telegram, facsimile, picture, and data transmission; and the broadcast of a standard time and standard frequency. Its actual lifetime in space has greatly exceeded the expected three years.

The DFH-2 Operational Satellite

China also has placed into geostationary orbit four operational communications satellites parked at 110.5° East (E), 103° E, 87.5° E, and 98° E above the equator, respectively. There are two kinds of these satellites: the DFH-2 (Dong Fang Hong-2), which is basically the same as the experimental satellites, and the four DFH-2A satellites, which are modified versions of the DFH-2.

Figure 27

An Intelsat IV (F-8) satellite relayed live television pictures of President Nixon's historic 1974 trip to China. At the time, China sought access to the Intelsat system, but claimed it would not join as long as Taiwan was a member. In 1974, eighty-seven countries were Intelsat members. (Courtesy of NASA)

The DFH-2 satellite was launched into geostationary orbit (103° E above the equator) on 1 February 1986. It carried two transponders and a domestic beam antenna. Its uplink and downlink communications frequencies were 6,225 to 6,425 megahertz and 4,000 to 4,200 megahertz, respectively. The antenna beam can cover the whole territory of China, thereby providing effective communications to outlying districts, such as the Xinjiang and Xizang Autonomous Regions, frontier stations, and islands. In April 1986, the DFH-2 satellite took over broadcasting services from the experimental communications satellite, and on 5 February 1987, it began providing fifteen channels for foreign program broadcasting from the Central People's Broadcasting Station. The number of channels available for foreign broadcasting increased to thirty on 30 September 1987. The DFH-2 satellite remained in continuous use until 8 July 1989.

Table 2

China's Launched Satellites up to August 1995

Launch Date	Satellite Name	Launch Vehicle	Brief Remarks
24 April 1970	Dong Fang Hong-1	Long March 1	Broadcast music "Dong Fang Hong"
3 March 1971	Shijian-1	Long March 1	Operated for eight years in orbit
26 July 1975	TTS-1	Fengbao-1	All onboard systems operated normally
26 November 1975	RRSS-1	Long March 2	Operated in orbit three days before landing
16 December 1975	TTS-2	Fengbao-1	All onboard systems operated normally
30 August 1976	TTS-3	Fengbao-1	All onboard systems operated normally
7 December 1976	RRSS-2	Long March 2	Operated three days and then landed
26 January 1978	RRSS-3	Long March 2	Operated three days and then landed
20 September 1981	Shijian-2 Shijian-2A Shijian-2B	Fengbao-1	Three satellites launched by one rocket for the first time
9 September 1982	RRSS-4	Long March 2	Operated five days and then landed
19 August 1983	RRSS-5	Long March 2	Operated five days and then landed
29 January 1984	Test-Satellite-1	Long March 3	Carried out communications, operational, and technical tests
8 April 1984	ECS	Long March 3	Stationed at 125° E above equator on 16 April
12 September 1984	RRSS-6	Long March 2	Operated five days and then landed

Table 2 (continued)

Launch Date	Satellite Name	Launch Vehicle	Brief Remarks
21 October 1985	RRSS-7	Long March 2	Operated five days for land survey and then landed
1 February 1986	OCS-1	Long March 3	Stationed at 103° E above equator on 20 February
6 October 1986	RRSS-8	Long March 2	Operated five days for land survey and then landed
5 August 1987	RRSS-9	Long March 2	Operated five days and then landed
9 September 1987	RRSS-10	Long March 2	Operated eight days and then landed
7 March 1988	OCS-2	Long March 3	Stationed at 87.5° E above equator on 23 March
5 August 1988	RRSS-11	Long March 2	Operated eight days and then landed
7 September 1988	Fengyun-1	Long March 4	Achieved predicted goal
22 December 1988	OCS-3	Long March 3	Stationed at 110.5° E above equator on 30 December
4 February 1990	OCS-4	Long March 3	Stationed at 98° E above equator on 14 February
3 September 1990	Fengyun-1 Daqi-1 Daqi-2	Long March 4	Transmitted cloud pictures to Earth and measured atmospheric density
5 October 1990	RRSS-12	Long March 2	Operated eight days and then landed
28 December 1991	OCS-5	Long March 3	Failed to achieve preset orbit because of launch vehicle failure
9 August 1992	RRSS-13	Long March 2D	Operated sixteen days and then landed
6 October 1992	RRSS-14	Long March 2C	Operated seven days and then landed

Table 2 (continued)

Launch Date	Satellite Name	Launch Vehicle	Brief Remarks
8 October 1993	RRSS-5	Long March 2C	Operated eight days, but failed to land because of malfunction
8 February 1994	Shijian-4	Long March 3A	Operated in a geosynchronous transfer orbit
3 July 1994	RRSS-16	Long March 2D	Operated fifteen days and then landed
30 November 1994	Dong Fang Hong-3	Long March 3A	Placed in quasi-geosynchronous orbit, but failed to station because fuel had drained

TTS—Technical Test Satellite
RRSS—Returnable Remote Sensing Satellite
ECS—Experimental Communication Satellite
OCS—Operational Communication Satellite

The DFH-2A Satellite Series

The first DFH-2A satellite was launched on 7 March 1988 and parked at 87.5° E above the equator. All four C-band transponders aboard the satellite have seen service. Transponders A and B relayed the first and second channels of the Central People's Broadcasting Station. Transponder C was used to transmit programs from Xizang television station and to provide special services for the Bank of China. Transponder D was used to transmit programs from the Yunnan, Guizou, and Xinjiang television stations. The power output of these transponders was 25 percent greater than that of the DFH-2. They were designed to last 4.5 years, or 50 percent longer. Moreover, the DFH-2A can provide 3,000 telephone channels or four television channels in contrast to the 1,000 telephone channels or two television channels provided by the DFH-2 satellite.

On 22 December 1988, China successfully launched the second DFH-2A satellite and parked it in a geostationary orbit at 110.5° E above the equator. Its four transponders relayed educational television programming over thirty channels and provided specialized communications services. On 4 February 1990, the third DFH-2A satellite was successfully launched and parked in an orbit at 98° E above the equator. Subsequently, the fourth DFH-2A satellite was launched on 28 December 1991, but unlike its successful predecessors, it did not achieve the desired orbit because of a firing failure of the launch vehicle's third stage. Nonetheless, the satellite entered an elliptical orbit using an onboard motor.

Figure 28

Launched in 1988, the first of China's DFH-2A series of satellites relayed television programming to key provincial capitals and provided special services for the Bank of China. (Courtesy of the Chinese Academy of Space Technology)

The DFH-3: The Second-Generation Communications Satellite

The second-generation DFH-3 communications satellite was a medium-capacity space-craft designed to take over from the first generation of satellites and to satisfy an increased demand for domestic communications capacity. It was to be put into orbit by a new launch vehicle, the Long March 3A. The beam of the DFH-3 communications antenna was designed to cover more than 90 percent of the territory of China. The DFH-3 satellite car-ried twenty-four C-band transponders, of which six were medium powered and were to be used to transmit television programming; the others were low powered for carrying tele-phone, telegraph, facsimile, and data transmissions. The DFH-3 was capable of relaying six channels of color television programming and 8,100 telephone channels simultane-ously. It was designed to last eight years, providing television and radio program and com-munications services to the entire country by the year 2000.

With the launch of the DFH-3 satellite, Chinese communications satellite technology will reach a new level. The DFH-3 was to be the first communications satellite geared to the needs of the whole society as well as to the needs of business and the state. For the most part, the Chinese Academy of Space Technology carried out the design and manu-facture of the satellite in China, although some foreign electronic devices and mechanical parts were purchased, and some components were produced in cooperation with the German firm MBB (Messerschmit-Bolkow-Blohm).

On 30 November 1994, the DFH-3 satellite was launched (about a year later than scheduled) into geosynchronous transfer orbit on a Long March 3A rocket. It was then injected into a quasi-geosynchronous orbit through maneuvers carried out by the onboard propulsion system. These maneuvers showed that the satellite's system design and technical concept were feasible and correct, but unfortunately, the satellite, on the verge of success, could not maintain its orbital position. Leakage in its attitude control thrusters had drained its fuel supply. Consequently, the DFH-3 did not become operative. The results of in-orbit tests indicated that the onboard subsystems, including the transponders, were all normal. The Chinese Academy of Space Technology is currently reviewing its systems and subsystems in the hopes of attempting another launch.

Benefits of China's Communications Satellites

Communications satellites have brought about many social and economic benefits to China. The application of communications satellites to the development of the country's telecommunications, television, and radio infrastructures has allowed China to leap over the traditional development stage and to realize countrywide coverage in a single step. Prior to the arrival of satellite communications, China's communications and broadcast services, especially its long-distance communications, were backward. Not only was the number of telephone channels small, but the quality of communications was poor. The poor quality and depressed quantity of telecommunications services severely limited the development of the Chinese economy: the speed of information transmission is one of the most important factors affecting economic, cultural, and educational development.

Since 1984, however, to effect a countrywide system of communications satellites, China has established a number of ground satellite stations in several cities, such as Beijing, Kunming (Yunnan Province capital), Urumqi (Xinjiang Autonomous Region capital), Lhasa (Xizang Autonomous Region capital), and Nansha island. Numerous diverse government agencies, such as posts and telecommunications, petroleum, mining, water conservation, electrical power, news, and the military, transmit most of their telephone calls, telegrams, facsimiles, data, tables, and pictures via communications satellites. The result is a significant improvement in communications, especially long-distance communications, and the overcoming of communications difficulties in remote districts. If China were to establish a high-quality radio and television broadcasting network covering up to 80 percent of the country using traditional microwave relay technology, the cost would be about 2 billion yuan ($370 million at the 1993 estimated price and exchange rate). The same services can be provided at a cost of only 1 billion yuan ($185 million at the 1993 estimated price and exchange rate) by a communications satellite system.

By the end of 1995, more than 30,000 receive-only stations had been built. The quality of television program transmissions has been enhanced and geographical coverage expanded. China has established more than 100 communications ground stations so far with antennas larger than five meters in diameter. In the field of educational television, the combined coverage rate achieved by the Central Television Station and local television stations was only 33 percent before 1983. The quality of transmission was low, and its effectiveness was poor because of geographical and climatic conditions. Subsequently, however, communications satellites began carrying two educational stations with a total of thirty hours of daily programming. By the end of 1995, China had more than 1,000 educational television receiving and transmitting stations, more than 10,000 receive-only stations, and more than 62,000 display points in the national educational system, representing a coverage rate of more than 83 percent. According to a poll conducted by the Chinese Educational Television Station, the total population receiving and watching educational television programs has reached 30 million.

As for long-distance telephony and telegraphy, if China were to establish a communications network linking all of its provincial and regional capitals by means of microwave relays and coaxial cable lines, the estimated cost would run into billions of yuan (nearly $1 billion). Use of a satellite communications system lowered this figure to only 500 million yuan ($93 million). By the end of 1995, China had more than 8,000 domestic and 25,000 direct international satellite channels available, as well as ten special satellite communications networks for specifics uses in the development of petroleum, mining, water conservation, and electrical power. At present, China is constructing a number of medium-sized satellite ground stations to add telephone and telegraph satellite channels. The number of very small aperture terminals countrywide has reached 35,000.

Satellite communications also has aided the field of finance. By using satellites for fund transfers, China's banks can reduce greatly the amount of money that is in transit from one point to another and that, as a result, cannot be used. The total amount of funds being transferred among Chinese banks at various levels is, according to one study, as much as 50 billion yuan ($9.3 billion) at any one moment, and the average time required for transfer is six days. By managing the transfer of funds by satellite, this figure can be reduced by as much 50 percent, so that an extra 25 billion yuan ($4.6 billion) would be available for six extra days, thus increasing the amount of working capital and greatly enhancing the availability of funds. The DFH-2A satellites now transmit financial data for the Bank of China. The bank's network is centered at the Beijing head office and includes 350 branch offices. The network not only helps modernize fund clearance and transfer transactions, but also provides financial management information services, telephone communications, business training, and teleconferencing.

For the management of railway traffic, the satellite system of communications and train dispatching has enabled a dramatic increase in the density of train traffic. The time interval between two trains in motion can be reduced from the present eight minutes to three minutes, thereby doubling rail transport capacity without much extra cost. In contrast, the construction of a duplicate railway line from Beijing to Shanghai to double transport capacity would cost 10.2 billion yuan ($1.89 billion), whereas building a satellite communications and dispatching system would cost only 1 billion yuan ($185 million). Preliminary studies to help select a suitable system are currently under way.

Future Prospects

Although it is a geographically enormous and populous country, China has a limited economic capability. As a result, the country finds it impossible to increase to a large extent its investment in space efforts. Therefore, the selection and planning of its communications satellite program, with an eye to its obvious social and economic benefits, have critical importance. China urgently needs to develop satellite communications and broadcasting capacity to offset its shortage of educational and communications services. Moreover, since the end of 1980s, the demand on communications satellites by both domestic and foreign users has increased rapidly. In particular, the recent appearance of the so-called information superhighway has increased further demand on communications satellites. Undoubtedly, the communications satellite will be one of the main thoroughfares of the future information superhighway.

For these reasons, during the 1990s, China undertook a large-scale development of its communications satellites. Future satellites will operate not only in the C-band, as do existing satellites, but also in the Ku- and L-bands to meet the requirements of educational television, fixed and mobile communications service, specialized data transmission, radio broadcasting, and television transmission. In addition, China plans to develop small, light, inexpensive satellites to meet the requirements of domestic and foreign users of small and medium scale.

At present, some satellites can broadcast directly to homes in several developed countries. The United States, Britain, France, Germany, and Japan have their own direct-television satellites, so that families can receive and watch television programs relayed directly by satellite. It is just a matter of time before China will develop direct-television broadcast satellites. Satellites also have unparalleled advantages over other media for mobile communication. In recent years, a variety of mobile satellite communications systems have been placed in high, medium, and low orbits or are in the planning stage. China, too, needs mobile communications satellites, and it is possible that a mobile communications satellite system will be developed in the not-too-distant future.

To meet the future demand for satellite bandwidth for the national high-speed information network, television broadcasting, and mobile communications, China will need a large number of satellite transponders by the year 2000. Public communications will require sixty to sixty-five transponders, including five to seven in Ku-band, while the special communications needs of various governmental agencies, banks, and large companies will require an additional twenty-five to thirty transponders, including twelve to fifteen in Ku-band. The country's remaining communications will require sixteen transponders, with four in Ku-band. It is anticipated that the number of users of mobile satellite communications will increase to between 200,000 and 300,000. In the field of television broadcasting, the Central Television Station, the television stations of thirty-one provinces, city-states, and autonomous regions, and the estimated four to six educational television channels will require a minimum of forty-four to forty-eight transponders. In total, China will require 145 to 153 transponders at least by the year 2000, including twenty-one to twenty-five Ku-band transponders with fifty-four to seventy-five megahertz of bandwidth each. To meet these requirements, China must build and launch six or seven satellites, if each satellite carries twenty-four transponders, or four to five satellites, if each carries thirty-six transponders.

The Next Generation of Communications Satellite

To satisfy these increasing communications needs, consolidate domestic markets, and penetrate international markets, China will focus on the development of satellites of higher quality and lower cost. The Chinese Academy of Space Technology, having developed and launched two generations of communications satellites, has begun reviewing its satellite experience and is studying foreign technology and management methods. The first step must be to standardize a common platform bus for a series of satellites. The platform is the base of a satellite. The use of a standard common platform can simplify the work of developing a new platform for each new satellite. A few standard platforms capable of satisfying the needs of various payloads will be developed. The main objectives of this approach are to reduce development time, to improve satellite quality, and to lower costs.

The Chinese space industry is currently developing four series of standard platforms, one of which is a geostationary orbit satellite platform to be used in communications satellites. The DJS-1 platform used by the DFH-3 communications satellite can support a payload weight of 150 to 170 kilograms, while its solar array is capable of generating 1,600 to 2,200 watts of electrical power to supply a payload electric power requirement of 900 to 1,000 watts. The DJS-1 platform also had 1,270 kilograms of available propellant for orbital maneuvers and an expected lifetime of eight to ten years.

The DJS-1 platform will be used to build additional medium-capacity communications satellites and to provide dual-band (Ku- and C-band) communications, mobile communications, data transmission, and other services. To satisfy the demand for large-capacity satellites, China plans to develop a third generation of communications satellite, one with large capacity and a large platform bus, dubbed DJS-2. The main capabilities of one option tentatively planned for the DJS-2 include such features as the ability to support a payload weight of between 400 and 500 kilograms, a solar array output of 4,500 to 6,000 watts, electric power available for payloads of 3,000 to 4,700 watts, 2,200 kilograms of propellant available for orbital maneuvering, and an operational lifetime of fifteen years. The DJS-2 common platform will be able to carry twelve Ku-band transponders of 100-watt output and twenty-four C-band transponders of sixteen-watt output, or fourteen to sixteen Ku-band transponders of 120-watt of output. It will serve to build large-capacity Ku and C dual-band communications satellites, Ku-band direct-broadcast satellites, and tracking and data relay satellites.

As payloads are the core of a satellite and the decisive factor in determining its uses and performances, and as the technical level of Chinese payloads is much lower than that of advanced countries, China must develop or acquire critical payload technologies on a priority basis. The main payloads of a communications satellite are transponders and antennas. In the field of transponders, China must first develop Ku-, L-, S-, and X-band transponders, dual-band (C- and Ku-band) transponders, higher powered transponders, and onboard processing techniques. In the field of antennas, China first must develop multibeam antennas, controllable spot-beam antennas, changeable-shaped-beam antennas, and a number of other antennas, as well as the techniques of onboard switching and intersatellite linkage.

While the Chinese Academy of Space Technology will carry out much of this developmental work, international collaboration must be an important part of those efforts. China should enlarge its range of cooperation with other countries and seek to collaborate at several levels in multiple fields. These efforts should include the importation of certain critical technologies that are of great importance, but such efforts require additional funding to develop and are difficult to achieve in the short term. China also should contract with foreign countries for some subsystems, while the design and integration of the satellite system will remain the responsibility of the Chinese Academy of Space Technology. Finally, China should carry out general programs of international cooperation for the joint development of new satellites, such as that undertaken with Brazil for the development of the China-Brazil Earth Resources Satellite Ziyuan-1 or the Sino satellite being developed jointly with French and German companies.

Chapter 16

No Free Launch: Designing the Indian National Satellite[1]

by Raman Srinivasan

The Indian National Satellite (INSAT) was the most advanced nonmilitary satellite ever launched anywhere. The satellite was and continues to be promoted by the Indian Space Research Organization as its flagship satellite and as being distinctly Indian in character. While most satellites fulfill a single, well-defined mission, INSAT was a multipurpose geostationary satellite. Its peculiar design arose partly from very unusual design constraints placed on it by India's insistence that the satellite carry at least four different payloads.

The most significant of the payloads on INSAT was a package that could receive television programs from selected stations in New Delhi and Ahmedabad and retransmit them to relay stations in Amritsar, Bombay, Madras, Gauhati, and Calcutta. Its importance arose from its special ability to transmit educational television programs directly to specially designed television sets owned communally by thousands of remote Indian villages. Many Indian leaders hoped that a significant fraction of India's somnolent villages thus would be awakened.

The second package was designed to provide telephone, facsimile, data, telegraph, video text, and other communications services among metropolitan areas. The third was a remote-sensing package built to survey the nation's resources and thus help in its development planning. The last payload was an ingenious meteorological system that not only transmitted pictures of cloud cover, but also collected weather information from several thousand unmanned data collection points on the ground; it served to trigger selected disaster-warning sirens in isolated coastal villages under the imminent threat of cyclones.

Thus, INSAT had the communications capacity of an Intelsat IV (the state of the art in communications satellites when INSAT was designed), a meteorological payload effectively equivalent to the Geostationary Operational Environmental Satellite (GOES)-A (then the most advanced of weather satellites), and a direct-broadcast television system akin to the Applications Technology Satellite (ATS)-6 (once again, the frontier of technology), all wrapped into one compact package. INSAT was a crowded Indian bus shot into space.

---

1.   Senior retired Indian officials who have read this text have advised the author that it contains materials "of a sensitive nature." Confidential sources may be found in two of the author's papers, "INSAT: The Politics of Appropriating High Technology" and "SITE: Traditionalizing Space Technology," typescript manuscripts, seminar papers in the author's possession. Unless indicated otherwise, this chapter is extracted from those two papers. Moreover, extensive interviews with Indian space program personnel form the basis of some of the assertions made here. The author wishes to thank his advisors, Raja Rao, his colleagues at the University of Pennsylvania, and the officials of the Indian space program for helping with both the research and the writing.

INSAT Historiography:
Questions, Sources, and Strategies

The history of INSAT was shaped by India's desire to "gatecrash" the glamorous world of modernism. When India gained independence in 1947, the moral authority of Gandhian nonviolence and the grandiose global vision of Nehru produced a heady euphoria. India not only wanted to join the party of the moderns, but insisted on wearing nothing but a loincloth, like Gandhi. Nowhere were these contradictory aims more apparent than in India's approach to its space program.

India made it clear, privately and through self-righteous declarations in international forums, that its plans for space technology were driven by "real needs" on land. Indians argued that technology in space was practically worthless without a vast array of other technological systems on the ground. INSAT sought to demonstrate the practical benefits of advanced technologies to the poorer nations. Its ground systems were designed to show that space technology was a technology appropriate for underdeveloped societies.[2]

INSAT could be equally renowned for the many meanings loaded into it. At one level, INSAT embodied the tensions and aspirations associated with decolonization. At another level, it illustrated the use of technology as an instrument of foreign policy. From the point of view of a third world country, it was also an example of using technology for economic development, social and cultural change, nation-state building, and the formation of a nonaligned bloc of third world nations politically equidistant from both the Soviet Union and the United States.

The past is a perishable resource, particularly so in the hot and humid climes of India. There are no depositories of documents, such as NASA's archives, for the Indian space program. The custodian of the satellite, the Indian Department of Space is a somewhat secretive quasi-governmental institution. In a society still largely oral rather than textual, crucial decisions are far less likely to be committed to paper than they are in the West. The culture of autonomy in these institutions encourages minimal documentation of decisions. Also, as is the case of good Hindus, the Department of Space cremates its dead files—and fairly regularly, too. Furthermore, that peculiar literary genre of the British Raj, annotations in official files, has declined, as has other institutions of the Raj in postcolonial India. Thus, researching INSAT poses many problems.

The sources for INSAT's history are scattered. This chapter relies extensively on field work, interviews, and archives in the West. Access to information was achieved in India using the data obtained in the West, such as limited access to the Indian Department of Space records.[3]

From Atomic Energy to Space Research

The Indian nuclear and space programs, although originally the private vision of a few scientists, quickly crystallized as high-priority programs of the new Indian state. A parliament, largely ignorant of science but lusting after advanced technology, approved the programs. An impoverished mass of mostly agrarian taxpayers funded the nuclear and space programs. These programs became planned efforts to appropriate alien technolo-

---

2.   Vikram A. Sarabhai, *Science Policy and National Development* (Delhi: Macmillan Company of India, 1974).

3.   The author would like to thank an anonymous worker in the Indian Space Research Organization for granting limited access to the records of the Department of Space. This person, of all the people the author met in India, understood history.

gies, paths to rapid industrialization, ways of catching up with the West, and symbols of a resurgent India.

Just as it is necessary to know Mahatma Gandhi to appreciate the history of independent India, an awareness of Homi J. Bhabha is crucial to the story of INSAT. By creating new institutions for the adaptation of advanced technologies, Bhabha laid the foundations of INSAT. His approach to high technology continued to shape the Indian space and nuclear programs for nearly four decades. To know and appreciate the life and work of Homi Bhabha is to understand not only INSAT, but also other high-technology enterprises of the Indian state.

Born to an influential Bombay Parsi family in 1909, the young Bhabha attended a private school established primarily for European children. He pursued a career in physics, starting with undergraduate and doctoral studies at Cambridge under the famed physicist Paul Dirac, as well as at the Cavendish Laboratory. He quickly earned a reputation as a brilliant theoretical physicist, and he was elected a Fellow of the Royal Society in 1939.[4]

On a short holiday when World War II erupted, Bhabha accepted a specially created readership in cosmic ray physics at the Indian Institute of Science in Bangalore, which was founded by his relatives, the Tata family. Bhabha decided to concentrate on the difficult task of creating new institutions embodying a new culture of science befitting modern India. The Tata Trusts, controlled by his close relatives, aided Bhabha in this task. In 1945, a generous and timely grant enabled him to create the Tata Institute of Fundamental Research at Bombay. There, amidst beautiful paintings and landscaped gardens, Bhabha planned how to shape the future of India.

A member of the Bombay Parsi elite by birth, Bhabha came into frequent contact with the leaders of the nationalist movement, such as Nehru and Gandhi. The independence movement received financial contributions from the merchant-princes of the Parsi community, and nationalist leaders were often their house guests. These informal contacts with the statesmen of India became important in helping the young Homi Bhabha direct India's high-technology ventures.

Just after independence, Bhabha quickly obtained formal approval to create the Atomic Energy Commission. Under its umbrella, he organized a vast empire of research. Convinced that even a backward country such as India could catch up with the West in an emerging field such as atomic energy, precisely because of the field's nascent character, Bhabha effectively welded Gandhian nonviolence and the rhetoric of the "peaceful uses of atomic energy" to Nehru's inspiring, if occasionally irritating, philosophy of political nonalignment.[5]

Although India began with atomic energy programs, it quickly diversified into space research. Propelling this move were both Indian and foreign influences. Bhabha's extensive travels in the West and his own interest in the physics of the upper atmosphere alerted him to the growing significance of space technology. He gradually expanded the

---

4.   The only scholarly work on Homi Jehangir Bhabha is a comparative study of Bhabha and Meghnad Saha. Robert S. Anderson, *Building Scientific Institutions in India: Saha and Bhabha* (Montreal: Centre for Developing-Area Studies, McGill University, 1975). The section on Bhabha also draws on: R.P. Kulkarni and V. Sarma, *Homi Bhabha: Father of Nuclear Science in India* (Bombay: Popular Prakashan, 1969); P.R. Pisharoty, *C. V. Raman* (New Delhi: Publications Division, Ministry of Information and Broadcasting, Government of India, 1982); George Greenstein, "A Gentleman of the Old School: Homi Bhabha and the Development of Science in India," *American Scholar* 61 (Summer 1992): 409–19; Shiv Visvanathan, *Organising for Science: The Making of an Industrial Research Laboratory* (Delhi: Oxford University Press, 1985); G. Venkatraman, *Journey into Light: Life and Science of C. V. Raman* (Bangalore, India: Indian Academy of Sciences in cooperation with Indian National Science Academy, distributed by Oxford University Press, 1988); as well as Homi Jehangir Bhabha, *Science and the Problems of Development* (Bombay: Atomic Energy Establishment of Bombay, 1966); the "Bhabha Report," the popular name for India (Republic) Electronics Committee, *Electronics in India Report* (Bombay: India Government Press, February 1966).

5.   For a history of India-U.S. relations after World War II, a surprisingly neglected research area, see H.W. Brands, *The Specter of Neutralism: The United States and the Emergence of the Third World, 1947–1960* (New York: Columbia University Press, 1989).

domain of his Department of Atomic Energy to encompass the upper atmosphere and thus, eventually, space. The Indian National Committee for Outer Space Research (INCOSPAR) was constituted in early 1962 under the umbrella of the Department of Atomic Energy. As a result, all space-related research was embedded in the technocratic Department of Atomic Energy until 1972, when an independent Department of Space was formed.

With the death of Homi Bhabha in an airplane accident, a new generation ascended to the throne. The generational transition, although nonviolent, produced discontinuities. In the area of large-scale science, Vikram Sarabhai succeeded Bhabha. Similar to Homi Bhabha, whom he had assisted in space research, Sarabhai was born into an elite family with a pronounced interest in social reform, the arts, and letters. His father, Ambalal Sarabhai, was one of the leading citizens of Ahmedabad, approximately 300 hundred miles north of Bombay. Ahmedabad, proudly proclaimed "the Manchester of India," had become a major textile center by the early twentieth century, thanks to the enterprise of a group of intricately connected Gujarati Jain families. Close ties grew between the Sarabhai family and national leaders. The Sarabhai family, described by a Rockefeller Foundation officer as the "Medicis of Ahmedabad," played a crucial role in postcolonial India.[6]

Vikram Sarabhai and his sisters, children of a wealthy Gujarati merchant, were educated at home by a carefully selected group of Indian and foreign educators inspired by Maria Montessori. As a child, Sarabhai met national leaders, such as Mahatma Gandhi, Jawaharlal Nehru, and Nehru's daughter Indira Gandhi, as family guests in the Sarabhai mansions.[7] At age seventeen, Sarabhai enrolled at St. John's College in Cambridge in 1936 and completed his natural science tripos in 1939. The onset of World War II forced him to return to India, where he continued to study physics at the Indian Institute of Science in Bangalore. There, the young Vikram Sarabhai found the opportunity to work closely with C.V. Raman and Homi Bhabha. After the end of the war, Sarabhai returned to Cambridge to finish his doctoral dissertation. On his return in 1947 to an independent India, Sarabhai, like Bhabha, persuaded charitable trusts controlled by his family and friends to endow a research institution near home in Ahmedabad, the Physical Research Laboratory. Barely twenty-eight years old, Sarabhai had embarked on an intense mission as a creator and cultivator of institutions.

Building Coalitions

Despite the frustrations and disillusionment of India during the 1960s, the early years of the Indian space program were euphoric. Vikram Sarabhai, a playful, Krishna-like successor to the solemn and remote Bhabha, ushered in a decade of naive technological enthusiasm in India with the formation of INCOSPAR in 1962. Sarabhai's personality definitely generated excitement with the space program. Only someone endowed with at least some of the attributes of Krishna could have built India's space program in such a traumatic decade.

Like Krishna, Sarabhai played several roles in his efforts to nurture the frail space program in its early years. He was a roving diplomat, teacher, strategist, friend, counselor, leader, and system-builder. Within India, his unusual combination of scientific eminence,

6.   Erik Erikson, *Gandhi's Truth: On the Origins of Militant Non-Violence* (New York: Norton, 1969), p. 298. For background on Ahmedabad, see Kenneth L. Gillion, *Ahmedabad: A Study in Indian Urban History* (Berkeley, CA: University of California Press, 1968).

7.   The section on Vikram Sarabhai is drawn from Chotubhai Bhatt, "Vikram Sarabhai," *Electronics India* 2 (January 1972): 35; Padmanabh Joshi, "Vikram Sarabhai: A Study in Innovative Leadership and Institution-Building," Ph.D. diss., Gujrat University at Ahmedabad, 1985.

aristocratic background, and disarming simplicity created a loving loyalty, often amounting to devotion, among those who knew him. Sarabhai's first attempts at technological evangelism within the Indian bureaucracy resulted in the Arvi Earth Station, the creation of which knit together a strong network of allies for the space program and set the pattern for the way in which INSAT would be constituted.[8]

The Arvi terminal, now a prominent landmark on the Pune-Nasik Road, stands as a testament to India's first success in space technology. Sarabhai had to persuade the rather conservative engineering bureaucracy entrenched in the Ministry of Communications to let him build satellite telecommunications terminals. The Overseas Communications Service had made plans to be connected to the international telephone network provided by Intelsat III, and it needed a ground station in India.

Because no one in India had built a ground terminal before, RCA was retained as a consultant and subcontractor for the electronics. The design of the antenna itself was based on the drawings of a similar antenna built by the U.S. firm Blaw Knox, but the engineering construction was done with the help of a Tata company, TELCO. The Arvi station was ready by October 1969, ahead of schedule. Sarabhai's ability to snatch the Arvi project from RCA resulted in saving India the equivalent of about $800,000 (in 1969 dollars) in foreign exchange and created a powerful profile for the space program within the bureaucracy. More importantly, it redefined the rules of the game. The space program acquired operational autonomy from the bureaucracy.

The sheer force of Sarabhai's personality subdued open dissent, and his reputation enabled him to slice through the bureaucratic jungle. Assured of loyal support at all levels, from the prime minister to the peon, Sarabhai, as did Bhabha, set out to secure cooperation from the spacefaring powers. He first turned to the United States. Unfortunately, NASA's pragmatic director of international programs, Arnold W. Frutkin, had heard it all before. Frutkin previously had been inoculated by none other than Homi Bhabha himself. Thus, Sarabhai's efforts to gain U.S. assistance met with polite but firm refusals.[9]

There was, however, one important exception. During the early 1960s, NASA was planning a series of advanced technology satellites known by the acronym ATS. Leonard Jaffe, NASA's director of communications, informed Frutkin of the need to field-test an ATS project, which involved the direct broadcast of television to receivers from a satellite. At the time, this technology was untested. The commercial and political advantages of a satellite system that could beam programs directly to television sets attracted NASA policy-makers.[10]

Frutkin and Jaffe examined a world atlas for a suitable site for the ATS experiment. The three countries that were large enough and close enough to the equator for testing a direct-broadcast satellite were Brazil, China, and India. Brazil proved uninteresting; the population was concentrated in a few cities, and conventional television broadcast technology was clearly a better solution. The People's Republic of China was out of the picture for political reasons. Therefore, India was the logical choice. It was densely populated, yet only Delhi had a television transmitter (a small one) left behind by a Dutch electronics company after a trade show.

8.   Kamla Chowdhry, "Vikram Sarabhai: Institution Builder," *Physics News* 3(1) (1972): 17; Kshitish Divalia, "Dr. Vikram Sarabhai: An Enterprising Industrialist," *Physics News* 3(1) (1972): 19; M. Sarabhai, *This Alone is True* (London: Meridian Books, 1952).

9.   Arnold Frutkin, interview, Washington, DC, 4 January 1989, NASA History Office, Washington, DC; Satish Dhawan, interview with author, Bangalore, India, 5 January 1990.

10.   Arnold Frutkin, interview, Washington, DC, 5 January 1989, NASA History Office; Leonard Jaffe, personal communication; James Wood, *Satellite Communications and DBS Systems* (Oxford: Oxford University Press, 1992); Michael E. Kinsley, *Outer Space and Inner Sanctums* (New York: John Wiley & Sons, 1976).

Frutkin and Jaffe calculated that it would be expensive to have a conventional television system covering the entire country; a satellite would provide a cheaper alternative. Apart from being free of technological encumbrances, India possessed other advantages. Politically, it was an ideal location to demonstrate the peaceful uses of U.S. space technology and to beat the Soviets in technological diplomacy. The potential for propaganda was immense; Frutkin vowed to exploit it to the fullest. His earlier experience with Bhabha had taught him to negotiate with India.[11]

The U.S. State Department, however, was not enthusiastic about India, having been frustrated many times in its crude attempts to win India over to the "free world." An embarrassing controversy over placing Voice of America transmitters in India remained fresh in its memory.[12] Having been turned down recently, the State Department was not about to ask India if it would let a U.S. satellite beam television programs directly into remote villages. The request had to come from India. Therefore, to spare the State Department further embarrassment, Frutkin arranged to have Sarabhai approach NASA. Sarabhai agreed gleefully.

He requested the use of an ATS satellite for a year to conduct a satellite instructional television experiment in India's villages. He saw a great opportunity to convince India of the need to invest heavily in space technology, a unique chance to learn the ground segment of a satellite system from the Americans, the possibility of baptizing a whole generation of Indian scientists and engineers, and a systems management lesson for an INSAT satellite. The Indian Department of Atomic Energy and NASA signed an agreement for the Satellite Instructional Television Experiment (SITE) in 1966.

The Satellite Instructional Television Experiment

SITE,[13] a massive experiment in social engineering designed jointly by NASA and the Indian Space Research Organization (ISRO), is a fantastic tale of technological cooperation between unfriendly democracies. Indian engineers placed television sets in 5,000 remote villages spread in six clusters across the subcontinent. Half of the televisions were further modified to receive programs directly from the ATS satellite, and each of which was equipped with a large, distinctive dish antenna that dominated the village landscape. ISRO technocrats, spurred by social engineering ambitions, devised a highly sophisticated computer program that chose villages specifically for their backwardness. Most villages were not electrified, and many could not be connected to the electric network within a year. Therefore, space technologists reengineered the television sets to adapt them to the rigors of rural life. Many were powered by solar energy and batteries. NASA wanted to test some new solar cells and encouraged the use of such television sets.[14]

For a year, from 1 August 1975 to 31 July 1976, hundreds and sometimes thousands of villagers gathered daily in front of each of these 5,000 television sets—placed outside like a processional deity of a temple—to watch educational television, which showed them

11.   R.S. Jakhu and R. Singal, "Satellite Technology and Education," *Annals of Air and Space Law* 6 (1981): 400–425.

12.   The United States wanted a Voice of America transmitter on Indian soil to counter communist propaganda in South and Southeast Asia. See James Tyson, *U.S. International Broadcasting and National Security* (New York: Ramapo Press, 1983).

13.   An experiment akin to SITE was conducted using radio technology from the mid-1940s and into the 1950s. However, SITE did not seem to have utilized the historical experience of either the educational radio experiments or a similar rural television experiment conducted in Europe during the early 1950s.

14.   Arbind Sinha, *Media and Rural Development* (New Delhi: Concept Publishing Company, 1985).

Figure 29
A technician with the Indian Space Research Organization (ISRO) stands next to a working model of the solid-state television set, designed with NASA assistance, for use in SITE (Satellite Instructional Television Experiment). The picture epitomizes the contrast between Indian traditional rural culture and the high-technology domain of satellite communications, as well as Indian technocratic hopes of using SITE to satisfy the country's social engineering ambitions. (Courtesy of NASA)

how to lead better lives and grow more food. During the day, the village school children watched science experiments on television. Not all the viewers were villagers. Often, engineers and bureaucrats watched. The American embassy in New Delhi had a SITE television set. In Sri Lanka, Arthur C. Clarke, the Jules Verne of satellite communications, was given a set to watch SITE from his home. Every major newspaper in the world wrote about SITE.

After a year, NASA parked the satellite in a new orbit away from India. Clarke pleaded forcefully with NASA to continue this revolutionary experiment beyond the stipulated one-year period. Many leftist journalists voiced the disappointment of villagers. Delegations of villagers trekked several miles to meet government officials. Hundreds of postcards petitioned the government to continue the program. Several of the anthropologists stationed in villages to study the effects of SITE stayed longer to conduct post-SITE evaluations, then returned home to write lengthy reports.[15]

SITE was, thanks to Vikram Sarabhai's foresight, a joint effort of All India Radio and Doordarshan (Indian Television), the Ministry of Telecommunications, the education and agriculture ministries, and ISRO. ISRO, of course, had the final responsibility for the project's execution. Although ISRO engineers were reasonably confident of being ready with the technologies for handling the ATS-6 ground segment, they necessarily had to delegate the task of producing programming to the Ministry of Telecommunications and, within it, to All India Radio and later Doordarshan.

All India Radio was an inertial bureaucracy totally unequipped to imagine the possibility of producing six hours of educational television every day for a year in four different languages. According to the agreement between the Department of Atomic Energy

15.   Arthur C. Clarke, *Ascent to Orbit: A Scientific Autobiography: The Writings of Arthur C. Clarke* (New York: John Wiley and Sons, 1984).

and NASA, the Indians had agreed to feed the satellite six hours of television program-ming for 365 days. To be fair to All India Radio, one must remember that the voracious appetite of ATS-6 amounted to almost twice the annual harvest of the extraordinarily active Indian commercial film industry. At least in the eyes of its Indian managers, SITE quickly transmogrified itself from a boon to a devilish nightmare of a bargain to be ful-filled by them.[16] As late as January 1975, less than six months before SITE was to go on air, only enough satellite fodder for one month was on hand.[17]

SITE provided a perfect opportunity for Indian engineers to acquire a wide variety of valuable technological knowledge—the sort of technological learning that occurs on any project. What marked SITE as an exceptional technological enterprise (and of importance to INSAT) was the way in which SITE affected those Indian engineers who went to the United States to help NASA prepare for SITE. The ISRO liaison in the United States between 1969 and 1973, Pramod Kale, had been recruited by Sarabhai to work in the Indian space program. Kale and other ISRO engineers working at NASA learned to design an advanced operational satellite, and Kale eventually became the project manager of INSAT.[18]

The phenomenal success of SITE in penetrating remote regions of rural India impressed everyone. It gave the Indian space program a level of state support that was oth-erwise unimaginable. SITE provided ISRO valuable technical expertise in building and managing the elaborate ground systems needed to utilize any satellite. It also enhanced the credibility of ISRO in the Indian scientific community and in the international space community. Indian scientists and engineers were more willing to work for ISRO. The recruitment of young professionals from elite engineering institutions increased for the first time after Sarabhai's death. SITE opened critical paths in the rapidly emerging European space programs. ISRO was accepted as an equal partner in several critical coop-erative programs managed by the European, French, and German space programs. The collaborations with European space agencies brought ISRO decisive technologies and much needed experience. The experience and confidence gained abroad fed into the design of INSAT. It also enabled India to practice its own technological diplomacy—one based on nonalignment.[19]

More important than all the technical expertise gained from SITE was the education that ISRO received from the Indian villages it had set out to instruct. SITE planted the seeds of social responsibility into the minds of India's elite space engineers. It etched India's social contexts into ISRO space technology. Exposure to the complex problems of India's villages tempered technocratic grandiosity. Although rural television promised rev-olutionary social changes, it also revealed the limitations of technological fixes.[20]

Unraveling INSAT

Homi Bhabha had begun scouting for a satellite for India as early as 1965. He was unwilling to purchase a ready-made satellite, because Indians would not learn anything about making satellites. Instead, he sought technical assistance in building a satellite and met several times with Arnold Frutkin to discuss U.S. assistance. Frutkin, however, saw no

16.   On All India Radio, see K.S. Mullick, *Tangled Tapes: The Inside Story of Indian Broadcasting* (New Delhi: Sterling Publishers, 1974).

17.   Arnold W. Frutkin to Yash Pal, letter, 30 January 1975, NASA History Office.

18.   M.S. Sridhar, "A Study of Indian Space Technologists," *Journal of Library and Information Science* 7(2) (1982): 146–58.

19.   G.A. Van Reeth and V.A. Hood, "Review of ISRO/ESA Cooperation," conference paper, Rome, 1983; Vijay Gupta, ed., *India and Non-alignment* (New Delhi: New Literature, 1983); H.W. Brands, *The Specter of Neutralism.*

20.   Vikram Sarabhai, *Science Policy and National Development* (Delhi: Macmillan Company of India, 1974); Vikram Sarabhai, *Management for Development* (Delhi: Vikas Publishing House, 1974).

practical benefits for NASA from teaching Indians to build satellites. Vikram Sarabhai encountered even greater obstacles—not only more resistance to technology transfer, but also systematic pressures from the United States and its allies, especially Britain, aimed at discouraging India's space program. Hesitantly, Sarabhai initiated planning for an Indian national satellite, to be called INSAT, in 1967.

Systematic thinking about INSAT began in mid-1968. A seventeen-member committee of engineers and scientists based at the Space Science and Technology Center in Thumba conducted a preliminary feasibility study. Ironically, their preliminary report did not consider the development of the payload or satellite. They had concluded that the first satellite had to be built in collaboration with other countries. They had not anticipated Western resistance to the transfer of space technology.[21] Thus, the first step in India's satellite program concentrated on the ground systems. This orientation continues to characterize the Indian space program today. The preliminary 1968 feasibility study explicitly focused on two tasks that the Indians could see themselves doing: (1) designing orbital parameters and ground stations and (2) developing those technologies for which no foreign assistance was available, such as propellants and rocket guidance.[22]

India's initial emphasis, then, was to be on the launch vehicle rather than the satellite. By 1968, negotiations with NASA to borrow the ATS-6 had generated considerable optimism about U.S. technological assistance. Several American aerospace firms, such as General Electric and Hughes Aerospace, had shown a keen desire to collaborate with India in building its first operational Indian satellite. However, these early efforts to build a launcher were unusually optimistic.

Sarabhai, as did Bhabha, found the United Nations an effective forum to argue for technological assistance. An international sounding rocket program conducted in India brought basic rocket technology from the United States, France, and the Soviet Union. However, scaling up to rockets capable of launching operational satellites was a daunting task. Sarabhai and his colleagues recognized the nature of the challenge: India would have to obtain launch services, at least initially, from either the United States or the Soviet Union and planned its satellites accordingly.

In 1970, Vikram Sarabhai announced plans for an Indian National Satellite (INSAT) at the Bombay National Electronics Conference. Attending the conference was an impressive collection of academics, industrialists, politicians, bureaucrats, researchers, scientists, and engineers, as well as a substantial crew of Indian journalists.[23] It was a perfect place to manufacture consensus for a national satellite. Sarabhai's presentation of the INSAT plans thrilled the crowd. Dissenting questioners were co-opted by well-informed "Sarabhai boys." Sarabhai's INSAT was primarily a direct-broadcast satellite meant to educate Indian villagers through television. To secure a wide range of support, Sarabhai indicated the possibility of including a number of special payloads to satisfy special clients. He also drew attention to the proposed SITE plan, which at the time was being worked out by a joint committee of Indian and U.S. engineers.

However, there was not much demand for telecommunications from the Postal and Telegraphs Department or for remote sensing from the Ministry of Defense. Weather information was not highly sought after either. Unlike most of the participants at the conference, the representatives of government bureaucracies were not excited by

21.   Anonymous, *Preliminary Feasibility Study Report* (Satellite Project) (Bombay: India Government Press, 1968), p. 8.

22.   Sreehari Rao and S.K. Sinha, "Orbit Determination for ISRO Satellite Missions," *Advances in Space Research* 5(2) (1985):147–153; Murray Stedman, *Exporting Arms: The Federal Arms Exports Administration 1935–1945* (New York: Kings Crown Press, 1947).

23.   Government of India, *Proceedings of the National Electronics Conference on Electronics* (Bombay: India Government Press, 1972).

Sarabhai's technological evangelism. It seemed, though, that they realized they could not pose any significant opposition publicly. The primary function of Sarabhai's INSAT was rural education through direct-broadcast television, making it explicitly a permanent replacement for SITE. If in the process INSAT could serve other needs, that was so much for the better.

Shortly after unveiling his INSAT plans, Vikram Sarabhai died, in December 1971. His death caused serious concern about the future of INSAT. Unlike SITE, which had NASA for a godfather, INSAT had no powerful patrons outside the Indian space program. Even more troubling was that no obvious successors to Sarabhai came forth from the Indian scientific community. At the same time, the U.S. Congress delayed the ATS program through budget cuts, giving time for India to revamp its space program. ISRO engineers committed to INSAT took pains to disconnect the fates of INSAT and SITE, which had been tied together implicitly, in the hope of ensuring INSAT's survival.

Toward the end of January 1972, the Indian government offered the leadership of the space program to Satish Dhawan. Dhawan had worked for Hindusthan Aeronautics Limited, India's state-owned aerospace industry, and he was familiar with Soviet aerospace technology. In appointing Dhawan, Prime Minister Indira Gandhi announced the creation of two important new organizations. One was the Department of Space. Dhawan convinced Gandhi to headquarter the Department of Space away from New Delhi, in Bangalore. The Department of Space was to be overseen by the other new institution, the Space Commission. At the same time, the space laboratories in Sarabhai's home town were consolidated into the Space Applications Center under the direction of Yash Pal of the Tata Institute of Fundamental Research. Also, the space laboratories in and around Trivandrum were consolidated as the Vikram Sarabhai Space Center. Brahm Prakash, recently retired from the Bhabha Atomic Research Center, was persuaded to direct its activities.[24]

Dhawan's leadership marked a significant change in the Indian space program. India could not withdraw from SITE without seriously jeopardizing its relations with NASA and the United States. Dhawan recognized that it had to make SITE a success. On the other hand, plans for INSAT could be shelved, especially because its advocates had separated it from SITE, and there were pressures from the Planning Commission to discard INSAT. The Ministry of Finance and the Planning Commission saw no cash flows in the original educational television version of INSAT. Fortunately, the fight over INSAT resulted in a compromise. No decision was made one way or the other. The government waited until SITE was completed to evaluate proposals for INSAT, even though it meant an expensive hiatus in the social revolution catalyzed by SITE.

Dhawan pushed hard to make SITE succeed. He saw the future of INSAT, if not ISRO, at stake. When SITE was hailed loudly as an unprecedented success by the Western media, thanks to NASA's self-interest, INSAT revived. Between 1970 and 1977, the Ministry of Telecommunications awakened to the necessity of satellite technologies. The Prime Minister's Office had also become aware of the potential revolutionary capabilities of direct-broadcast satellite television. ISRO personnel, fired by Sarabhai's vision, succeeded in creating a small, independent rural television station based in Pij, Gujrat. Nonetheless, the Prime Minister's Office began to see the spread of satellite television of the SITE variety to be socially disruptive. It wanted nothing more than a gradual social revolution.[25]

---

24.   K.A.V. Pandalai, "Aerospace Personalities: Prof. Satish Dhawan," *Aeronautical Society of India Newsletter* 3(8) (August 1986): 2–12; Satish Dhawan, *Prospects for a Space Industry in India* (Bangalore, India: Patiala Technical Education Trust, 1983).

25.   Satish Dhawan, interview with author, Bangalore, India, 4 January 1990.

Much had changed by the time the INSAT concept revived in 1977. Dhawan, while well respected, did not command the influence that Sarabhai had. Back-of-the-envelope calculations, on net revenues generated by rural education through satellite television, did not hold any weight without Sarabhai. The Planning Commission, in collaboration with the Prime Minister's Office, sought to ground INSAT's direct-broadcast capability. Also, the Ministry of Finance made it clear that INSAT had to be a fully operational and revenue-generating satellite for it to receive funding. This implied that INSAT had to be a collaborative venture involving the Department of Space (directly under the purview of the prime minister), the gargantuan Postal and Telegraphs Department under the Ministry of Communications, the Meteorological Department of the Ministry of Tourism and Civil Aviation, and Doordarshan and All India Radio of the Ministry of Information and Broadcasting. Suddenly, INSAT had five clients, but they were all ossified bureaucracies dating from the British Raj into which had transmigrated a thoroughly Indian soul. They had little commitment to Sarabhai's vision of INSAT as an educational tool. Paradoxically, All India Radio and Doordarshan were anxious *not* to be saddled with the direct-broadcast segment. They feared the chore of feeding the demon satellite. The government revealed no intention of freeing broadcasting from its total control. In addition, maintaining several hundred thousand village television sets, as called for in the INSAT plans, was a logistical nightmare for any government bureaucracy. Thus, the market for INSAT changed radically between 1970 and 1977.

Many idealistic ISRO engineers, fresh from the experience of SITE, realized that the original INSAT might be hijacked. They refused to consider any INSAT design that did not include a direct-broadcast television payload. To placate them, the INSAT design committee, now a real working group with engineers from the Ministry of Telecommunications, agreed to install a television transmitter comparable to the one used in SITE. However, the Planning Commission refused to sanction money for ground equipment, such as television sets and antennas. Such a compromise served only to save face. The activists in ISRO wanted to be prepared for an opportunity to renew SITE. On the dark side, this decision to include a SITE capability imposed tough design constraints on INSAT. For instance, it increased the power needed to keep the satellite alive.

Why did Indian space technologists prefer not to design several single-purpose satellites? Each such satellite would have been fairly simple and built with proven, off-the-shelf technology. Building several satellites also would have spurred the serial production of an Indian space platform. The official answer to this obvious question is the economics of satellite launching. Because Indian rockets were incapable of launching a satellite into geostationary orbit, India had to purchase foreign launch services. "There are no free launches," NASA told Dhawan. The Soviets did not launch geostationary satellites; their launch pads were too far north to achieve an equatorial orbit. The European Space Agency offered free launches, but only on experimental rockets. INSAT could not be risked on experimental rockets. Rough calculations showed that the launch costs of a multipurpose satellite would be lower than the cost of launching three or four smaller spacecraft.

The technological patrimony of SITE especially influenced INSAT. The success of SITE, in a sense, blinded INSAT designers. The activist culture of the Indian space program did not permit leisure introspection. The shortage of skilled personnel ensured minimal dissent on technological alternatives. The United States—and Canada to some extent—lured away a significant number of elite Indian engineers and scientists, stunting the growth of a healthy scientific community.

Moreover, ISRO progressively lost the ability to define the satellite. When Sarabhai succeeded Bhabha, he continued to follow his mentor's strategy without responding to changes in the geopolitical context. India did not command the same moral authority it had even in the early 1950s. Also, Sarabhai was not fully aware of the motivations prompting NASA and the United States on SITE. He could not foresee that SITE committed the Indian space program to an expensive detour that would ultimately freeze unwieldy features, such as a television transmitter, into the second generation of INSAT satellites. After Sarabhai's death, the Indian space program had to compromise more to keep its allies. Dhawan had even less room to negotiate than Sarabhai. In the meantime, the space organization had grown into a large bureaucracy. In real terms, budgets did not grow at a healthy rate. Dhawan sought to balance carefully the diverse needs of his various allies, while preserving a demanding coalition.

Not all of the payloads on INSAT had powerful patrons fighting for them. The Meteorological Department was not really interested in the satellite. Its clients were small farmers who did not know how to lobby for their share of technology. Yet, the designers of INSAT added the meteorological subsystems. In doing so, they made the satellite even more difficult technologically. To predict weather, INSAT designers placed a camera on the satellite. Thus, the satellite could send pictures of cloud movements that could be used to predict weather patterns. The camera required that the satellite be ten times more stable in space than it needed to be for the other missions. This added another dimension to INSAT's technological complexity. Why was the weather subsystem not left out? Who would have complained that INSAT did not predict the weather? A few people on the INSAT design committee experienced a deeply felt responsibility to the Indian farmer and stood their ground in design committee deliberations. India's efforts to get INSAT built required it to harness several payloads, to yoke several interests, and even to create new constituencies. INSAT became a crowded Indian bus.

Conclusion

Indians do not enjoy concluding stories. For complex epics such as the Ramayana and Mahabharata, conclusions are often beginnings. In 1947, an independent Indian state came into being. Barely two years after independence, food shortages forced the government of India to beg from the West. The United States, especially under the Johnson administration, cynically sought to use India's food crisis to further its own Cold War agenda. The uneasiness between the two countries is evident in the history of India's nuclear and space programs, especially when they are seen in the light of decolonization.

The history of INSAT is also a case study in the emergence of a nation-state in South Asia. Nascent science and technology institutions learned to deal, trade, and negotiate with the West under the leadership of Bhabha and Sarabhai, both of whom came from merchant-industrialist families. Doing business with the rest of the world, a skill not in great demand during centuries of foreign oppression, was what INSAT taught some Indians.

The rise of a professional, middle-class leadership in the Indian space and nuclear programs may be read as an indicator of the growing integration of the Indian nation-state into the family of nations. The politics of accommodation vividly illustrated by INSAT is at some level a comforting sign. The politics of foreign aid and the geopolitics of INSAT show India that nothing comes free.

Chapter 17

Footprints to the Future, Shadows of the Past: Toward a History of Communications Satellites in Asia

by Brian Shoesmith

"If a new means of communication makes its appearance, who are its patrons? If new knowledge is produced, who controls it and for what ends?"

—Brian Stock

"Satellites already have inspired one revolution in Asia. In the space of four years since Star TV was launched in Hong Kong on the Asiasat 1 spacecraft, satellite and cable have transformed the face of broadcasting in the region."

—Brian Jeffries[1]

Writing a history of satellites in Asia is like building a house on quicksand. For a start, there are contextual problems with the term Asia. Questions that quickly spring to mind are: Precisely whose Asia are we talking about? Which Asia are we speaking of? Then there is the technology itself. Satellites are a comparatively recent addition to the repertoire of available communications media, and their status remains largely unresolved as other communications technologies compete for markets. Furthermore, conditions surrounding satellites in Asia are volatile: more satellites are launched; new regulations on their uses are announced by national governments; and new strategic alliances are forged among governments, satellite providers, and commercial broadcasters, thereby clouding the issues.

In short, identifying the major trends and contributions to the evolving mediascape of Asia is difficult because things change, quite literally, over night. Consequently, the following analysis of the forces shaping the current Asian communications satellite environment is necessarily provisional. One should bear in mind that contingency seems to be the overwhelming determinant of the direction the communications industry has taken in Asia. Nevertheless, it is appropriate that a history of satellites in Asia be begun—if for no other reason than that, within a rapidly changing scene, there is a need to chart the foundations of what has become a powerful regional force—before they, too, are clouded by the forces of change.

To understand how powerful an influence satellites have had, it is necessary to outline briefly the mass media systems of Asia that have been challenged by the new technology. The following discussion is a "broad-brush" approach. Clearly not all Asian media systems have developed along identical lines, nor have they been confronted with the same regulatory systems or subjected to the same levels of political censorship.

---

1.   Brian Stock, *Listening for the Text: On the Uses of the Past* (Baltimore: The Johns Hopkins University Press, 1990), p. 21; Brian Jeffries, "The Sky's the Limit," *Far Eastern Economic Review* 158 (1995): 49.

Nevertheless, one can say that almost without exception the modern mass media in Asia were introduced into the region by the colonial powers. Initially, it was the press, followed by radio then television, which formed the postcolonial Asian media inherited by the nationalist regimes and melded into indigenous ideological systems. The media became indissolubly linked to localized concepts of development, being assigned an entirely different sociopolitical function than that of the Western media. The media in Asia have developed within national ideological frameworks since at least the 1950s, and they are best exemplified by the Indonesian state ideology of *pancasila*.[2]

The significance of satellites lies in their potential to precisely and comprehensively subvert prior national systems of control of the media. The material presence of satellites in Asia has forced the various Asian governments to rethink their policies on broadcasting and communications. The corollary of this is that the political will of the Asian states to retain some vestige of control over their mediascape also has forced Western transnational broadcasters to reevaluate their broadcasting practices and ideologies.[3]

A second feature of satellites requiring elucidation is their scope. When Apstar 2 finally gets into orbit, it will have a footprint that covers the whole Eurasian landmass, part of East Africa, and the western Pacific rim. All political boundaries and topographical impedimenta will be transcended, fulfilling a trend begun in the Victorian era with the introduction of telegraphy.[4] The geopolitical consequences inherent in this situation were recognized at the beginning of this century with the extension of telegraphy in the service of empire. As Halford J. Mackinder wrote in 1904: "For the first time we can perceive something of the real proportion of features and events on the stage of the whole world, and may seek a formula which shall express certain aspects . . . of geographical causation in universal history." For the first time in history, there was a potential for an "empire of the world" dominated by whoever controlled the pivot area or "heartland" of Eurasia.[5]

Satellites are more than artifacts of universal history, however. They also have profound commercial and cultural attributes. Since Rupert Murdoch purchased Star TV in 1993 from the Hong Kong company Hutchinson Whampoa,[6] a number of other European and U.S. communications companies, such as Pearson and NBC, have entered the domain of Asian satellite broadcasting.[7] There is a view, generated largely by Asia's booming economies, that Asia holds a vast market of consumers waiting to be serviced by an endless supply of consumer goods. In many respects, this view echoes that of the Manchester school of nineteenth-century England, which viewed China similarly as a vast market for their products and which influenced British foreign policy accordingly.[8]

---

2.   The five principles of pancasila are: belief in a single supreme God; a just and civilized humanity; national unity; democracy, led by the wisdom of consensus among representatives; and social justice for the people of Indonesia. Edward Janner Sinaga, "Indonesia," in Achal Mehra, ed., *Press Systems in ASEAN States* (Singapore: Asian Mass Communication Research and Information Centre (AMIC), 1989), pp. 27–39, esp. p. 27.

3.   John Sinclair, "The Business of International Broadcasting: Cultural Bridges and Barriers," paper read at Communications Research Forum, Sydney, 19–20 October 1995; Will Atkins, " 'Friendly and Useful': Rupert Murdoch and the Politics of Television in Southeast Asia, 1993–95," *Media International Australia* 77 (August 1995): 54–64.

4.   James W. Carey, *Communication as Culture: Essays on Media and Society* (London: Unwin Hyman, 1989), pp. 201–23; Daniel Czitrom, *Media and the American Mind: From Morse to McLuhan* (Chapel Hill, NC: University of North Carolina Press, 1982), pp. 11–12.

5.   Cited in Stephen Kern, *The Culture of Time and Space, 1880–1918* (Cambridge, MA: Harvard University Press, 1983), p. 228.

6.   Atkins, " 'Friendly and Useful'," pp. 54–64.

7.   Nick Masters, "UK's Pearson secures a significant stake in TVB," *Televisionasia*, March 1995, p. 8.

8.   Mary Clabaugh Wright, *The Last Stand of Chinese Conservatism: The T'ung-Chih Restoration, 1862–1874* (Stanford, CA: Stanford University Press, 1957), pp. 23–33.

Technological issues also must be addressed. Three in particular stand out. First, the ability to develop and maintain a satellite infrastructure implies a powerful economic base and technological sophistication. Second, access to satellite technology also implies a complicated set of international relations. In reality, only the United States at the end of the Cold War had the necessary economic power to develop an autonomous satellite program. All other programs were hybrids combining, in varying degrees, local technology with imported technologies. Thus, the disastrous Apstar 2 launch in January 1995 combined a satellite developed by the American company Hughes Aerospace with a Chinese Long March rocket. The hybrid technology approach first developed in India where locally fabricated satellites were combined with either U.S., Russian, or French rockets throughout the 1960s and 1970s. Australia adopted similar strategies in the mid-1980s with the launch of AUSSAT. Europe has formed its own consortium to produce communications satellites and at the same time make the technology available to others who wish to put satellites into orbit. Thus, one can suggest that communications satellites are the product of complex global technological systems that include Asia structurally as a partner, unlike previous world communications systems that included Asia only as a client.

Finally, there is the question of the relationship between technology and culture. Social scientists still have a tendency to characterize explanatory systems of change that feature technology as deterministic.[9] Unfortunately, the specter of technological determinism will not disappear, irrespective of whether it is couched in terms of "hard," "soft," or technological momentum when the relationship between technology and culture is discussed. The heat generated by discussions of technological determinism clarifies nothing. Regarding communications technologies, it would be preferable to use the Innisian term "bias."[10] Harold A. Innis placed communications technologies at the core of cultural activity and argued, persuasively, that all such technologies have either a temporal or spatial bias. These arguments are well rehearsed and require little elaboration here,[11] except to say that communications satellites have the most pronounced spatial bias of any communications technology yet devised. Apstar 2 is the exemplary case.

If we view communications satellites simply as artifacts, we can only formulate a partial account of their history. There is no doubt that technical developments in rocketry, solid state circuitry, digitization, and miniaturization all have played crucial roles in the unfolding history, and any comprehensive account has to accord them due significance. According to Carolyn Marvin, though, "the early history of electric media is less the evolution of technical efficiencies in communication than a series of arenas for negotiating issues crucial to the conduct of social life,"[12] and "[W]hen audiences become organized around these uses, the history of the new medium begins."[13] With these factors in mind, this discussion proposes a history of communications satellites in Asia that has three distinct stages—all of them linked to the development of television as a medium in Asia.

Stage one lasts approximately from 1962 until the late 1980s and is characterized by a perception that satellites should be harnessed for development purposes, placing them securely in the public sector sphere of broadcasting history. Stage two is a short, transitional period spanning the end of the 1980s and is characterized by a response among Asian governments to the end of the Cold War. Stage three begins in 1991 and is

---

9.   Merrit Roe Smith and Leo Marx, eds., *Does Technology Drive History? The Dilemma of Technological Determinism* (Cambridge, MA: MIT Press, 1994), *passim.*

10.   Harold A. Innis, *The Bias of Communication* (Toronto: University of Toronto Press, 1991), *passim.*

11.   William Melody, Liorar Salter, and Paul Heyer, eds., *Culture, Communication and Dependency: The Tradition of H.A. Innis* (Norwood, NJ: Ablex, 1981), *passim;* David Crowley and Paul Heyer, *Communication and History*, first ed. (New York: Longmans, 1991), *passim.*

12.   Carolyn Marvin, *When Old Technologies Were New: Thinking About Electric Communication in the Late Nineteenth Century* (New York: Oxford University Press, 1988), p. 4.

13.   *Ibid.,* p. 5.

characterized by the dominance of commercial forces in Asian broadcasting. In making these distinctions, they are not posited as exclusive. They are clearly not exclusive, as the transformation of development communication from a didactic practice to a postmodern application demonstrates.[14] Moreover, public broadcasting retains a strong presence in Asia. In short, what are described here are dominant trends.

In the Public Sphere

Asia officially entered the space age in 1965 with the founding of the Centre for Training and Research in Satellite Communication for Developing Countries in Ahmedabad, India, under the direction of Vikram A. Sarabhai, although research had begun earlier in 1962.[15] This discussion of satellite communications in Asia during in the period 1965–1988 is organized around three themes: (1) the carryover of an ideology of public broadcasting into satellite communications that was consonant with Asian political ideals based on notions of development and nationalism; (2) a view that satellites were an expensive and scarce resource whose use should be limited to development communication; and (3) problems of regulation by international bodies that emphasized a "first-come, first-serve" basis—principals that were contentious because, even at the earliest stages, they were perceived as disadvantaging the developing nations. The period is also characterized by a shift away from India toward China as the principal space-oriented nation in Asia. The shift is emblematic of China's changing economic and political status in Asia, signifying changing geopolitical realities that will inform this discussion.

Sarabhai was both visionary and persuasive[16] and had a clear agenda for satellite communications in India. That agenda was to serve the educational needs of India's rural population rather than the entertainment wants of the urban masses. He forcefully articulated a developmental ideology that has governed India's use of satellites until very recently. It must be acknowledged, however, that broadcasting in India at the time was not only limited, but highly bureaucratized, and it was aimed principally at education and information, placing a very low priority on entertainment. Thus, both terrestrial and satellite communications in India were locked into the public sphere and organized accordingly.

It was not just domestic policy that shaped Indian space policy in this period. Under the leadership of Jawaharlal Nehru, India became the leading exponent of nonalignment in international affairs and thus sought aid and advice from both sides in the Cold War. This policy applied to satellite technology as much as any other aspect of Indian *realpolitik*. Between 1967 and 1979, the Indian Space Research Organization (ISRO) cooperated with NASA, the Soviet Union, and the Franco-German Symphonie program on space projects. In 1967, ISRO negotiated with NASA for access to its Applications Technology Satellite (ATS)-6 spacecraft for one year to conduct the Satellite Instructional Television Experiment (SITE). In 1975, India launched its Aryabhata research satellite using a Soviet rocket in 1975, and later, during 1977–1979, India conducted the Symphonie Telecommunications Experimental Project (STEP) with the Franco-German Symphonie satellite administration. The outcome of these collaborations was that India had the most advanced space program in Asia by the end of the 1970s. This effort probably for the most part dissipated because it was too oriented toward development and too heavily bureaucratized. The experience of SITE exemplifies the situation that developed.

---

14.   Pat Howard, "The Confrontation of Modern and Traditional Knowledge Systems in Development," *Canadian Journal of Communication* 19 (1994), http://edie.cprost.sfu.ca/cjc/cjc-info.html.

15.   Heather Hudson, *Communication Satellites: Their Development and Impact* (New York: The Free Press, 1990), pp. 202–07.

16.   *Ibid.*, p. 202.

In 1967, engineers from ISRO visited NASA to study the ATS-6 satellite to assess whether its capabilities suited their needs for developing SITE, a rural-based instructional network in India. The manner in which SITE was set up, organized, operated (1975–1976), and evaluated is now well known.[17] What SITE revealed was a set of fissures within the Indian state apparatus. By placing responsibility for SITE in the hands of the ISRO, the government unwittingly turned it into a site of conflict over who "owned" the public sphere. Doordarshan (the Indian national television broadcaster) felt it had sole responsibility for broadcasting; officials from the Ministry of Telecommunications argued that they controlled telephony. The other administrative units, such as health, education, agriculture, and family planning, for whose programs the technology was specifically designed to facilitate, felt that the experiment was a drain on their respective resources.[18] Its problems were even more acute, however; the program had no concept of an audience—neither its needs nor its desires.

SITE serviced 2,400 remote villages, each of which was supplied with battery-driven television sets under the care of a village guardian. Specially designed programs were broadcast and then evaluated by teams of academic sociologists, psychologists, and developmental theorists. In retrospect, SITE was a classic example of top-down planning and execution divorced from the needs of its clientele. Leela Rao points out that the custodian of the television set was the single most important component of the chain of command. Quite frequently, these individuals were teachers who, as the state's chief functionary in the village, had a range of activities competing for their time and attention.[19] In most cases, they placed television last in their priorities. Moreover, although SITE's ideology was vigorously anti-entertainment, it was the popular Hindi films screened on Sunday nights that attracted the largest audiences and that significantly provided the programs audiences remembered best.[20] The net effect of SITE was to lock India into a terrestrial system of broadcasting and telecommunications that only ceased to dominate in the 1990s.

India continued to develop expertise in space and satellite technology through the 1970s and 1980s. However, it was expertise largely directed toward servicing the Indian bureaucracy rather than the population at large. It was also expertise that was rigorously localized. India manufactured the components used in SITE—a fact determined by India's economic policy of self-sufficiency. During 1981–1982, STEP involved a satellite "designed and fabricated in India"[21] and launched on an Ariane rocket. The expertise thus acquired was translated into the INSAT project, which included four Indian satellites launched between 1982 and 1985. The problem of the excess technical capacity created by those launches was solved only by the move away from the public sphere and developmental ideology of the 1960s and 1970s to the marketplace ideology that currently prevails. Nevertheless, television expanded exponentially throughout India, as INSAT became operational, and created a massive rural and urban audience for entertainment programs. Television also became the site for increasing political argument and intellectual dispute, as India moved away from a command to a market economy.[22]

17.   Leela Rao, "Medium and the Message: an Indian experience," in Neville Jayaweera and Sarath Anunugama, eds., *Rethinking Development Communication* (Singapore: AMIC, 1987), pp. 176–90; Srinivas R. Melkote, *Communication for Development in the Third World: Theory and Practice* (New Delhi: Sage, 1991), *passim;* Hudson, *Communication Satellites*, pp. 202–04.

18.   Hudson, *Communication Satellites*, p. 203.

19.   Rao, "Medium and the Message," p. 188.

20.   *Ibid.*, pp. 188–89.

21.   Hudson, *Communication Satellites*, p. 204.

22.   Geoffrey W. Reeves, *Indian Television: The State, Privatisation and the Struggle for Autonomy*, Occasional Paper 3 (Perth: Centre for Asian Communication and Media Studies, 1994).

The second country in Asia to acquire a regional communications satellite was Indonesia, which launched Palapa 1 in 1976. The name Palapa is highly significant in the lexicon of Indonesian political ideology; it has symbolized harmony and unity to Indonesian people since at least the thirteenth century. The introduction of Palapa 1, according to Philip Kitely, was "driven by political and ideological imperatives concerned with nation building and circulation of ideas of national culture."[23] Indonesia is an archipelagic nation formed in 1945 and composed of more than 300 ethnic groups with as many languages and dialects. Bahasa Indonesian was invented as a language in the 1920s and adopted during the 1930s as the national language by the nationalists, who consciously used the mass media and education to establish it on a national footing. However, Bahasa has always coexisted in tension with local and regional languages, and it could never be guaranteed preeminence in the face of tradition and custom until the advent of satellite broadcasting. Technocrats who came to prominence in the administration of Indonesia during the 1970s realized that the spatial and cultural bias of satellites served their ideological ends well and advocated the introduction of communications satellites during the early 1970s. In 1976, Indonesia became the third country in the world, after Canada and the United States,[24] to install a geostationary satellite for domestic communications purposes.

The introduction of Palapa 1 accelerated the growth of television and telecommunications in Indonesia. The government funded the construction of an additional eleven ground stations, which distributed the state-owned TVRI channel throughout Indonesia. Moreover, private capital constructed television stations on the outer islands, giving rise to a commercial regional television system.[25] However, these private regional stations were prohibited, by law, from broadcasting nationally, which was the preserve of the state-owned TVRI channel. The public nature of television in Indonesia was enhanced in 1981 when advertising was banned. Up until that time, TVRI had carried advertising, which covered more than 90 percent of its production costs by 1976–1977.[26] The ban on advertising marked a particular stage in developmental thinking in Indonesia, which emphasized industrial development within a framework of "balanced growth."[27]

The particular convergence of satellites with state-directed patterns of growth had two unintended consequences for communication in Indonesia. First, the use of satellites to distribute information meant that audiences were no longer dependent on terrestrial services to receive images, messages, and meanings. From the early 1980s, a proliferation of privately owned parabolic dishes in urban centers had a profound effect on television watching in Indonesia.[28] The possession of a parabola effectively meant that viewers were no longer dependent on locally produced television for information and entertainment; they could tune in to any provider that was beaming in their direction. CNNI (Cable News

23.   Philip Kitely, "Fine Tuning Control: Commercial Television in Indonesia," *Continuum: The Australian Journal of Media and Culture* 8 (1994): 102–23, esp. 103.

24.   There is confusion on this issue. The Soviet Union is usually written out of the equation. The countries are the United States, Canada, and Indonesia.

25.   Kitely, "Fine Tuning Control," p. 104.

26.   *Ibid.*

27.   Howard Dick, James J. Fox, and Jamie Mackie, eds., *Balanced Development: East Java in the New Order* (London: Oxford University Press, 1993), p. 13.

28.   Brian Shoesmith and Hart Cohen, "Cultural Values of Media and Asian Audiences: Local Responses to Global Media," forthcoming in *Asian Journal of Communications*. These comments also are based on research conducted in East Java during 1994–1995 on the impact of satellite broadcasting on the middle classes of Surabaya with the author's colleagues Hart Cohen (University of Western Sydney at Nepean), Basis Susilo, Andarini Susanto, Rachmah Ida, and Emy Susanto (Faculty of Social and Political Sciences, Univeritas Airlangga, Surabaya).

Figure 30

Indonesian delegation in special support room, Johnson Space Center, Houston, during the deployment of the Palapa 3 communications satellite by the Space Shuttle (STS-7). Palapa 3 was the second of two communications satellites placed in orbit during the STS-7 mission. (Courtesy of NASA, photo no. 83-HC-406)

Network International) thus gained a foothold in the middle-class, tertiary-educated Indonesian audience. The ban on advertisements also had an impact on Indonesian television insofar as it "lessened TVRI's appeal."[29] This in turn encouraged viewers to sample foreign language programming—meaning that, almost by default, Indonesia acquired an "open skies" satellite-broadcasting policy that has been quite at odds with the communications policies of its neighbors.

The second important consequence was that Indonesia became a regional distributor of television programming. Intelsat's initial approval for Palapa forbade international broadcasting, but the Asian Broadcasting Union began using the service for program exchange. By 1981, Thailand and Pakistan booked time on the satellite for their own domestic programming because it was cheaper than Intelsat.[30] Indonesia became both a domestic and international supplier of satellite broadcasting within a very short time period—something that has developed considerably as Indonesia has improved its satellite capacity. ATV, the Australian international television service, has rented a transponder from Palapa since its inception in 1993.[31]

During the mid-1980s, three events occurred in Indonesia that radically altered the national mediascape—all of which were related to both satellite communication and domestic politics. First, in 1986, advertising was reintroduced to Indonesian television. Second, in 1989, RCTI (Rajawah Citra Televisi Indonesia), the first commercially owned

29.   Kitely, "Fine Tuning Control," p. 105.
30.   AMIC, *Satellite Technology: The Communication Equaliser: An AMIC Compilation* (Singapore: AMIC, 1985), p. 76.
31.   Brian Shoesmith, "Asia in Their Shadow: Asia and Satellites," *Southeast Asian Journal of Social Science* 22 (1994): 125–43.

television station, was established in Jakarta. Third, in 1993, the Indonesian government lifted the bans preventing the commercial regional broadcasters from broadcasting nationally. Subsequently, RCTI rapidly became Indonesia's most popular television station.

While the state retained ultimate control over broadcasting, the dominance of commercial television over the state system in Indonesia is a manifestation of what Philip Kitely has called "patrimonial relations."[32] RCTI is controlled by one of President Suharto's sons, TPI (the educational broadcaster) is run by a daughter, and the owners of the other commercial stations have strong links to Suharto family interests.

The other Asian nation in which notions of the public sphere and the public good shaped attitudes toward satellite communication is China, of course. China is discussed more extensively below, but two points should be made here. Compared with India and Indonesia, China entered the realm of satellite broadcasting relatively late, launching its first geostationary satellite in 1983. It can be argued that this delay reflected certain attitudes toward communication and technology within the Chinese ideological framework between 1949 and 1977, in which communication was geared around pervasiveness with minimum technology—namely, the press, radio, and loudspeaker.[33] Television, as a late addition to the repertoire of propaganda tools available to the Communist Party of China, never acquired the same status for propaganda as the press or radio, both of which resonated with Maoist principles about the mass line. Indeed, Chinese authorities have tended to see television largely in terms of entertainment rather than political ideology—a view shared by the audience. Consequently, the provision of communications satellites to distribute television was not a high priority of the Chinese government. In the period following the introduction of the Deng economic reforms in 1977, however, television underwent a surge in terms of both hardware and interest among audiences. Instead of being a scarce resource, television became commonplace throughout China and built up expectations among the audience that the government found difficult to meet.[34]

As in India and Indonesia, television in China underwent dramatic changes during the 1980s, with the introduction of advertising, access to foreign programming, and the presence of transborder communication through Star TV. Nonetheless, the evolution of television from the public sphere to the marketplace in China is less pronounced than in India or Indonesia, because the Chinese state has applied a unique set of conditions on its development.

Transitions

No one event can be identified as the causal factor in bringing about the change from public sphere broadcasting to commercially dominated broadcasting in Asia. A combination of exogenous and exogamous factors are at play. Similar to the rest of the world, public sphere broadcasting in Asia came under attack for economic and ideological reasons.[35] Moreover, Asia became enmeshed in international broadcasting events through Intelsat and the various nongovernmental organizations responsible for the regulation of global broadcasting, such as the allocation of orbit slots.[36]

---

32.   Kitely, "Fine Tuning Control," p. 116.

33.   A. Doak Barnett, Jr., "China: Some Generalizations, Hypotheses, and Questions for Research," in *Mass Communication Review Yearbook*, Vol. 1 (London: Sage, 1980), pp. 620–27.

34.   Wang Handong, "Chinese Television in the 1990s," in Brian Shoesmith, ed., *Three Aspects of the Chinese Media* (Perth: Centre for Asian Communication, Media and Cultural Studies, 1996); James Lull, *China Turned On: Television, Reform and Resistance* (London: Routledge, 1991).

35.   Jonathan Karp, "TV Times," *Far Eastern Economic Review* 157 (15 December 1994): 56–60.

36.   Hudson, *Communication Satellites*, pp. 251–62.

The shifts in the changing mediascapes of India, Indonesia, and China outlined above are in line with trends in international communication. Running parallel with these trends is the remarkable economic transformation that has gripped Asia since the mid-1980s, especially in urban centers. This economic transformation has led to significant changes in consumption habits, some would argue, fueled by television. The internal dynamic of countries such as China and Indonesia, which will have the first and fifth largest economies, respectively, in terms of gross domestic product by the year 2020,[37] contribute greatly toward changing technological patterns. Significantly, however, these changes emerged and developed as the Cold War ended, suggesting that the respective governments and peoples took stock of the situation and then developed new social and cultural practices that access to television encapsulated. Throughout Asia during the late 1980s, television became a common item in the domestic setting, creating audiences who developed both expectations about what they wished to see and sophisticated ways of looking at what they saw. Returning to Carolyn Marvin, the history of the medium begins "when audiences become organized around . . . uses."[38]

The Level Playing Field

Much of what has happened in Asia between 1990 and 1995 regarding satellite communications is too recent and too complex for detailed analysis in a work of this scope. However, it is argued here that the transformation of communications in Asia associated with the introduction of satellite transmission is unprecedented. So powerful is the perceived influence of satellite broadcasting that we find Asian governments adopting contradictory positions regarding their regulation. National governments variously have invented ideological systems to combat the cultural pollution that satellites are alleged to bring in their wake, and at the same time they have competed fiercely to enter the satellite age: all self-respecting Asian countries now have a satellite or at least rent transponder space.

Furthermore, satellite relations are no longer conducted on a government-to-government basis. When Rupert Murdoch visits an Asian country, he is virtually accorded "head-of-nation" status. When he visited India in 1993, it was alleged that all that was missing was the twenty-one-gun salute. Murdoch was the most obvious example of this trend, but as other powerful commercial interests enter the satellite field, an analysis of their operations becomes as much a question of market relations as international relations between governments. Complicating this even further are the nongovernmental organizations, such as the Asian Broadcasting Union, the International Telecommunications Union, and UNESCO, which play a regulatory role. Thus, contemporary conditions surrounding satellite communications in Asia involve three levels at the very least: commercial enterprises (both indigenous and transnational), the nation-state, and nongovernmental organizations. These relationships are explored here with reference to three aspects of the current situation—namely, the introduction of Star TV into Asia, recent developments in China, and telecommunications.

Although it was previously argued that there is no one dominant causal event shaping recent developments in satellite communications in Asia, there can be no doubt also that the launch of Asiasat 1 in 1990 radically altered the communications equation in the region. It dramatically extended the broadcast footprint covering virtually all of the

37. "War of the Worlds: A Survey of the Global Economy," *Economist* 333 (October 1994): 1–44.
38. Marvin, *When Old Technologies Were New*, p. 5.

western Pacific rim, Central Asia, and South Asia, providing a potential audience of three billion persons to advertisers. It also was privately owned rather than a state initiative. The principals were CTI (China), Hutchinson Whampoa (Hong Kong), and Cable and Wireless (Britain). Star (Satellite Television Asia Region) TV was created shortly after— mainly through the initiative of Richard Li, son of Li Ka-shang, a Hong Kong property developer, a billionaire, and a principal of Hutchinson Whampoa.

Around the same period as the launch of Asiasat 1, China (ChinaSat 3), Indonesia (Palapa B2R), Japan, and India all launched communications satellites, presenting Star TV with massive competition at the national level. Consequently, Star TV had to invent Pan-Asian broadcasting. It did so by developing strategies whereby it piggybacked existing services, such as BBCWST (BBC World Service Television), MTV, and ESPN, while at the same time developing its own production profile. There was a degree of ambivalence toward Star TV on the part of the television industry in Asia and the various nation-states. As a new concept, it faced considerable criticism and misunderstanding. For example, the Chinese were always suspicious of transborder broadcasting for ideological reasons. The fact that Star TV carried BBCWST, which offended Chinese authorities on a number of occasions, fueled their suspicions. Hong Kong, the base for Star TV, was reluctant to let them broadcast into the colony. The greatest enthusiasm for Star TV came from the transnational advertising industry, which grasped at the possibility of creating a vast Pan-Asian consumer market. However, what transformed Star TV from a marginal operation in global broadcasting was News Corporation's (Murdoch's) decision to purchase a 64-percent controlling interest in Star TV for $550 million (U.S. dollars) in 1993. Murdoch quietly purchased the remaining stock in January 1995, thereby gaining total control of the broadcasting company, although the original partners retained control of the satellite, and Richard Li kept the Hong Kong broadcast license.[39]

Murdoch's pronouncement that "the advancements in the technology of telecommunications have proved an unambiguous threat to totalitarian regimes everywhere"[40] highlights the potential for ideological conflict over satellite broadcasting in Asia. China immediately banned its citizens from watching Star and other transborder television programs. As John Sinclair pointed out, "the announcement, which was designed for consumption by Western shareholders of News Corporation, seemed a fair characterization of the attitude with which national leaders in the erstwhile Third World regarded this new technology in the 1970s and 1980s. Yet it has proved hollow. . . ."[41] Under Murdoch, Star TV continued a Pan-Asian approach built around English as the major linguistic medium and his American Fox Television product, but the loss of the Chinese market made that strategy problematic. Aggressive marketing in India and the eventual purchase of 49.9 percent of Zee TV in December 1994 partially compensated for the loss of China. However, Star TV also continued to attract political and cultural criticism from other Asian states. Malaysian Prime Minister Mohamed Mahathir proved the most forceful critic; he suggested that Murdoch was seeking to impose Western values on Asia and thereby corrupt and pollute Asian societies, while at the same time controlling Asian broadcasting.[42] Such criticism was damaging and led Star to develop more "culturally sensitive" programming. Meanwhile, Murdoch practiced shuttle diplomacy, meeting with Malaysian, Indian, Japanese, and Indonesian leaders seeking to establish Star TV in their nations. Star also

39.   Atkins, " 'Friendly and Useful'," pp. 54–64.
40.   Stephen Hutcheon, "Murdoch Makes Amends," *The Age* 27 (January 1995), p. 11.
41.   John Sinclair, "The Business of International Broadcasting: Cultural Bridges and Barriers," paper read at the Communications Research Forum, Sydney, 19–20 October 1995, p. 3.
42.   Brian Shoesmith, "Asia in Their Shadow: Asia and Satellites," *Southeast Asian Journal of Social Science* 22 (1994): 125–43; Atkins, " 'Friendly and Useful'," pp. 54–64.

dropped BBCWST from its northeast Asian transponder in an effort to mollify the Chinese authorities and at the same time promised "noncontroversial" programming on its northeast service.[43]

Underlying much of the discussion about satellite communications are assumptions about globalism and the demise of the nation-state. Marjorie Ferguson has questioned these assumptions and shown them to be fallacious.[44] The Asian experience tends to support Ferguson's contention that arguments about globalism are predicated on the desire of capitalism to maximize its profitability. The Economist[45] observed that Murdoch had been savaged economically by his experience with Sky television in Britain and that there was no economic nor commercial evidence to suggest that Star TV would perform any better in the short to mid-term. Indeed, it was announced in the fall of 1995 that Star TV would lose $100 million in the next year,[46] significantly higher than previously published figures. It is clear that if Star TV is to succeed financially, it cannot afford to offend the leaders of the Asian states, and it has begun to develop programming strategies accordingly. Consequently, critics see Star as becoming more anodyne and commercially oriented, seeking security for its investment rather than blazing a trail for democracy in Asia.[47]

The point here is that Star TV, through its praxis, has thrown into doubt all of the prognoses about the impending demise of the nation-state. On the contrary, what the Star TV experience has demonstrated is that while transborder broadcasting via satellite has had a profound effect on the way terrestrial mediascapes are organized, it in no way undermines national sovereignty unless the government has lost the "mandate of heaven." China's role in this unfolding history is a case in point.

China has a comprehensive domestic satellite system that was established in 1983 with the launch of ChinaSat 1. China, including Tibet, is serviced by satellite television; even the most remote areas have access to a minimum of four Central China Television (CCTV) channels. In the larger urban areas, audiences can access up to twenty-two channels. In addition to this domestic service, China, if it chooses, may access a full range of global and regional satellite services. For example, the Indonesian Palapa C series satellites include southern China within their footprint. Despite the fact that there are an estimated 40,000 parabolic dishes in China, none of these services are accessed except under the most stringent of conditions. The Chinese polity always has assumed that it has total control over communications within its boundaries, irrespective of its political allegiances, and the present decaying regime always has resisted transborder incursions.

Surprisingly, the 1993 bans on relaying Star TV signals throughout China have been successful for two reasons. First, when the government issued new regulations that limited satellite reception to three categories of receivers (foreign enclaves, luxury hotels, and selected educational institutions),[48] it had the support of the Chinese television industry. Since 1985, the Chinese television industry has been owned by the state, but it has operated along commercial lines. The industry was unhappy that much needed advertising revenue was being denied by transborder broadcasting. Second, since the late 1980s, the Chinese television industry had become professionalized. Although ideologues still hold

43.   Geetikar Pathania, "Ambivalence in STAR-ry Eyed Land: Doordarshan and the Satellite Television Challenge," Sagar 1 (1994), http://asnic.utexas.edu/asnic/sagar/sagar.main.html.

44.   Marjorie Ferguson, "The Mythology About Globalization," European Journal of Communication 1 (1992): 69–93.

45.   "Murdoch's Asian Bet," The Economist 328 (31 July 1993): 13–14.

46.   Deborah Brewster, "$1.34 bn Profit a News Record," The Australian, 23 August 1995, p. 42.

47.   Pathania, "Ambivalence in STAR-ry Eyed Land."

48.   Joseph Man Chan, "National Responses and Accessibility to STAR TV in Asia," Journal of Communication 44 (1994): 112–31.

key positions within the Chinese media administration, the numerous local television stations that form the Chinese hybrid television infrastructure of cable, microwave, and satellite are staffed by people driven by professional concerns. Chinese programs are now slickly produced and marketed.

As television became woven into the fabric of contemporary Chinese life,[49] the authorities understood that they had to continue providing a service similar to that which had developed between 1980 and 1993, when foreign programming had been freely available. The Chinese government opted for a fully developed cable system, believing that it provided greater opportunity for the effortless control of the circulation of meanings.[50] More than 5,000 cable operations exist in China. Many of these are Industrial Community Television operations set up during the 1970s and early 1980s to enhance the spread of CCTV throughout China, while others are clearly illegal operations. The largest cable operators, such as those in Shanghai, Beijing, and Wuhan, have an estimated 700,000, 600,000, and 300,000 subscribers, respectively.

By attempting to create an environment that protects their communications integrity and political sovereignty, the Chinese authorities have created a highly contradictory situation for their regime. Satellites transcend space and create highly centralized institutions, as well as particular textual communities[51] consonant with Chinese political theories that have emphasized centralized authority as the preferred model. In contrast, cable television tends to lean toward regionalism, both in terms of pragmatics and as a bias. In China, this problem is best exemplified in language. As W.J.F. Jenner pointed out,[52] when we talk about Chinese dialects, we are accepting a centralized view of the world. What passes for dialects in China are mutually unintelligible languages bound together by a universalizing script. Until now, no other medium of communication has been sufficiently powerful to challenge this particular hegemony. Television may be the exception. A system based essentially along regional lines presents centralist ideologies with particular problems, not the least of which is a tendency for local languages to be heard more frequently. This fact has not escaped the attention of the Chinese Ministry of Radio, Film, and Television, which has decided to reintroduce the concept of "one region, one network," in which domestic satellites combine with cable to ensure the central authorities a degree of control, as the prevailing broadcasting model for China in the 1990s. This suggests that in a mediascape increasingly shaped by new communications technologies, nation-states may survive the trend toward globalism for particularistic cultural reasons, but may not survive the forces of localism. In many respects, the nation-states of Asia, as creations of colonial powers in an earlier stage of globalism, are better equipped to confront the issues of globalism than the problems associated with the premodern articulated in localistic demands for ethnopolitical autonomy.

One may be forgiven for thinking from this account that communications satellites are concerned almost exclusively with problems of transborder television. This is clearly not the case; communications satellites are capable of much more than merely broadcasting television signals. Indeed, it may be that the role satellites have in the global capital system—facilitating the instantaneous global electronic transfer of trillions of dollars on a daily basis—is more culturally transformative than the broadcast of soap operas.

---

49.   James Lull, *China Turned On: Television, Reform and Resistance* (London: Routledge, 1991), *passim.*

50.   These comments are based on research conducted in Wuhan, People's Republic of China, in 1995 with the author's colleague Wang Handong (Wuhan University).

51.   On textual communities, see Stock, *Listening for the Text, passim;* Marvin, *When Old Technologies Were New, passim.*

52.   W.J.F. Jenner, *The Tyranny of History: The Roots of China's Crisis* (London: Penguin, 1994), *passim.*

Furthermore, their increasing role in expanding telephony also may have radical implications for Asian cultures. It is to this aspect of communications satellites that this discussion turns.

With the exceptions of Singapore and Hong Kong, terrestrial telephone services in Asia are antiquated and generally inefficient. For example, it can take up to two years to have a telephone installed in China, following normal channels. Communications satellites have altered this situation dramatically. Handheld telephony has developed into an area of intense activity and competition throughout Asia. What we are witnessing is a convergence of Western technology and local capital to establish satellite-driven technologies that bypass older systems of telephony. Three examples will illustrate the significance of this trend.

Asia Cellular Satellite Systems (known as ACeS) is based in Jakarta, financed by Indonesian, Filipino, and Thai capital, and uses American technology (Lockheed Martin satellites and rockets). It will service India, Bangladesh, Burma, China, Korea, Japan, Hong Kong, Indochina, Thailand, Malaysia, and Burma. The Afro-Asian Satellite Company is financed with Indian money and serviced by the Hughes Space and Communication International. Asia Pacific Mobile Telecom is jointly owned by China (75 percent) and Singapore (25 percent); Hughes, Lockheed Martin, and Space Systems/Lorel are all bidding to supply the technology. Moreover, we are not discussing geostationary satellites here; the companies involved in expanding telephony in Asia are exploring medium and low satellite orbits as options for providing services.[53] This suggests that the technological lag of the past between East and West in most cases has dissipated.

Parabolic dishes and handheld telephones are largely the domain of the Asian middle classes, but their potential to alter endogamous cultural relations may be of greater significance than television's capacity to effect changes, which has attracted so much attention. The systems structured around telephony are more intensely personal and immediate than television; they impact directly on individual temporal and spatial regimes in unprecedented and novel ways.[54] By contrast, Asia has been imbricated in various economic world systems since the Roman era, because its products have been in high demand on world markets. Asian cultures have learned to deal effectively with foreign economic and cultural incursions. Global economic systems are not new phenomena; it is the conditions under which they operate to establish economic ascendancy that has changed. In his classic account of the Canadian fur industry, Harold Innis observed that the "First People of North America" became enmeshed in an international economic system with the sale of their first fur to a trader.[55] Communications satellites in Asia, to a degree, register a set of changing economic power relations where technological transfer is being replaced by technological dialogue—for example, Hughes Space, Lockheed Martin, and Space Systems/Lorel bid for Asian business. Their success in this arena becomes crucial to their success as commercial operators, which can be viewed as radically different from the Asian engineers who went to NASA during the 1960s to learn about space technology.

---

53. Faith Keenan, "Looks Who's Talking: Newcomers Face Huge Risks as Well as Vast Potential Markets," *Far Eastern Economic Review* 158 (1995): 38–46.

54. The author's postgraduate student Kenneth Staples is currently researching the mobile phone in Perth. One of his findings is how people rethink their ideas about place when they begin using a mobile phone.

55. Harold A. Innis, *The Fur Trade of Canada: An Introduction to Canadian Economic History*, rev. ed. (Toronto: University of Toronto Press, 1970), *passim*.

However, it is the impact made by technologies at the microlevel that is ultimately more important. Few media analysts in the West accept any longer the "bullet theory of communications" that grants the sender almost total control in the communicative process. Rather, the receiver is perceived as the most important element in the communications chain. This perception has led to notions of the active audience becoming dominant in thinking about the relationship among text, audience, and the institutions of communications. This notion has been articulated clearly in respect to television viewing, but not so clearly in respect to the new telecommunications technologies. Nevertheless, there may be a degree of correlation between the active television audience and the active user of telecommunications. Furthermore, if we accept that the Western audience is an active producer of meaning, we also must accept that Asian audiences are active in the same way. Any other approach is fraught with dangers. Yet, much of the discussion around the supposed impact of the new communications technologies on Asian audiences and receivers is couched in terms of the passive. It is alleged that television introduces concepts and ideas that traduce Asian cultures, corrupt the young, and pollute the moral environment. These arguments are probably peddled not because the speakers believe them, but because they believe that the messages the technologies make available to the mass audiences, apparently without effort, challenge the established bases of power. Herein lies the terrain of the history of communications in Asia that has yet to be written.

Conclusion

Asia has been part of the global satellite communications scene since the 1960s, when Intelsat parked its second satellite over the Pacific Ocean. However, the major distinguishing feature of satellite communications in the region has been the way in which it has accelerated from a scarce resource to one of abundance. Between 1966 and 1996, the number of satellites serving Asia has increased from one to forty-six.[56] It is too early to pronounce what effect they have had, although it can be suggested that whatever the outcome, their impact will be profound and more pronounced at the domestic than the international level.

It can also be suggested that all technologies have a contradictory potential[57]; they do not behave in the way they are expected to act. For example, it is now clear that the new communications technologies, including satellites, have not killed off the newspaper as was predicted. On the contrary, the press in Asia is booming, especially at the regional level, in the local vernaculars. Two examples will suffice. In India, the local vernacular press has mobilized the new technologies to its own advantages and expanded dramatically.[58] In Surabaya, East Java, the *Jawa Pos* has increased its circulation from approximately 50,000 to more than 350,000 in five years. It is a dedicated exponent of using modern communications technology in production.[59] So much for the forces of homogenization and standardization supposedly inherent in the technology!

---

56.   "Regional Satellites," *AsiaPacific Satellite* 1 (March 1995): 18–23.
57.   Daniel Drache, "Introduction," in Harold A. Innis, ed., *Staples, Markets, and Cultural Change: Selected Essays* (Kingston, Ont.: McGill/Queen's University Press, 1995), pp. xii–lix.
58.   Hamish McDonald, "Paper Tigers," *Far Eastern Economic Review* 158 (1995): 24–26, esp. 26.
59.   See note 3. The *Jawa Pos* employs thirty-eight journalists in Surabaya to produce a major regional newspaper. To achieve impressive results, the paper has invested heavily in modern technology, including renting a transponder to communicate between Jakarta and Surabaya.

On a more cautious note, it is clear that communications satellites in Asia are still confronted by technical and cultural problems. Technically, there is the issue of orbit allocation, which is also a geopolitical issue, as well as the problem of congestion. Economically, there is the problem of oversupply; forty-odd satellites with an increasing number of more powerful transponders is an issue to be explored in its own right. Culturally, there is the problem of enunciation: whose voice is to be heard? Malaysian Prime Minister Mahathir has called for an Islamic satellite to provide services to Arabia, South Asia, and Southeast Asia. The most obvious question here is: Whose version of Islam will prevail? That of Iran or Saudi Arabia? These are essentially political problems, but problems of cultural identity and national sovereignty remain to be resolved as well. What is suggested here is that the presence of communications satellites in the Asian context is highly significant, but problematic.

In constructing a partial history of the present, one must be conscious of what is not said. In this context, Japan has hardly been mentioned and the role of radio ignored. However, in analyzing communications satellites in Asia within a historical perspective, this discussion has tried to suggest both a trajectory of development for satellites and a way of looking at the relationship between technology and culture that escapes the strictures of technological determinism.

Chapter 18

From Shortwave and Scatter to Satellite: Cuba's International Communications

by José Altshuler

Early in the sixteenth century, the Spanish conquistadors perceived that the island of Cuba was a most important strategic outpost in the New World—one that required rapid and reliable communications with Spain. These communications had to be essentially by ship; hence, it was very slow for centuries, until the first telegraph service by submarine cable between Havana and Florida was established in 1867.[1] Long before, the Spanish colonial authorities had rejected the implementation of a similar project for fear that it would serve the cause of the island's annexation to the United States. In different historical contexts since the 1850s to this day, Cuba's insularity and its proximity to the United States have strongly influenced the development of its international communications.

By the end of Spanish rule in Cuba, in 1898, the island was linked telegraphically to the rest of the world by submarine cables owned by U.S., British, and French companies. Early in the twentieth century, some spark radio telegraph stations went into operation in Cuba, which already was a virtual protectorate of the United States. These spark radiotelegraph stations served mainly, but not exclusively, to communicate with ships sailing in the Gulf of Mexico and the Caribbean Sea. The first two of them belonged to U.S. companies, but in 1909, some stations owned by the Cuban state were installed.[2] They all operated on long waves or medium waves, following the international standard practice at the time.

Radio broadcasting came to Cuba in 1922 and expanded rapidly on a commercial basis, with twenty-nine medium-wave stations operating in 1923 and sixty-one in 1933. By the end of 1933, moreover, the first shortwave commercial broadcasting station was installed. A few other low-power shortwave transmitters went into operation afterwards for the purpose of serving the interior of the island, but they were not effective enough, and most of them ceased to operate after some time. Only a few shortwave low-power broadcasting stations remained in operation in Cuba in the 1950s, each of which was dedicated to transmitting on shortwaves the ordinary commercial programs broadcast by an associated medium-wave station serving a national audience. As for international point-to-point shortwave radio communications services, practically all of them were in the hands of private companies, the most important of which belonged to International Telephone and Telegraph (ITT).

---

1. Thanks to the transatlantic telegraph cable link established between America and Europe in 1866, the Havana-Florida submarine cable made it possible for Havana and Madrid to exchange messages almost instantly, in comparison with the sixteen to twenty-one days taken by the courier steamships then traveling between Havana and the Spanish ports of Cádiz and Vigo.

2. José Altshuler, "Cuba," in Asociación Hispanoamericana de Centros de Investigación y Empresas de Telecomunicaciones (AHCIET), *Las Telecomunicaciones en Hispanoamérica: pasado, presente y futuro* (Madrid: AHCIET, 1993), pp. 73–88.

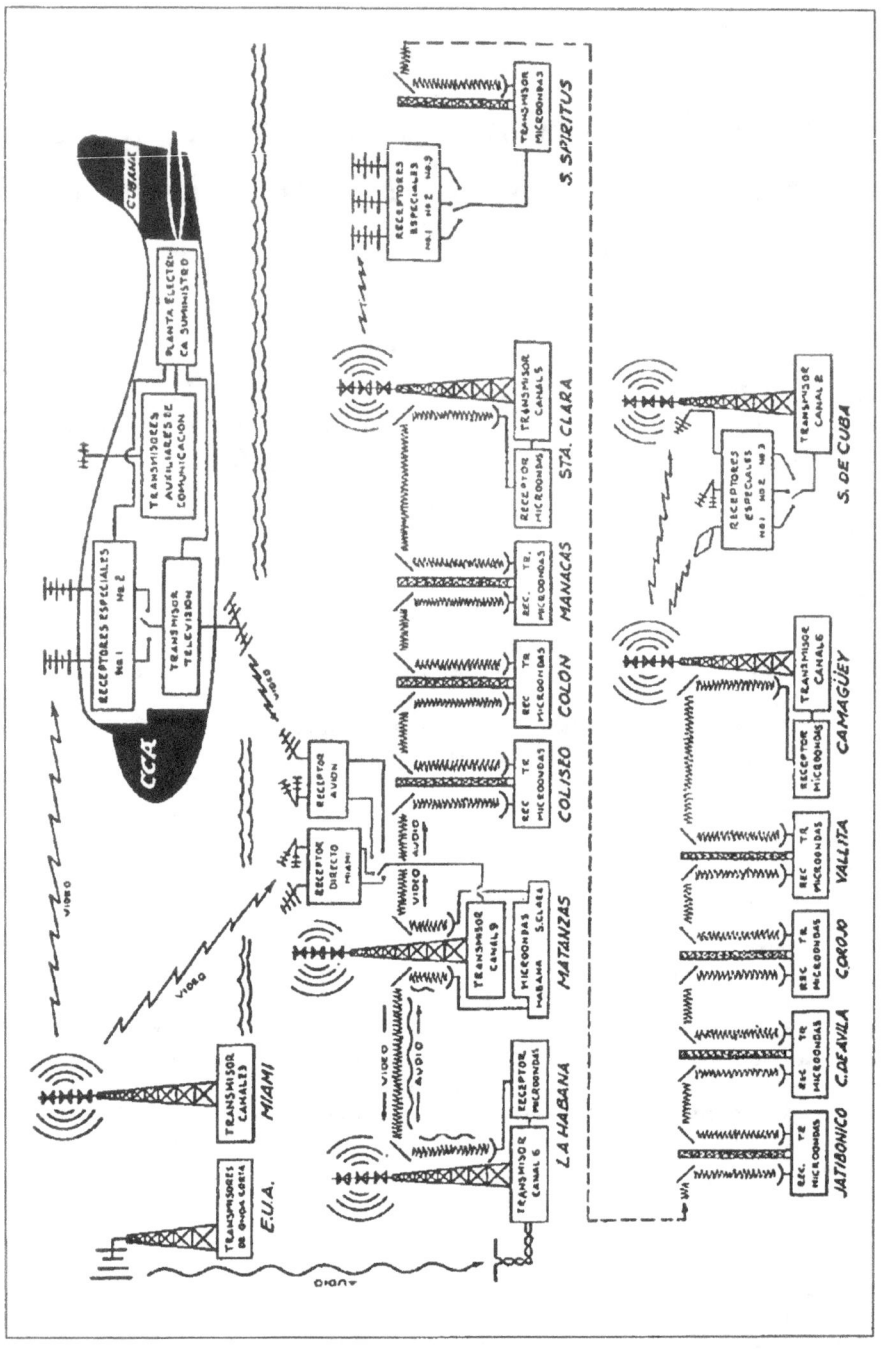

Figure 31

Airborne relay system used in 1955 to transmit the World Series television programs to Cuba. (Courtesy of José Altshuler)

After the introduction of television to Cuba in 1950, the possibility of receiving "live" television programs from the United States came under serious consideration. Cuban engineers and technicians met the challenge by using an airplane flying for three to five hours over the sea near the island. Equipment aboard the airplane received video signals from television stations in Miami and retransmitted them to a station belonging to a Cuban television network with coverage over a large part of the country (audio signals were transmitted separately by cable and shortwave).[3] In this way, the games of Major League Baseball's World Series were watched by Cuban fans during 1955 and 1956. By the same method, scenes from a cabaret show taking place in Havana in 1955 were transmitted to Florida and broadcast live on television in the United States.

A stable broadband communications link via tropospheric scatter, the first of its kind ever designed and used to transmit commercial television, was implemented in September 1957 between Guanabo, near Havana, and Florida City. It had a designed capacity of 120 telephone circuits, in addition to one monochrome television channel.[4] The Cuban station belonged to ITT, and the station in the United States was the property of American Telephone and Telegraph (AT&T). Both companies already had associated with one another to set up a telephone link between Havana and Key West in 1921. The link utilized the longest and most deeply submerged submarine cables ever applied in the world at the time. ITT and AT&T again collaborated in 1950 to link the same locations by the first deep-sea telephone cables with submerged repeaters.[5]

Shortwave Radio Development in Cuba After 1959

After the downfall of the dictatorship that had been in power in Cuba during the previous years, a popular government was established early in 1959. The avowed purpose of the new regime was to accomplish deep social, economic, and political changes in the interest of the country in general and of the needy in particular. As the events that followed led to more and more radical measures that affected important corporate interests, both national and transnational, the new Cuban government realized that the country must have, as soon as possible, its own means of international shortwave radio communications, including broadcasting. This measure was seen as an important way to avoid complete informational isolation by powerful unfriendly interests, especially in the case of a foreign military attack. Such a point of view was akin to the one prevailing at the close of World War I among high-ranking U.S. government officials, who believed that "foreign ownership of any part of American telecommunications would prove dangerous in any future war."[6]

---

3.   Enrique Valdés Pagés, "La Serie Mundial por T.V. en Cuba," *Ingeniería Eléctrica* 1(1) (1957): 15–18.

4.   Maurice Deloraine, *Des ondes et des hommes: Jeunesse des Télécommunications et de l'ITT* (Paris: Flammarion, 1974), pp. 70–71; "El Enlace Radioeléctrico por Difusión Troposférica entre Cuba y Norteamérica," *Ingeniería Eléctrica* 3(1) (1959): 15–19.

5.   These were two cables (213 and 232 kilometers long) laid between Havana and Key West with three flexible repeaters each, working at a maximum depth of approximately 1,900 meters. They were actually put to work as a prototype for a similar transatlantic telephone link inaugurated a few years later in early 1957, which could carry thirty-six telephone channels. Archie Clow, "First Transatlantic Telephone Cable," *Science News* 44 (1957): 59–68. Each one of the Havana-Key West cables was used to transmit, simultaneously in one direction, twenty-three telephone channels and twelve voice-frequency telegraph channels.

6.   According to Robert Sobel, *ITT: The Management of Opportunity* (New York: New York Times Book Co., 1982), pp. 34–35, this was the belief of Assistant Secretary of the Navy, Franklin D. Roosevelt. While President Wilson feared that the British might develop "a veritable monopoly in the transmission of intelligence throughout the world," Admiral William Bullard urged a small group of businesspeople "to retain in American hands the complete control of radio communication, not only in the United States, but in Central and South America as well, [thus outlining] a policy of wireless doctrine not unlike that of [the] greater Monroe Doctrine," as reported by Gen. J.G. Harbord, "Radio in the World War and the Organization of an American-owned Transoceanic Radio Service," in Anton de Haas, ed., *The Radio Industry: The Story of its Development* (Chicago and New York: A.W. Shaw, 1928), pp. 67–96.

Not too long afterward, a shortwave transmitting radio station was designed by the Cuban Ministry of Communications. It included a 100-kilowatt Swiss-made broadcast transmitter, as well as a twenty- to thirty-kilowatt German-made point-to-point communications transmitter, complemented by a British-equipped receiving station. The transmitting station was built near Havana in record time and was operative a few weeks before the middle of April 1961, when Cuban airports were subjected to enemy bombing in preparation for the Bay of Pigs invasion, which took place immediately afterwards. The shortwave station was used quite successfully to counteract hostile propaganda and to announce the defeat of the invaders, thus contradicting the totally different picture reported by international press agencies.

Direct shortwave radio communications circuits were established for the first time during the 1960s by the Ministry of Communications to support telegraph, telephone, facsimile, and telex international services with Prague, Moscow, Mexico City, Berlin, Paris, Algiers, Shanghai, and other distant places. Radio communications services with the United States continued to be handled by the older companies, as were communications with the United States via the Guanabo-Florida City scatter link. Prensa Latina, the Cuban international press agency, developed its own facilities utilizing Chinese and U.S. transmitting and receiving equipment, and the agency established shortwave press circuits with its offices in several countries.

The country's new interest in the development of shortwave radio communications required adequate training of technical personnel on the characteristics of the ionospheric propagation of radio signals. Pertinent training material was prepared in the early 1960s, first by the Ministry of Communications and afterwards by the University of Havana, for courses taken there by electrical engineering students. A program of ionospheric research also started at the Institute of Geophysics of the Cuban Academy of Sciences, backed by important scientific research institutes of the Soviet Union.

After a decade of operation and progressive expansion of shortwave radio services in Cuba, however, it became clear that additional measures would have to be taken to meet the demands of growing traffic. In addition, the prospect of getting rid of the vagaries of the ionosphere as a transmitting medium, which was incapable of accommodating broadband channels, made the idea of setting up a ground station for international communications via satellite seem exceedingly attractive.

The Beginning of Satellite Communications in Cuba

Less than three weeks after the launch of the Intelsat Early Bird geostationary satellite on 6 April 1965, which established broadband transoceanic communication between North America and Europe, the Soviet Union put into orbit its first communications satellite, Molniya 1. Its orbit was a very elongated ellipse, strongly inclined with respect to the equator and with its apogee above the Northern Hemisphere, so that it spent most of the time over the territory of the former Soviet Union. Suitable tracking of the satellite by ground stations made it possible to set up an experimental link between Moscow and Vladivostok lasting eight to ten hours.

In November 1965, representatives of the Cuban government attended a meeting in Moscow, whose purpose was to discuss a Soviet proposal to use its launch facilities and basic know-how to implement a comprehensive program of scientific cooperation among socialist countries "for the peaceful study and use of outer space." This was to be known later as the Intercosmos Program.[7] Following subsequent discussions held just a few weeks

---

7.   José Altshuler and Kiril Serafimov, *La hora del espacio* (Havana: Editorial Científico-Técnica, 1991), pp. 255–84.

after the Soviet Union placed a second Molniya 1 satellite into orbit, it became quite clear that an international communications system based on a few essentially similar satellites could be used to establish a permanent communications link between Cuba and Eastern Europe. "In comparison with shortwave links," remarked Soviet Minister of Communications N.D. Psurtsev, "such a system would be acceptable and not too expensive, and under extraordinary conditions it might come to provide the only possible transmission channel."[8] Needless to say, the October 1962 missile crisis had made it dramatically clear to all how important it was for Cuba to have reliable long-distance communications facilities under certain "extraordinary conditions," quite apart from the convenience that a satellite link offered to establish broadband channels capable of transmitting color television programs and other kinds of information.

At the time, however, space research was far from being regarded as a priority at all for the newly founded national research institutions, and only the Cuban Ministry of Communications took the initiative to fully support the country's participation in the Intercosmos Program. Accordingly, when the Cuban government created a national agency for the coordination of the peaceful study and use of outer space in 1966, to facilitate the country's participation in the Intercosmos Program, the agency was placed under the patronage of the Ministry of Communications. This activity spurred the country's early introduction to space research, which was to reach its highest point when some twenty experiments prepared by Cuban scientists, in the framework of the Intercosmos Program, were performed in space during the flight of a Cuban cosmonaut in September 1980.[9]

Still, the main interest of the Ministry of Communications was to establish a broadband communications link via satellite between Cuba and other countries, especially those with which it was closely associated at the time. That possibility came within reach by the end of 1971, when work in the area of space communications, again within the context of Intercosmos, led to the creation of Intersputnik, an intergovernmental organization intended to meet the need for long-distance telephone and telegraph circuits, as well as exchanges of radio and television programs, among different countries through the use of communications satellites.[10]

A bilateral agreement between the Cuban and Soviet Ministries of Communications soon materialized, and construction of an Intersputnik receiving and transmitting ground station began in Cuba. This station was essentially similar to standard ones used in the Soviet Orbita communications system that utilized the Molniya 1 satellites. Construction of the so-called Caribe ground station began in February 1972. It was able to receive live from Moscow the first color television programs on 7 November 1973. On 1 January 1974, live images of the celebration then taking place in Havana were transmitted to Moscow, and they were retransmitted from there to other European countries via the Intervidenie system.

The Caribe station, provided with a twelve-meter-diameter tracking parabolic antenna, was erected in a valley near the town of Jaruco, approximately forty kilometers east of Havana. It was designed to work with Intersputnik satellites, which were initially of the Molniya 2 type.[11] The station operated around four and six gigahertz and was capable of

8.   Roberto Díaz Martín, "Génesis de las comunicaciones espaciales en Cuba," *Orbita* 6 (1984): 32–40.

9.   José Altshuler, "Space Activities in Cuba," in the proceedings of the Study Week on the Impact of Space Exploration on Mankind, 1–5 October 1984, Vatican City, 1986, pp. 207–17. The patronage of the Cuban space commission was handed over to the Academy of Sciences in 1974.

10.   Ralph Chipman, ed., *The World in Space* (Englewood Cliffs, NJ: Prentice-Hall, 1982), pp. 584–86.

11.   The Molniya 2 satellites were put into a very elongated elliptic orbit, inclined about sixty-five degrees with respect to the equator, with its apogee 40,000 kilometers above the Northern Hemisphere and its perigee 500 kilometers above the Southern Hemisphere, making a complete revolution around the Earth in about twelve hours. Three satellites simultaneously in orbit were used to obtain around-the-clock service. Actually, the Molniya 2 was an improved version of the Molniya 1 type suitable for international use. Its working frequencies were about 900 megahertz. It continued to be used for a long time for regional coverage.

simultaneously handling one television channel and sixty telephone channels. This system was used to communicate with the Soviet Union and Eastern Europe and, through them, also with France, Italy, and Spain.

However, broadband communications with countries in Latin America, Western Europe, and other important regions of the world through Intersputnik was difficult. The system had limited coverage. This fact became an obvious hindrance when Havana was chosen by the nonaligned countries as the site of their sixth summit, which took place in September 1979. As a palliative for the occasion, a Standard B Intelsat ground station was installed close to the original Caribe station.

During the early 1980s, Intersputnik shifted from Molniya to Statsionar geostationary communications satellites, which the organization leased from the Soviet Union. These satellites operated in the same frequency bands as the old Molniyas (six gigahertz uplink, four gigahertz downlink), but their transmission capacity was much greater. One of them was placed at longitude 53° E (Indian Ocean), and the other was at longitude 14° W (Atlantic Ocean),[12] which gave Cuba the possibility of delivering television programs to large regions abroad, especially in Africa, beginning in 1986.

In the meantime, telecommunications between Cuba and the United States continued to depend basically on the use of submarine cables and tropospheric scatter technologies. When the deterioration of the old Cuba-Florida cables made their operation too troublesome and costly, they were replaced in November 1989 by one with submerged repeaters between West Palm Beach and Cojímar, a small town approximately ten kilometers east of Havana.[13] However, the newly laid cable, which was capable of carrying 144 telephone circuits, remained idle for a long time. No mutually satisfactory business agreement could be reached under the pressure of the embargo conditions imposed by the U.S. government on Cuba. As a result, only the eighty-nine telephone circuits provided by the deteriorated tropospheric scatter system were available for direct communications between the two countries. Furthermore, even those circuits disappeared suddenly in August 1992, when Hurricane Andrew seriously damaged the Florida City station, and its owner, AT&T, subsequently decided to abstain from repairing it.

The Current Situation

The current tense relations between the United States and Cuba have made bilateral negotiations on the replacement of the old telecommunications links stagnate for more than two years. Nonetheless, despite the political situation, an agreement was reached to operate future links on an equitable basis from an economic point of view.[14] In November 1994, telephone and other telecommunications services were restored in part through the AT&T-owned Cojímar-West Palm Beach submarine cable, but mostly by Intelsat communications satellites, and this continues to be the case to this day.[15]

---

12.   Chipman, *The World in Space*, p. 585.

13.   Apparently, this cable came from a suitable recovered length of a submarine telephone cable laid in the late 1960s across the Atlantic Ocean. All of the Cuba-Florida telephone cables belonged to the Cuban-American Telephone & Telegraph Company, jointly owned equally by ITT and AT&T. In the early 1990s, AT&T bought out ITT and became the company's sole owner.

14.   The Torricelli bill, approved in 1992 by the U.S. Congress, authorized payment to Cuba of its share in the telecommunications business established with the United States.

15.   A Standard A Intelsat unit had been added to the Caribe ground station in 1991, to be able to effect an efficient television transmission for Intelsat users of the eleventh Pan-American Games, which took place that year in Cuba.

Meanwhile, Cuba's information exchanges with the rest of the world also have continued to expand using both the Intersputnik and the Intelsat satellite systems. The current international context is quite different from that which existed when the Caribe ground station was inaugurated more than twenty years ago. From today's perspective, the Caribe station can be viewed as a working monument to the best that the space age can offer to a small island nation struggling to develop under arduous and daunting circumstances.

Chapter 19

The Recent History of Satellite Communications in Cuba

by Roberto Díaz-Martín

This chapter discusses the development of satellite communications in Cuba—beginning with the first ground station, named Caribe 1, which was installed because of Cuba's early participation in the Intercosmos Program[1]—up to the present day. Satellite communications in Cuba can be said to have begun in 1965.[2] In that year, the Soviet Union invited several countries to attend a meeting held in Moscow between 15 and 20 November 1965 to address "space research and the peaceful uses of outer space." At this conference, participating countries were invited to join a new "Collaborative Program for Research and Peaceful Uses of the Outer Space." The program proposed to extend to collaborating countries the satellite communications know-how and facilities already available in the Soviet Union. At this meeting, the Soviet minister of communications offered to provide reliable direct communications between Cuba and the European continent, using the existing operational Molniya communications satellite system. Subsequently, Cuba joined another new program known later as Intercosmos, as well as its successor organization, called Intersputnik.

Within a relatively short time period, the Caribe 1 ground station was erected in 1974, in the town of Jaruco near Havana, so that Cuba could participate in these satellite communications programs initiated by the Soviet Union. The site selected for the ground station was perfect from a geographical point of view: a small valley surrounded by hills protected the station from undesirable city noise and from electromagnetic interference. The Caribe station was Cuba's crucial link to the Intersputnik satellite system.

An intergovernmental agreement established Intersputnik on 15 November 1971, and the signatory nations confirmed the agreement on 12 July 1972. Cuba has been an Intersputnik member nation since the organization's beginning. The Caribe ground station, completed in 1974, became only the second such station outside the territory of the Soviet Union. The Intersputnik system utilized three Molniya 1-class satellites in elliptical orbits. The station's initial capacity was one television channel and twenty-four analog telephone circuits (expanded later to sixty). In 1978, the Intersputnik system switched to geostationary satellites of the Horizont, and later Statsionar, series. This change eliminated interruptions caused by the need to switch from one satellite to another so as to maintain a continuous communications link.

1. José Altshuler, "El primer sputnik, un cuarto de siglo después," *Orbita* 1(1) (1983): 1–20; Secretary of UNISPACE '82, *Cooperación intergubernamental multilateral en las actividades espaciales: Documento de antecedentes,* U.N. document A/CONF.101/BP/10 (New York: United Nations, 1981).

2. Roberto Díaz Martín, "Las comunicaciones espaciales en rápido desarrollo," *Orbita* 2(1) (1984): 32–40; R. Díaz Martín, "Génesis de las comunicaciones espaciales en Cuba," *Orbita* 6(1) (1986): 7–15; Pedro Luis Torres, "Informe de la Conferencia de Moscu sobre la investigacion y utilización del espacio cósmico con fines pacíficos," 1965, conference paper, personal communication.

Figure 32

The Caribe satellite communications ground station near Havana, Cuba, equipped for working with both Intersputnik and Intelsat satellites. (Courtesy of José Altshuler)

The first television program received by the station, and rebroadcast nationwide later because of the time difference between Europe and Cuba, was the Moscow commemorative parade of 7 November 1974. Some weeks later, the Havana commemorative parade of 1 January 1975 was transmitted to Europe from Havana. The Cuban public also enjoyed the television transmission of the Montreal Olympic Games in 1976. To exchange television programs among Cuba, the Soviet Union, and other socialist countries, it was necessary to overcome the handicap of different television broadcast standards. The television standard used in Cuba was NTSC, while the Soviet Union and the majority of Eastern European countries used SECAM. This difficulty was resolved in 1975 through the use of camera tubes that carried out the conversion optically. Later, an electronic standard converter replaced this analog system.

Selection of a Color Television Standard

One of the key factors affecting the operation of satellite communications in Cuba was the selection of a standard for color television broadcasts. The availability of regular color television programming through the Caribe station encouraged the introduction of color television broadcasting, which was not generally available at the time because of the lack of a single television standard. A large number of NTSC black-and-white television receivers were in use throughout Cuba. However, the color television receivers available from Eastern Europe were all SECAM receivers. Financial support to replace all the NTSC black-and-white receivers with NTSC color sets was not available. Moreover, complicating the switch from NTSC to SECAM was the difference in electrical power standards. The Cuban electrical power system universally supplied 110 volts at sixty hertz, while SECAM receivers required 220 volts at fifty hertz, the European standard. Voltage differences aside, adopting the SECAM standard would cause screen images to flicker, because of SECAM's fifty-hertz time base.

Color television arrived in Cuba in 1958. Havana's Channel 12 aired in color using some RCA telemovie equipment capable of processing color television images. After 1959, however, Channel 12 no longer broadcast color programming. Many years later, once the Caribe station began operation, the technical staff of the Cuban Institute of Radio and Television began to reactivate the old Channel 12 color television equipment and to redesign and adapt new SECAM studio equipment obtained from the Soviet Union. These efforts demonstrated the possibility of transmitting in color using the NTSC standard through the existing broadcasting network. Other tests were carried out with the SECAM color standard with acceptable results.

Selection of a color standard for Cuba was conditioned not only by technical factors, but also by economical, social, and political ones. The final government decision reached at the end of 1974 was to stick to the national color television standard (the NTSC standard). This was the standard used by most countries in the Caribbean geographical area. Using a different standard would isolate Cuba from its neighbors. Once the government made this decision, the country rapidly acquired color remote control equipment from NEC in Japan, while existing Soviet equipment was duly modified for color transmission. The first standard color transmission in Cuba via the Caribe ground station took place in 1975.

Cuba Joins Intelsat and the Global News Network

In 1979, the Sixth Summit of Nonaligned Countries took place in Cuba. The summit required coverage of all the communications needs of the member countries, including worldwide telephone and television services. Most of the participating countries were located geographically around the Pacific, Indian, and South Atlantic Oceans—areas not completely covered by Intersputnik. Yet, Intersputnik was Cuba's communications satellite system. As a result, Cuba had to find an appropriate technical solution that would permit it to offer telecommunications services to countries outside the Intersputnik footprint. One proposal was to install additional equipment in Algeria, in the Sahara Desert, to retransmit to areas around the Indian, South Atlantic, and Pacific Oceans. This proposal was discarded in favor of installing a new "Standard B" Intelsat ground station in Cuba. Thus, in 1979, in a record time of three months, Cuba became an Intelsat user. The new ground station had twenty-four telephone circuits and one television channel. It provided communications services satisfactorily during the Summit of Nonaligned Countries, thanks to the joint operation of both the Intersputnik and Intelsat stations.

Because of satellite communications, Cuba was able to participate early on in the international exchange of television news. In 1977, supported by the International Organization for Radio and Television, Cuba entered into television news exchanges with the Intervision Network, televised from the network's center in Prague. Every day at 3:00 p.m. European Central Time, the broadcast exchange took place during approximately one hour. A leased common audio channel provided simultaneous sound programming to audiences scattered around the globe, from Managua, Nicaragua, to Ulan Bator, Mongolia, and Hanoi, Vietnam. The Intervision Network continued to broadcast until the disappearance of the International Organization for Radio and Television, when Eurovision and EUTELSAT picked up the service. All participants shared in the expenses of this service.

Cuba's satellite communications capability also benefited groups of interested users located in Europe and other regions of the world. For this purpose, Cubavision was produced and transmitted regularly through a leased transponder on a Statsionar satellite. The program included news from Cuba, as well as entertainment material.

Space Communications Research and Special Services

Cuba participated in the International Space Communications Work Group of the Intercosmos Program. Within its framework, the Cubans conducted research to study radio propagation in the tropics in the ten- to thirty-gigahertz frequency range. Existing data from radio telescopes, which had conducted geophysical studies of the Sun's impact on the Earth's ionosphere, were compiled and analyzed. Cuban researchers conducted additional studies using equipment obtained through, and with the backing of, the Intercosmos Program.

Beginning in 1981, Cuba launched a project to develop the Central National Science and Technology Library. The objective of the library, inaugurated in June 1988 in the Capitol Building, was to promote the expansion of information services to the entire country. As part of the library project, Cuba instituted national and international networks to provide computer-assisted exchanges of information services. The Caribe ground station provided these networks with communications support.[3]

As an island nation, Cuba has a close relationship with the sea. Political and economic reasons have pushed Cuba to broaden its merchant and fishing fleet. Cuban vessels now cover large distances to maintain the country's needs. For this reason, maritime communications, especially maritime satellite communications, has taken on greater interest and importance. Therefore, on 27 July 1989, Cuba joined Inmarsat to modernize its maritime communications capabilities.

Conclusion

The Caribe ground station has been the subject of continuous technical improvement. The Intersputnik system featured time division multiple access and a single carrier per channel—features that made it compatible with the Intelsat system. The Intelsat B ground station increased its telephone capacity to eighty circuits, and the Intersputnik station expanded to 144 telephone circuits. However, to meet the needs of the constantly growing numbers of foreign visitors, businesspeople, and tourists, Cuba had to enlarge its telephone and data communications facilities even more.

Consequently, a new Standard A Intelsat station, named Caribe 2, became operational in 1991. This new ground station, equipped with the latest digital technology, provided the necessary capacity to cover present and future island needs. This station satisfactorily covered the eleventh Caribbean and Pan-American Games. In addition, small-capacity ground stations were installed in outlying areas to serve the needs of remote zones and the small islands of the Cuban archipelago.

---

3.   R. Díaz Martín, "Actividades Espaciales en Cuba, Diseminación de la información via satelites," paper read at Taller de Naciones Unidas sobre Sistemas Regionales de Información Espacial, Lima, Peru, 24–28 October 1988, copy in United Nations library.

PART III:

The Unfolding of the World System

Applications

Chapter 20

Project SHARE and the Development of Global Satellite Communications

by Joseph N. Pelton

The past is often prologue, and the history of satellite communications is no exception. The two earliest practical concepts of satellite communication were offered by Hermann Noordung, the Polish writer who used the pen name Potocnik, and Arthur C. Clarke. In articles published during the 1930s and 1940s, they formed the first specific image of what a communications satellite might be and what it might do. Certainly others, such as H.G. Wells, Everett Edward Hale, and Jules Verne proposed earlier concepts in fanciful fiction, but Noordung and Clarke had specific ideas based on scientific and even quasi-engineering principles. These first "practical" views included the idea of geosynchronous orbit, the use of radio-relay technology, and the belief that these space stations would be large, manned contrivances that would have on-board personnel replacing radio tubes and repairing electronic equipment.

The subsequent invention of the transistor, reliable solid-state circuitry, and high-speed computers served to alter the ultimate reality of communications satellites that materialized during the late 1950s and early 1960s. These technological breakthroughs allowed satellites to be smaller, cheaper, more reliable, and, most critically, unmanned remote-controlled versions of those proposed by Noordung and Clarke. The initial concept that these communications satellites essentially would be radio broadcast towers in the sky endured, however. Clarke's earliest vision, in fact, was a very convenient one for telecommunications carriers and broadcasters. These entities, which frequently were monopolies with huge investments in terrestrial networks, very much wanted communications satellites to become an adjunct or extension of their large and expensive terrestrial switching systems. At the dawn of the age of space communications, telecommunications moved through the key hubs of Paris, London, New York, and a few other key urban centers of the world, such as Tokyo and Rome. Thus, the early history of satellite communications was a story of the technical triumph of electronic and rocketry technology, as well as a political struggle to retain a global hierarchy for the dominant communications carriers of the day.

There was nothing sinister or even conspiratorial about this situation. The concentration of traffic through hierarchical switches for many decades was considered the ultimate network strategy. Suddenly, the network architecture dictated by wire and terrestrial switches was confronted with a new option—that of the communications satellite. The satellite did not depend on physical wire connection, but rather wireless links through free space. Network planners, however, saw satellites as long-distance trunk lines that simply augmented the terrestrial telecommunications network. Because the first ground stations were of necessity large, expensive, labor-intensive operations ($10 million each and staffed by forty people operating over a twenty-four-hour shift), the number of points where satellites and wire networks could link together were small in number. Even so, the new communications satellites immediately began to make a difference. Developing countries could build a national ground station to link via satellite not with one or two major world centers, but with dozens of places. Chile could call Colombia or Brazil directly without going through New York. Nigeria could contact Ghana or Senegal without going through London or Paris.

Thus, by 1969, satellite communications, which began in large part with Intelsat's launch of Early Bird in 1965, had affected the form and structure of global telecommunications in a decisive manner. Chile, for instance, was the first country to initiate direct satellite communications in South America. Santiago Astrain, Intelsat's first director general, was the head of Entel Chile during the middle to late 1960s. For an investment equivalent to the purchase of a Boeing 707 jetliner, his country could enjoy global communications. Suddenly, we could connect directly with a growing family of nations, rather than transit through a handful of major powers.

The existence of national telecommunications monopolies and the high cost of large, high-gain ground stations, however, served to limit this "satellite revolution." Comsat's chief scientist, Sid Metzger, who then headed the team responsible for the technical design characteristics of Intelsat's ground stations, conducted a number of ongoing systems studies, which showed that the best "balance" of capital investment between Intelsat satellites and ground stations strongly favored the big and expensive thirty-meter Intelsat "A" stations. The key assumptions of these studies (which were indeed quite valid at the time) included the following:

- The power of satellites will be largely limited, and greatly increasing their power will be quite expensive.

- Ground stations generally will be international gateways, and most countries will have no more than one or two.

- To the extent that the Intelsat system must have smaller, low-gain antennas for special reasons, such as in the case of a shipborne antenna to support a manned space mission, the satellite access charge should have a rate adjustment factor to charge much more for access.

This type of design engineering and charging policy remained in place through the 1960s, but as the next decade approached, the new and dramatically more powerful Intelsat IV satellite was being built by the Hughes Aircraft Company. This satellite, with about sixteen times more capacity than the Early Bird satellite, permitted new approaches to satellite communication. In an unusual move, Comsat organized a special study team, called the Intelsat IV Charging Policy group, within its system manager's organization. Instead of being an engineer, the head of this group was the director of the firm's International Division, John A. Johnson, who also chaired the Intelsat Interim Communications Satellite Committee, whose eighteen members from around the world decided overall management policy. That group recommended a new policy: a new family of ground stations of different sizes and greatly reduced rate adjustment factors for smaller, cheaper stations. The new policy envisioned the use of satellites for regional, national, and international communications services, as well as services to smaller terminals in rural and remote regions. Experiments supported by NASA and Canada through their joint Communications Technology Satellite program in such locations as the Brazilian Amazon supported the technical feasibility of such a concept.[1]

---

1.   It was the author's first important assignment at Comsat to head the Intelsat IV Charging Policy study in 1970, with a team of eight members drawn from the procurement, finance, legal, technical, operations, and international divisions of Comsat.

The recommendations of the Intelsat IV Charging Policy group report were quite controversial. Some recommendations involving the use of excess Intelsat capacity for domestic communications had to wait approximately two years for Algeria to formally request this service from the Intelsat Interim Communications Satellite Committee in 1973. As it turned out, the United States also directly benefited from this policy. The committee decided that traffic between the U.S. mainland and Hawaii and Alaska could be considered under bulk lease for "domestic service."

As the "decentralization" of the satellite network went forward with the approval of the Intelsat Interim Communications Satellite Committee in 1971, during the following years Intelsat designed and built bigger and more powerful satellites, authorized smaller and smaller ground stations, and offered a full range of national, regional, and international communications services. Still, there was a feeling that much more could be done. By the early 1980s, Intelsat began to define a new type of international service that could use the Intelsat V and V-A satellites to go directly to customer premises. This became the Intelsat Business Service. Intelsat even defined a service for rural and remote services known as VISTA, as well as a microterminal-based service for medium-rate data called Internet, which could operate via a sixty-five-centimeter "desktop" dish about the size of a pizza pan.

This burst of new services occurred in close parallel with the 1983–1984 announcement of plans for competitive services. Critics claimed that Intelsat needed competition to develop and deliver new and innovative services. The fact that Intelsat's twentieth birthday was coming up on 20 August 1984 suggested to a number of people that Intelsat should do something particularly noteworthy and innovative to mark that upcoming anniversary.

The Beginnings of Project SHARE

This author was named the Intelsat director of strategic policy in 1983 and, in that capacity, proposed an activity to promote access to educational and health services in rural and remote areas via satellite. Because Intelsat did have spare satellite capacity for emergency restoration, it was possible to think of offering the space segment for free on the understanding that any such free demonstrations would be preempted in the case of a failure of regular commercial service.

It was not easy to convert a vague idea into a clear specific proposal acceptable to Intelsat management, Intelsat signatories from around the world, and the Intelsat Board of Governors. This Board of Governors replaced the old Intelsat Interim Communications Satellite Committee in 1974 under the newly negotiated Intelsat Definitive Agreements. It required a lot of hard work and a lot of cooperation among a number of people to make this vague idea truly workable. Marcel Perras of Canada, who was a former board chairman and at that time Intelsat's director of business planning, came up with the name: Project SHARE. SHARE was the acronym for Satellites for Health and Rural Education in English, but Intelsat operated with three written languages: English, Spanish, and French. Those involved could not get the same acronym or meaning to work in the other two languages. Finally it was simply decided that the name would have to be Project SHARE in English, Projet SHARE in French, and Proyecto SHARE in Spanish. After test-marketing the idea, it turned out that the English word "share" was rather universally understood around the world.

The advantages to Intelsat of making free satellite capacity available to test rural and remote health and educational services were not hard to understand. These were potential new markets. The extension of ground stations into more remote territories

Figure 33

Project SHARE envisioned making free Intelsat satellite capacity available to furnish health and educational services to rural and remote regions. The use of satellites to link remote peoples appeared much earlier, however, as this NASA drawing dating from the early 1960s illustrates. (Courtesy of NASA)

expanded the ever-increasing interconnectivity of the Intelsat system. The capacity was not being used for other purposes. Most importantly, it was a very good and exciting thing to do.

The twentieth anniversary celebration took on an important substantive dimension that went beyond throwing a big party. The rest of the world could celebrate also. The Intelsat signatories around the world had a somewhat different perspective. They needed to make available ground station equipment and terrestrial telecommunications systems to make the tests and demonstrations work. How would they be able to pick and choose? How much additional effort would this require? Would it interfere with paying commercial customers? If they allowed free service to start, how could they gracefully end the service when the tests ended?

Here, the Intelsat Board of Governors and the individual signatories were extremely helpful in working out a detailed procedure that limited requests in terms of scope and duration and that eliminated trivial or frivolous proposals. As a result, Project SHARE activities were meaningful, focused, and oriented toward participants who had a serious interest and intent to implement a test program. It took months of drafting and coordinating these procedures for final approval by the Board of Governors, but eventually the formal authorization was in place.

Another serious problem, however, had to be addressed. Intelsat knew much about satellite communications, but very little about rural education or health care. Again, willing help was found in the international community. Intelsat approached the International Institute of Communications, headquartered in London, for help. This nonprofit organization, with a global span of academic, professional, and industrial members, agreed to help form an international advisory board to assist in assessing incoming proposals and even to help generate proposals from relevant global organizations, such as the Pan-American Health Organization, the World Health Organization, and the open university systems around the world. This advisory committee proved invaluable to the planning, implementation, and evaluation of ongoing Project SHARE activities.

After much hard work, the Board of Governors gave its approval, the application forms were printed, the Intelsat signatories' ground rules for participation were in place, and the stage was set for the launch of the program. Something special in terms of publicity and fanfare seemed appropriate. So, those involved turned to Charles Schulz of "Peanuts" fame. Research indicated that the comic strip "Peanuts" was perhaps the most widely read and recognized cartoon in the world. Because education and health care for children were key objectives of Project SHARE, United Features and Charles Schulz were contacted to ascertain whether he might donate a distinctive and appealing logo—one that would be appealing to a broad audience rather than just scientists and engineers. In characteristically generous fashion, he agreed. Within a matter of weeks, we had a Project SHARE logo featuring Snoopy atop a doghouse, which also served as the main body of an Intelsat V satellite.

The "launch" of Project SHARE subsequently occurred in August 1984 at the International Club in Washington, D.C., amid much fanfare. Snoopy presented an educational degree certificate to the Intelsat's director general. As it turned out, it was a very hot day, and the diminutive actress inside the Snoopy costume went to her changing room and promptly fainted. She was quickly revived. The press in attendance gave Project SHARE enthusiastic support. The framework was in place, the project was announced, and the advisory committee was on alert, but there were no projects. The people involved waited patiently for a couple of months and then began to wait impatiently. Finally, proposals were recruited on a very active basis.

A Few Notable Projects

The turning point came in the form of a delegation of the Post and Telecommunications Ministry from the People's Republic of China. They requested the use of an Intelsat transponder (or a full television channel) for a several-month test of a new National Television University. After some discussion, and a couple planning and coordination trips, the first major Project SHARE activity was launched with a signing ceremony at Intelsat headquarters in August 1985.

The first phase involved distribution of educational programs developed by Central China Television and the Ministry of Education. Initially, thirty hours of programming were distributed to some forty regional television receivers with a test audience of several thousand students and more than 100 instructors. The tests were so successful that the Chinese signatory requested a three-month extension, which was promptly granted. By the end of the second trial, China formally leased a transponder from Intelsat and began implementing a full-scale electronic university network across most of rural and remote

China, as well as in many other locations. As of 1996, this network is still very much in operation. Approximately 90,000 terminals have been built and deployed in every part of China, and the number of students now exceeds 3 million. Eventually, a target of more than 10 million teachers and students will be served by this satellite network started under Project SHARE. This is the world's largest single educational network, and even if nothing else arose from Project SHARE, one could claim that this program alone made it extremely worthwhile.

Fortunately, many other projects quickly followed. Everything started to come together. The advisory committee began to bring forward key projects. Signatories began to recognize from the experience in China that Project SHARE could attract new customers and favorable publicity. The early successes tended to breed new opportunities and innovative suggestions. The case of Dr. Max House of the Memorial Hospital of Nova Scotia and the Canadian signatory Teleglobe Canada was a key case in point.

In the 1970s, Max House started a highly innovative project to bring health and medical assistance to remote areas of Canada. His initial mission was to extend the treatment coverage of Memorial Hospital to drilling rigs off the coast of Nova Scotia. Using communications links, House helped assist with the broken limbs, diagnostic tests, and medical emergencies of workers on those remote off-shore towers. In light of this positive practical experience, House began to consult with hospital and clinic officials in East Africa and the Caribbean regarding the provision of their medical treatment needs via satellite links. Various suppliers of specialized medical terminal equipment and videoconferencing codecs (that is, encoders/decoders), such as Colorado Video, donated approximately $300,000 in equipment that could be used at remote medical facilities to help with medical testing and training.

In parallel, officials at Teleglobe Canada and Intelsat developed plans to set up the satellite links. The plan was to operate the Canada-to-Kenya medical link seven hours a day and then switch to a link between Canada and the Caribbean for another seven-hour shift. The link was designed for multiple purposes. The most frequent use was to be for training nurses and doctors on a variety of medical techniques. The link also could serve to relay medical information and records, such as the results of EKGs, EEGs, x-rays, sonograms, and blood serum or urine tests.

Intelsat approved the request, and the Canadian signatory implemented it without delay. The service operated for more than one year. In the case of the Caribbean link, the existing network serving the entire region, known as the UDIWITE network, was able to reach dozens of clinics and hospitals. During the test period, the network was able to provide effective new training to paramedics, nurses, and doctors, as well as diagnoses of many rare medical conditions that could not have been made at the remote locations. The narrowband link also was highly efficient. The use of slow-scan video over the equivalent voice channel was found to be adequate for most needs. When higher resolution and color images were required, air courier systems were employed instead.

Over a period of nearly two years, Project SHARE covered a large number of projects and involved more than 100 different countries. The intent of the project was to develop specific educational and health services for remote areas that were geared to specific needs and suited technically and financially to implementation on a practical basis. The range of projects was geographically, technically, and functionally diverse.

The largest single project in terms of coverage was the so-called "Day of Five Billion" global television production created by CNN. This show included footage from dozens of countries around the world and addressed the educational and environmental issues arising from the Earth's total population reaching 5 billion people. Its satellite distribution network reached more than 150 countries and showed the potential of developing countries being able to supply, not just receive, programming.

Conclusion

Rather than attempting to summarize or characterize the many dozens of programs that were tested, demonstrated, or evaluated during Project SHARE, it is more important to focus on what was learned from the collective experience and the trends it helped create for the future. In terms of key conclusions, the final review of Project SHARE seemed to suggest the following:

- The programs that were the most successful in the test and demonstration stage, and that went on to operational implementation, were typically ones in which the participating country developed its own programming and designed its own projects.

- The reverse was also true—that is, success levels were much lower when developed countries produced programs to demonstrate in developing countries.

- The appeal of video-based services was strong and broadly based in the tests and demonstrations, but the economic realities of narrowband voice and data services were reflected in the actual operational implementations in virtually all cases, with the exception of China.

- The importance of using small, low-cost ground equipment in rural and remote locations for a wide range of social services was shown in locations in Africa, the Caribbean, South America, and Asia. In short, the established model of large ground stations and large streams of satellite traffic were not well suited to the increasing demands for distributed networks and rural and remote services.

During the five years that followed Project SHARE, which ended in 1986, a revolution in satellite communications occurred. A host of new services sprang forth. These included customer-premise ground stations with apertures three meters in diameter (known as "very small aperture terminals") and even microterminal dishes 65 centimeters in diameter. Direct broadcasting also emerged. New commercial applications, especially the demands of oil companies, banks, and insurance companies, were the primary drivers behind the creation of these new satellite services and the demand for smaller ground terminals. Nevertheless, the vision of the future embodied by Project SHARE tests and demonstrations also aided and speeded this process. Their impact was large. Project SHARE exposed dozens of heads of state, ministers, and high political leaders to the potential of remote satellite educational and health services. Doctors, teachers, politicians, United Nations officials, broadcasters, and telecommunications experts around the world learned how to work together to create new and even remarkably innovative capabilities. The unique scope and reach of Project SHARE was truly a phenomenon. In some ways, the world of modern satellite communications did change the way we think about services to remote areas in an increasingly smaller and interconnected world.

Commercial telephone, data, and television requirements played a fundamental role in the rapid development of satellite communications and in the utilization of the 2,000-fold increase in satellite capacity from Early Bird in 1965 to Intelsat VI twenty years later. Nevertheless, the unique requirements for innovative social applications of satellite technology in developing countries and remote areas of the world strengthened, and possibly even accelerated, that trend. Key innovations expedited by the Project SHARE initiative included the bulk procurement program for low-cost Intelsat very small aperture terminals known as VISTA ground stations. Such bulk procurement programs clearly advanced the cause of delivering educational and health services to remote areas. The development of spread-spectrum microterminals for remote data services again promoted satellite-based services in very rural and isolated areas. This started as technology for national satellite services by the Equatorial Communications Corporation, and it was very much the personal mission of former Stanford University professor Edwin Parker. In short, there are numerous examples of how business and entertainment requirements drove satellite and ground-station technology forward, but likewise there are important examples from Project SHARE, in which social and educational requirements helped shape the technical, operational, and market agenda as well.

Ultimately, only about one Project SHARE program in ten made it past the demonstration stage into operational service. This result is in a way still remarkable, given the project's highly experimental nature and the lack of a firm funding base for tests and demonstrations at the program's outset. As anniversary celebrations go, Project SHARE must be considered a significant success, for it created a new framework for international cooperation in providing educational and health services to remote regions, and it spawned dozens of projects that affected nearly 100 countries and millions of people.

Chapter 21

The Evolution of Mobile Satellite Communications

by Edward J. Martin

The development of mobile satellite systems and institutions traces its roots to the close of World War II and proceeds along a torturous path with a number of false starts and successes in the ensuing years. Gaining an understanding of how we arrived at where we are today is worthy of some scholarly research. This chapter, though, does not pretend to produce a well-documented historical review; rather, it is an attempt to sketch the outline of events over the years relying almost entirely on the author's personal recollections, having participated in many of those events. Many other important developments, particularly in Europe and in Japan, are omitted because of a lack of adequate personal recollection.

Because memory is this discussion's main source, many errors in facts, dates, and other details may be found herein. It is hoped, nonetheless, that the general coverage of events is both accurate and interesting enough to stimulate some conclusions on why things went right when they went right, as well as what errors were made when they did not go right. The various attempts to achieve mobile satellite capability in the pre-Inmarsat years are not very well known, but they contain interesting lessons for the policy-maker and entrepreneur alike. Therefore, this discussion dwells more heavily on that era, assuming that more recent events are far better known.

The decade and a half that has just passed may be called in the future the "golden age" of Inmarsat—a period when a unique international political and commercial institution enjoyed great success in developing a global system of satellite communications on the sea, in the air, and on the ground. This success was born out of the convergence of many, largely uncoordinated initiatives that placed the necessary technical, financial, and political resources in the hands of a chosen instrument, the Inmarsat organization. Inmarsat's great success can be measured by its persistent 30-percent annual growth rates in both revenues and customer base. Today, regional and domestic mobile satellite systems are blossoming not only in North America, but also in Asia and the Pacific. Moreover, after three decades of commitment by commercial satellite operators to the geostationary orbit, a number of private entities now plan to offer global mobile services in competition with Inmarsat using constellations of satellites in low- or intermediate-altitude orbits. Also, other private entities are planning geostationary orbit systems for regional coverage, primarily over land areas, from a number of orbit locations around the globe. Whereas the first commercial mobile terminals, which were designed for shipboard use, weighed up to a ton and cost $50,000 to $75,000, technological advances and higher powered satellites today are driving down size and cost. Now, one can buy a satellite phone that fits into a briefcase for under $10,000. Within a few years, the phone will fit in your pocket and cost well under $5,000. The markets of the future will be measured in the millions of users, as compared to the thousands or tens of thousands today. This chapter attempts to outline the major events that set the stage for the establishment of Inmarsat and the factors that have, in more recent years, led to a proliferation of competing systems.

Mobile Communications Before Sputnik

Between the end of World War II and the dawn of the space age, the U.S. military engaged in an intensive search for a reliable means of air-to-ground communications. The U.S. Department of Defense required reliable control of its airborne nuclear forces and its interceptor fleets, which might have to shoot down enemy bombers threatening U.S. population centers. To maintain a credible counterthreat, Strategic Air Command bombers would be deployed toward the North Pole to await a signal to continue or to return home. Reliable contact with these aircraft, as well as with airborne command posts, was absolutely imperative. It was essential to extend reliable communications ranges well beyond the line of sight. High-frequency communications offered long-range but highly unreliable performance, especially in the polar regions.

Although Arthur C. Clarke's classic *Wireless World* article appeared after the war, recognition of the possibilities offered by the satellite relaying of air-to-ground communications did not emerge until several years later. Clarke's concept, which involved the use of manned television broadcast stations operating more than 20,000 miles in space, was more the stuff of science fiction than a practical alternative. Therefore, U.S. military research efforts concentrated on terrestrial technologies. Much effort went into improving the reliability of high-frequency air-to-ground communications, but no fully satisfactory solution emerged.

At the end of World War II, little was known about the ability of radio waves to penetrate the atmosphere. The war, however, had spawned a powerful new tool: radar. The U.S. Army's Project Diana succeeded in detecting radar signals bounced off the Moon's surface in 1946. Ten years later, British scientists using the powerful Jodrell Bank radar telescope discovered and measured the Faraday rotation of a very high-frequency (VHF) radar signal reflected from the lunar surface. This type of information was to become vital to the planners of VHF mobile satellite systems several years later.

During the late 1950s, U.S. Air Force scientists began to explore the prospect of using the Moon for communications. Unknown to them, though, the U.S. Navy already was heavily engaged in lunar communications work, with the idea of eavesdropping on line-of-sight communications half way around the world in the Soviet Union. The Navy started the design and construction of a 600-foot-diameter steerable antenna for this purpose, but the project failed because the cost was unreasonable; in any case, spy satellites were on the horizon. Nonetheless, the Navy realized a lunar-relay communications system between Hawaii and the U.S. east coast—a system about which little has been published. The U.S. Navy also placed large antennas aboard two ships for Moon bounce communications on the high seas, but the large antennas and small signals associated with this technique offered no promise for widespread future use.

During the 1950s, the U.S. military also undertook tests to assess the potential of using radio wave scattering from irregularities in the atmosphere to generate usable signals on a reliable basis. An Air Force Cambridge Research Laboratories research team demonstrated that tropospheric scattering could support communications out to several hundred miles. Scattering from the ionosphere extended that possibility out to well over 1,000 miles. An elaborate test bed was established on a C-135 aircraft (more commonly recognized as a Boeing 707). Directive transmit antennas were fitted in the nose and tail, each fed by a multikilowatt transmitter; two parasitic arrays were fitted to the wing tips to receive signals. The ground station had an array of six diversity antennas to receive. The goal of this investment in equipment was to establish a two-way teletype link between the ground and aircraft. The test program was coming to a close just as Sputnik heralded the opening of the space age in October 1957, and it became clear that better alternatives to this brute force scatter technology were available.

The Dawn of the Space Age

By 1959, the U.S. Air Force had studied a number of alternative space technologies for improved communications and had concluded that active electronics in orbit were not practical, at least for a while. Instead, they favored the use of passive satellites— that is, radar reflectors—in orbit. NASA, the fledgling space agency, launched into orbit the two Echo balloons, the first artificial satellites visible to the naked eye. If rocket technology could launch those balloons, the military reasoned, then the same technology could destroy the satellite as well.

MIT's Lincoln Laboratory devised an ingenious response to that dilemma: Project West Ford. Commonly known as Project Needles, this program involved deploying a belt of tiny dipole "needles" in orbit about the planet. If two antennas were directed at the same belt location, they could establish a reliable link by scattering energy from the dipoles. One such belt was successfully deployed. Some of the signal processing research carried out for atmospheric scattering now was applied to a space system. Various communications configurations, including ground-to-submarine, were successfully tested. Lincoln Laboratory induced the Air Force Cambridge Research Laboratories to explore the feasibility of communicating with an airborne terminal using the West Ford belt. The ensuing study program resulted in a set of design parameters that included a two-foot (0.6-meter) tracking X-band dish mounted under a radome on the top of an aircraft's fuselage, a ten-kilowatt transmitter, and a maser receiver with a noise temperature of fifty degrees Kelvin. Subsequently, the studies were transferred to the Air Force's Wright Air Development Center at Wright-Patterson Air Force Base, Ohio, for implementation, and an airborne X-band system eventually came into service, but without the West Ford belt, whose usefulness was ended by the advent of the active repeater satellite.

NASA introduced the active satellite era with the launch of the experimental Telstar and Relay satellites—programs carried out in cooperation with Bell Telephone Laboratories and RCA's Astro Division, respectively. No mobile applications were foreseen at that stage because of the low power of the early satellites, but these satellites showed conclusively that electronics could operate reliably in space. In what may arguably have been its most important early decision, NASA decided to support the Hughes plan to put a satellite in the "Arthur Clarke" or geosynchronous orbit. Although Syncom was not expected to have any relevance to mobile communications, Roland Boucher, a Hughes engineer, showed otherwise.

The Syncom communications transponder used a two-watt L-band transmitter, which was far too weak for mobile communications applications. However, as Boucher demonstrated, Syncom's VHF telemetry and command system could be fooled into relaying a communications message. Subsequently, a Pan Am Airlines 707 cargo airplane, fitted with a set of experimental electronics, flew out of Hong Kong in early 1965. The pilot pointed his aircraft toward the Pacific Syncom satellite and relayed a message to the ground station at Camp Roberts, California, sending the first satellite message from a regularly scheduled airliner in flight. Thus was born the first initiative for commercial mobile satellite communications.

Meanwhile, NASA had started its Applications Technology Satellite (ATS) series. Two of these, ATS-1 and ATS-3, incorporated VHF transponders that enabled the airlines and others to conduct extensive test programs. These satellites were so successful that NASA was under constant pressure to employ them in operational roles. The ATS-5 had an L-band transponder, but this regrettably provided only limited test potential because its attitude control system, a gravity-gradient stabilization experiment, failed.

The United States and Europe had great interest in using geosynchronous satellites in a satellite-to-aircraft ranging system for position determination—an interest spurred largely by the increase in North Atlantic air traffic. Because of the range errors produced by the ionosphere on the satellite-to-ground path at VHF frequencies, many in the field believed that the VHF band was unsuitable for position fixing and that the L-band was the only sound option. However, Roy Anderson, a General Electric engineer, demonstrated that differential ranging techniques using fixed reference stations could reduce most of these errors. Anderson's scheme, of course, was a progenitor to the differential techniques now in use with the Global Positioning System. In any case, NASA and the Federal Aviation Administration (FAA) agreed with their European colleagues that the aeronautical system of the future would involve aircraft-to-ground communications and ranging in the L-band. Finding little support from the U.S. airlines and FAA operations, NASA turned to the European Space Research Organization (ESRO). The ensuing cooperative effort involved the deployment of two L-band satellites over the Atlantic Ocean for communications and ranging and sought to achieve closer spacings for more efficient but safe traffic flow. The airlines now feared that a new expensive system would result without their participation and that they eventually would be required to pay for it through user fees.

The Rise and Fall of Aerosat, 1963–1975

The Boucher Syncom demonstration fired the imagination of the airline industry, particularly Frank White of the Air Transport Association (ATA), Ben McLeod of Pan Am, and Cap Petrie of ARINC Radio. Quickly, ATA, ARINC, Pan Am, Boeing, Bendix Radio, and Hughes Aircraft formed an informal industry cooperative in consultation with NASA. The communications satellite firm Comsat was now in operation, and the first commercial satellite was already under construction. In mid-1964, ATA, on behalf of the group, asked Comsat about the feasibility of a VHF aircraft voice link via a geosynchronous satellite. Comsat gave an enthusiastic affirmative reply and joined the group. To assess the experimental performance of VHF satellite-to-aircraft links, Hughes presented Comsat a proposal to piggyback an experimental VHF transponder on the third flight model of the Early Bird satellite at an incremental program price of $250,000. NASA dissuaded Comsat from this course, however, for the reason that it had just decided to deploy a much more powerful VHF package on its ATS spacecraft.

The FAA would be an important player in any planning for aeronautical service, but the agency did not have an internally agreed policy. The air traffic service staff was especially interested because of the explosive growth of traffic in the North Atlantic and Pacific corridors, where high frequency then was the only available long-range communications technique. The lack of reliable communications precluded any major reduction in the standards for separating aircraft on those routes. The FAA research and development staff, on the other hand, seemed totally engaged in designing and implementing five-year plans for developing satellite applications and had no interest in near-term applications. In 1965, Comsat proposed to the FAA a VHF aeronautical satellite service in the Atlantic region. The proposal was extremely sketchy, so Comsat was asked to put together a more comprehensive plan.

Meanwhile, Comsat also was planning a follow-on to Early Bird with cost-sharing arrangements that set a useful precedent for the establishment of the first commercial

Figure 34
The first commercial mobile communications satellite, Marisat, in 1975, built by Hughes for Comsat and used by both U.S. Navy and merchant marine ships. (Courtesy of Hughes Space and Communications Company)

mobile system, Marisat, several years later. The follow-on, designated by Hughes as HS-303A and later known as Intelsat II, devoted 75 percent of its capacity to connect NASA's ground network in support of the Apollo program. The balance was available to extend the coverage of Intelsat communications prior to the establishment of the "global system." Along with its proposal to build these satellites, Hughes offered two unsolicited options, the HS-303-B and HS-303-B-E. The HS-303-B satellite was a VHF version of Intelsat II, equipped with VHF hardware developed for NASA and two ground-to-air channels. The HS-303-B-E version simply doubled the size of the solar array to double the power and, therefore, the channel capacity to four channels. The price for four satellites, enough for global deployment, was just under $10 million.

Comsat did not evaluate these options; rather, it was considering a novel concept: "cross-strapped" communications transponders that used VHF only for the mobile links and C-band for the ground feeder links. Comsat elaborated on this plan in a 1967 proposal to the FAA and incorporated in it a number of technical schemes later used in the Marisat and Inmarsat programs. Comsat then solicited industry bids. It soon became apparent that the change in satellite design from a simple VHF repeater to the C-band/VHF "cross strap" sharply increased the price, and the proposal failed under the weight of the additional costs.

ARINC Radio officials then suggested that Comsat reissue the proposal to them, because that firm already operated the airline high seas communications facilities serving the FAA for air traffic control and the airlines for operational control. The cost burden was still too high, however. ARINC asked Comsat to return to the simpler all-VHF design. As a result, Comsat gave ARINC a design proposal for a four-channel all-VHF satellite in 1968. FAA operational staff were enthusiastic, but they wanted to initiate operation in the Pacific region for a number of reasons, including the very long flight routes over the Pacific Ocean with highly unreliable high-frequency communications, as well as the difficulty experienced in coordinating the use of VHF frequencies with the European aviation community. This design looked promising. The airlines, led by Pan Am, persuaded Boeing to incorporate VHF satellite capability on the world's first jumbo jet, the B-747. All early 747s were wired for VHF satellite communications and carried a VHF slot-dipole antenna mounted under a radome on the fuselage just behind the familiar nose hump. All of this effort came to naught, however, as policy issues at both the national and international levels intruded.

Europe viewed these U.S. efforts with alarm. The United States was pursuing satellite plans with global implications, but with very little international consultation. Comsat, having established its dominance in the development of Intelsat, now appeared to be targeting a new conquest in the field of mobile communications. Moreover, the Europeans felt that VHF, while a feasible near-term solution, did not seem to be the way to proceed in the long run because of a perceived shortage of assignable VHF frequencies as well as propagation anomalies along the North Atlantic aviation routes caused by ionospheric effects. Perhaps even more conclusively, European industry would not have enough time to catch up to the United States and play a meaningful role in space system development. These and other factors pointed the Europeans in the direction of a move into a higher frequency range—namely, the L-band—and a slowing down of progress. The FAA research and development staff, contrary to the agency's air traffic service, supported the Europeans' "go slow" and L-band viewpoint. Given the FAA's lack of a consistent in-house policy, one could hardly expect the United States as a whole to be in tune.

As the 1960s drew to a close, the U.S. government, looking for a way out of the VHF versus L-band dispute, referred the aeronautical satellite question to the Interagency Group on

International Aviation (IGIA). Led by the White House Office of Telecommunication Policy, the IGIA tried to draw up a national plan for aeronautical services.

Comsat submitted a draft plan for the Office of Telecommunication Policy to present to the IGIA. The plan called for the development of two hybrid satellites with both VHF and L-band capability to be deployed over the Pacific Ocean. An early operational capability would be provided at VHF, as long-range development of L-band capability advanced. The plan was well received as a useful compromise, and Comsat prepared an implementation proposal. VHF communications services would be leased to ARINC on behalf of the airlines, while an advanced test program of L-band applications was to be conducted through a spot-beam L-band transponder leased to the FAA. Only one satellite was needed to provide communications services; the second satellite could be deployed as an option to permit two-satellite ranging tests to be performed in both frequency bands.

Comsat's proposal set off alarm bells in Europe and at NASA. It was a clear threat to the joint venture of NASA and ESRO. Pierre Langereux, a leading French aviation pundit, called the hybrid scheme "dangerous." Finally, in early 1971, the White House intervened, issuing an Office of Telecommunication Policy statement after extensive consultation with industry and government agencies. The statement declared that the U.S. government would support only L-band applications, that the FAA—not NASA—was the appropriate agency to manage any satellite effort, and that the space segment should be furnished from commercial sources, if practicable. Thus, Aerosat was born. The work being carried out by NASA in partnership with ESRO was, in effect, turned over to the FAA.

Although practically no one was completely happy with the Office of Telecommunication Policy statement, at least it established some clearer ground rules. Canada now joined what had become the FAA-ESRO enterprise, making it a tripartite venture. Dissension was far from over, however. FAA management concluded that the procurement of service on a commercial basis would not be practicable, and the agency attempted to shut out Comsat and any other potential service providers. The White House again had to intervene. This time, Henry Kissinger spoke for the Nixon administration and confirmed its commitment to commercial procurement. Finally, a two-tier arrangement was agreed on for Aerosat: the FAA, Canada, and ESRO were the operator users of the system on one level, while the FAA would be replaced by a U.S. service provider on the second level.

The next question was how the U.S. operating entity would be selected. Comsat was not the only U.S. company that expressed interest. The United States told the Europeans that it would support any qualified U.S. entity for that role and that ESRO itself could make that decision. This news was brought to ESRO headquarters in Paris by a member of the ESRO staff, who reported that he was not taken seriously at first. That the U.S. government would allow the Europeans to select a U.S. partner was considered to be a hilarious joke.

Once ESRO realized that this was in fact U.S. policy, that space agency had to design a selection process. It decided on competitive bidding. A briefing for bidders took place in Paris; Comsat and the three U.S. international record carriers—RCA, ITT, and Western Union International—attended. The request for proposals that ensued consisted of a series of questions generally along the lines: "Would the U.S. company agree to this?" or "Would the U.S. company agree to that?" The questions were intended to be answered by essays, but Comsat concluded that they were, for the most part, simply multiple choice requiring either a "yes" or "no." Comsat submitted a proposal to ESRO that consisted heavily of one-word paragraphs. Confronted with Comsat's proposal, ESRO had difficulty in using its carefully created proposal evaluation scoring system. As later reported, they were confident that simple "yes" or "no" answers deserved either 100 percent or 0 percent in the scoring, but they were not always sure which. Comsat, in any case, won the contract, and the Aerosat project was launched.

The parties established two Aerosat bodies: one represented the customers, the FAA, ESRO, and the Canadian government, and the other represented the providers, Comsat, ESRO, and the Canadian government. Also, the parties agreed on the creation of a joint program office located in the United States with a European director, if a U.S. prime contractor were selected, or located in Europe with a U.S. director, if the contractor were a European company. Until the satellite contract was awarded, the interim joint program office was set up in 1975 at the European Space Technology Center at Noordwijk, The Netherlands. An international proposal evaluation board under Canadian leadership evaluated the responses to the request for proposals, and General Electric won the contract. Signed in 1975, the contract required a "Notice to Proceed." Although General Electric received an earnest money check for $1,000 (U.S. dollars), the "notice" was not forthcoming.

It should be noted that up to this point, the users of the air space—namely, the airlines—had had no voice in any of the planning. If ever there were a recipe for disaster, this was it, and a disaster it became. Outraged over the process, the airlines lobbied the U.S. Congress, which eliminated the FAA's funding share of the program and, instead, appropriated $1 million to study what really should be done. The dream of a satellite system completely devoted to aviation died then and there, never to be revived. At that point, maritime applications were to take center stage.

All during the flurry of activities during the mid-1960s, Comsat attempted to involve Intelsat in the provision of aeronautical services. Intelsat seemed to be an appropriate international mechanism, because the provision of aeronautical services had been included in the Intelsat charter, although under the more stringent requirements of "specialized services." During the earliest discussions between Comsat and the FAA, the Intelsat governing body formed a special advisory committee on aeronautical matters, chaired by the FAA, to provide a forum for aviation authorities from the various member countries to discuss the topic. Within the first five minutes of the committee's first meeting, the U.S. chairman managed to offend the French delegation, and the situation never recovered. The committee never met again. Nonetheless, over objections from the French, Intelsat agreed to authorize Comsat, as Intelsat manager, to conduct an aeronautical satellite systems engineering study. Comsat awarded the contract to Philco Ford, which did what all have acknowledged as an outstanding job. Unfortunately, Intelsat interest was lost forever. Comsat, however, used the results of the study to stimulate FAA interest in its hybrid scheme, until the Office of Telecommunication Policy torpedoed that program, as mentioned above.

Inmarsat Is Conceived, 1962–1979

Meanwhile, the maritime community was moving slowly but surely toward a mobile communications satellite system. The maritime arm of the United Nations, the International Maritime Consultative Organization, started to examine the use of satellite techniques during the early 1960s in its radio communications subcommittee. Captain Charles Dorian of the United States chaired the subcommittee; Captain Yuri Atserov of the Soviet Union was its vice chairman; and good representation from the world's major seafaring nations rounded out its membership.

Although maritime satellite developments lacked the excitement and urgency of the aeronautical initiatives, they showed far greater promise of reaching fruition. Among other advantages, the Soviet Union had a significant voice in the proceedings; in Intelsat, the Soviet Union would never get a seat at the table because it lacked the level of international traffic necessary to generate a sufficiently large investment share. The Soviet Union suggested the establishment of a new international satellite organization, separate from Intelsat, and even proposed a name: "Inmarsat" (International Maritime Satellite Organization).

The work of the International Maritime Consultative Organization extended over approximately 10 years and relied on, for the most part, a panel of experts drawn from the various member nations. By 1973, the organization agreed to convene an international conference to "decide on the principle of setting up an international maritime satellite system." The conference met three times during 1975 and 1976. At its final session, the conference adopted two agreements: a "convention" among governments and an "operating agreement" among entities designated by governments to finance and operate the system. This two-tiered organization borrowed heavily from the Intelsat experience, but it exhibited many simplifications and improvements. No signatory could vote an investment share greater than 25 percent, for example, unless it first offered the excess share for sale to other signatories. This provision was a thinly veiled effort to prevent the United States from having a totally dominant voice in Inmarsat, as had been the case in the earliest days of Intelsat.

The purpose of the new Inmarsat was defined originally to: "make provision for the space segment necessary for improving maritime communications, thereby assisting in improving distress and safety of life at sea communications, efficiency and management of ships, maritime public correspondence services and radio determination capabilities." Later, its charter was broadened to include aeronautical services. Further amendments to include land mobile services are awaiting ratification. Also of significance was Recommendation 4 of the conference; it encouraged the early study of the possible use of multipurpose satellites to provide both maritime and aeronautical services. This measure clearly recognized the potential economic benefits of meeting maritime and aeronautical requirements through a common space system—a prospective benefit now well recognized after the Aerosat debacle.

The Inmarsat agreements were open for signature in 1976. They provided that investment shares in Inmarsat would be determined by actual use of the space segment, much as was the case for Intelsat. At the outset, however, it was necessary for signatories to bid for a share in the investment. Two or three years later, after the establishment of a global satellite system, Inmarsat would redetermine investment share in proportion to actual use. However, for Inmarsat to come into existence, initial investment commitments of at least 95 percent of total capital outlay was required; if the 95-percent figure were not achieved by July 1979, the enterprise would fail. Two years into this interval, the necessary 95-percent commitment had not been achieved, and the United States had yet to designate an operating entity. Finally, in 1978, after a most acrimonious debate, Congress selected Comsat as the U.S. signatory to Inmarsat under conditions specified by the Maritime Satellite Act of 1978.

Comsat, as the U.S. operating entity, closed the investment gap by offering to subscribe to all of the unsubscribed initial investment shares and thereby guaranteed the establishment of Inmarsat. Europe met this news with mixed emotions. Some feared that Comsat's huge investment share would allow it to dominate Inmarsat during its critical start-up period. The 25-percent voting limit did not seem to totally allay European fears. If initial investment in Inmarsat was oversubscribed, then all shares would be reduced pro rata to reach 100 percent. The European bloc decided to increase its bids to drive Comsat's share down to a level acceptable to Europe. The bidding war was on.

As the deadline for Inmarsat investment commitment neared at midnight on Sunday, 15 July 1979, signatory representatives gathered at the London headquarters of the

International Maritime Consultative Organization on Piccadilly Lane. The Comsat delegation waited until thirty minutes before the deadline to arrive at the final session. Among those in the room were the Chinese delegation ably assisted by British Post Office representatives, as well as a huge mechanical calculator to help in following any changes in investment bids to interpret the impact on China's share.

By midnight, the bidding had risen to a total of close to 300 percent. Comsat made a final offer at approximately ten seconds before the final bell, immediately followed by a joint European offer. This offer contained a formula by which the amount of their investment would increase to whatever level was required to offset any increase in the U.S. offer, after calculating the pro rata adjustment. Comsat objected to the bid as improper; only numerical values, not formulas, should be accepted. The next day, at the very first meeting of the Inmarsat Council in Brighton, England, Comsat filed the first and, to date, the only signatory dispute. The dispute quickly reached a negotiated resolution, thereby removing the threat of lengthy arbitration procedures and saving Inmarsat's credibility from severe damage.

Throughout the events that transpired from Sunday night through Monday morning, when the first Inmarsat Council met, no one pointed out the sad news to the Chinese delegation. The pro rata adjustment for the artificially high investment bids had forced their share down too low to qualify for a seat on the Inmarsat Council. Not realizing this, the Chinese delegation showed up on opening day in Brighton to take their seat. Some quick quiet diplomatic conversation was required, and the Chinese disappeared the following day. Thus was Inmarsat launched in a most colorful fashion!

Marisat: The United States Takes the Lead

Spectrum allocation for mobile satellites was always a critical issue. At the first International Telecommunication Union Space Conference in 1963, maritime mobile services received only a footnote in a small piece of VHF spectrum for the development of space techniques. The more organized aviation community succeeded in having bands allocated to aeronautical space services all the way from VHF to the Ku-band. It was not until the next Space Conference in 1971 that a meaningful frequency arrangement was made to the maritime services: a piece of the L-band spectrum was wrested away from the powerful aviation lobby. With the aim of facilitating the sharing of satellites by both aeronautical and maritime mobile services, the Space Conference placed their satellite allocations in adjacent bands. A single translation frequency, 101.5 megahertz, thus could serve both services. This allocation was significant, because the initiative to establish a dedicated aeronautical satellite system was about to self-destruct.

For the first time, the maritime community now had a meaningful place in the radio frequency spectrum to plan its services. Almost immediately after this frequency allocation, the first high-profile demonstration of satellite capability using a commercial satellite took place. Comsat, Intelsat, and Cunard Line cooperated in a live test program on the high seas. An eight-foot (2.4-meter) World War II radar converted to C-band and installed on the *Queen Elizabeth 2* served for this demonstration of maritime satellite communication. Two entire transponders of an Intelsat IV satellite were required for a voice link, one for each direction of transmission, but the test was a public relations success that hastened efforts to develop a commercial system.

During the 1970s, as Inmarsat was coming into existence, an important parallel activity had begun in the United States. In early 1972, the U.S. Navy made Comsat an urgent request for satellite services to meet fleet communications requirements for ultrahigh-frequency (UHF) capacity during the period between the expected failure of the first-generation experimental satellites, TACSAT and LES-6, and the deployment of the first operational system, FLEETSAT. Initially, the Navy contacted Comsat staff with Navy backgrounds, Commander Burt Edelson and Captain Bill Wood, who in turn referred them to the Comsat project office set up to develop domestic and mobile applications. Comsat recognized the Navy inquiry as an opportunity to introduce a commercial mobile service.

The concept of a hybrid satellite for two different services, as proposed a few years earlier by Comsat for aeronautical applications, was adapted for this program. A system was designed to share satellite power in a flexible manner between Navy UHF transponders and an L-band system. Each Marisat satellite had up to 75 percent of its capacity available at UHF for use by the U.S. government; the remainder was L-band. In June 1972, Comsat presented an unsolicited proposal to the Navy for a three-channel UHF service—specifically, one wideband channel and two narrowband channels, each of which could be activated independently by ground command. The L-band system had three levels of L-band transmit power, so that commercial L-band service could increase as the Navy withdrew its UHF service.

The proposal required the Navy to commit to all three UHF channels on two satellites—one over the Atlantic Ocean and one over the Pacific—for a period of three years, by the end of which time, it was hoped, growth in L-band service would take up the revenue slack. The size of the financial commitment required for the UHF service, $69.8 million, was more than the Navy budget could support; the proposal was not accepted. After a difficult internal evaluation of market risk, Comsat decided to offer a revised proposal in December 1972. At that point, the Navy was required to commit to a wideband channel in each of the two satellites, but for only two years, although with options to extend both time and capacity, and the initial monetary commitment was reduced to $27.9 million, an offer sufficiently attractive for the Navy to proceed.

The U.S. regulatory process did pose a brief threat. Word of Comsat's efforts leaked out, and the U.S. international record carriers asked the Federal Communications Commission (FCC) to insist that they have an opportunity to compete. The Navy promptly issued a request for proposals for the UHF service and required a response within a few weeks. The carriers complained that they did not have enough time, so the Navy provided a one-week extension. Time was a critical factor because the planned service date was July 1974. Still, Comsat was the only bidder and won the contract.

The FCC issued many instructions to Comsat about the program, including the requirement that other service providers be afforded an opportunity to share in system ownership. AT&T decided not to participate on the grounds that the limited ship-to-shore L-band telephone capability would not be large enough to support a viable business. The three international record carriers, though, did join in; RCA opted for an 8-percent investment share, Western Union International had a 3.41-percent share, and ITT had a 2.3-percent share. Comsat formed a subsidiary called Comsat General to serve as the Comsat investor and system manager for the Marisat joint venture—a partnership among the four companies owning the system.

From the outset, Comsat had concluded that a sole-source contract with a satellite manufacturer would be essential to meet the July 1974 target and that Hughes was the only credible supplier. By May 1973, Comsat and Hughes had concluded a contract for three Marisat hybrid satellites at a price just over $40 million. In taking this marketing risk, Comsat remained confident that once the Navy started using the service, it would want all channels for an indefinitely long time. This judgment proved more than correct; the Marisat UHF service still continues 20 years later.

Figure 35
The first commercial ship installation of a mobile communications satellite terminal aboard the ESSO Wilhelmshaven *in 1976. The antenna is sheltered inside the radome perched atop the ship's antenna mast.* (Courtesy of Scott Woods)

In February 1973, immediately after the Navy contract award, Comsat announced the Marisat plan at an international mobile satellite symposium in London. The European community was in a state of shock. The European nominal game plan for maritime satellite services involved a fairly orderly set of assumptions. The International Maritime Consultative Organization was on the verge of convening a conference to set the stage for the establishment of Inmarsat; European governments were investing heavily in their industry to put it into a position to supply the Inmarsat space segment; and the process of designing a communications system was well under way. Western Europe had to decide whether to cooperate with this U.S. initiative or to try to suppress it. Many decided on suppression.

The satellite design had to be completely determined before any significant work began to select a communications design. The satellite had a shore-to-ship C-band-to-L-band transponder and a ship-to-shore L-band-to-C-band transponder, both four-megahertz wide. Comsat assembled a small engineering design team that was led by David Lipke and included Dan Swearingen and Tom Calvit, aided by Dick McClure from Comsat Laboratories. Their challenge was to devise a system to utilize the L-band transponders for ship-to-shore communications and to have the system ready for service in eighteen months. Although faced with a daunting task, the team put together a careful blend of digital and analog technologies for a combined telephony and telex system.

The limited capability of the satellite's L-band system required a careful evaluation of shipboard antenna size. If it were too large, it would cost too much, and no one would install it; if it were too small, the system's channel capacity would be too low to be economical. Europeans believed that a practical shipboard antenna would have to have a very low gain. In the United States, fortunately, Scientific Atlanta had invested internal funds on the development of a low-cost stable platform capable of supporting a one-meter L-band antenna. Encouraged by this work, Comsat chose the higher gain antenna and, in a competitive procurement, ordered 200 units from Scientific Atlanta to seed the market.

The ship-based station, initially called simply Marisat, is known better now as Inmarsat-A, the workhorse of the Inmarsat system. Today, twenty years later, it still accounts for about 90 percent of Inmarsat's revenues. In Europe, work continued on an all-digital system, which the Europeans referred to as the "Forward Looking System." The European system was intended to replace Marisat when Inmarsat began operating; however, Marisat's technical success put off conversion to an all-digital system to the remote future.

After overcoming some satellite development problems, caused in no small measure by the pressure of the schedule, the Marisat system inaugurated commercial service in the Atlantic and Pacific regions through its stations in Southbury, Connecticut, and Santa Paula, California, in the summer of 1976. Pleased with the UHF service, the Navy asked that the third satellite, then held on the ground as a spare, be launched over the Indian Ocean to complete the global service. This request presented Comsat an opportunity to expand L-band service into the Indian Ocean region. Comsat negotiated an agreement with KDD (Kokusai, Denshin, Denwa), the Japanese international carrier, which allowed it to establish a third Marisat station at Yamaguchi, Japan. In November 1978, Marisat service began for ships in the Indian Ocean region through the Yamaguchi station, thus extending Marisat to global coverage.

Inmarsat Opens for Business, 1979–1982

After the establishment of Inmarsat in 1979, a number of start-up issues arose, such as the selection of a director general and the location of a headquarters site, but the overriding issue was the procurement of satellite capacity to begin operations as soon as possible. During the years just before 1979, the debate already had begun. The European national space agencies had invested large subsidies in the development of the Marecs satellites, a maritime version of the European Communications Satellite. Under the European plan, Inmarsat would procure four Marecs satellites (one spare) and deploy them over the three ocean regions. The United States feared that the size of the market would not support the cost of these dedicated satellites. Within Intelsat, another initiative was under consideration with strong U.S. support. It consisted of an L-band piggyback transponder placed on a number of Intelsat V satellites. A compromise "three plus three" proposal was put forth: the Inmarsat space segment would comprise three Marecs and three Intelsat V satellites. The proposal assumed that the Marisat satellites would die conveniently in 1981 after their nominal five-year lives were up, and that they would play no role in Inmarsat. The Marisat concept was far from dead, however.

In 1978, the U.S. Navy, still worried about the delays in its FLEETSAT program, decided to seek another commercial lease of UHF capacity. In a most unusual procurement, Comsat found that its only competitor for the follow-on Navy business was Hughes, which also was its satellite supplier. The Comsat approach to the Navy requirements was to design a Marisat II satellite with the same benefits to all as the existing Marisat. The Soviet Union and Western Europe alike were distressed that the threat of this "Military/Commercial Hybrid," as it was called, had not gone away. In December 1978, at a plenary session of the Inmarsat Preparatory Committee—the international organization created to prepare the way for the establishment of Inmarsat—Comsat put forward its Marisat II plan. The opposition from representatives of the approximately forty Inmarsat countries would have been unanimous, except for the warm support accorded the hybrid by the U.S. State Department. The Marisat concept was not to go away completely, however.

In one of its very first actions, the Inmarsat Council appointed a special working group to develop plans for establishing its first space segment. Three options were on the table: the purchase of Marecs satellites; a possible lease of L-band transponders on Intelsat V; and a possible lease of the L-band capacity on the three Marisat satellites already in orbit. Europeans supported their "three plus three" (Intelsat plus Marecs) solution, with the assertion that the Marisats, with lifetimes of only five years, would be dead soon anyway. The United States argued that the cost of three dedicated Marecs might entail too large an investment for a limited market and that the Marisat satellites showed great likelihood of surviving for a very long time.

Inmarsat issued a request for proposals to all three suppliers. The United States took the position that all proposals should be on a lease basis to allow for fair comparisons. The European Space Agency (ESA) notified the Inmarsat Council that a Marecs lease was not possible. In a tough weekend negotiating session, Comsat, pointing to the procurement rules in the Inmarsat agreements, argued that if the purchase of Marecs satellites were to be considered, then the procurement must be delayed to allow all potential hardware suppliers in the world to compete. The following Monday morning, the Inmarsat Council learned that ESA had made arrangements with a European bank to finance a lease deal. ESA furnished an amended offer, and Inmarsat negotiated contracts with all three suppliers.

In addition to procuring its own space segment, Inmarsat had to arrange for all the details of transferring the operation of a global system from the Marisat Joint Venture to Inmarsat. Inmarsat and Comsat staff, in a spirit of mutual cooperation, worked out the myriad technical, operational, and administrative details. After considering the possibility of making the transition on an ocean-by-ocean basis, the parties decided to switch the whole world at the same moment. Thus, Inmarsat launched global operations on 1 February 1982, only thirty-one months after its establishment.

Inmarsat had hoped that the first Marecs satellite, with its higher capacity, would be available for service over the Atlantic Ocean at the time of transition, but program delays, plus some early operational hiccups in orbit, served to delay its entry into service. As a result, all three Marisat satellites were required for the transition. Marisat made three key contributions to the rapid development of Inmarsat: a global constellation of three satellites, an initial customer base of 1,000 ships, and an operating communications network.

The Golden Era of Inmarsat, 1982–1995

After entry into service, Inmarsat was able to concentrate on developing the market and planning for its next generation of satellites. The internationalization of the system removed those forces suppressing growth in the Marisat era, and Inmarsat began to enjoy annual growth rates of 30 percent in both customer base and revenues. A number of related issues immediately arose in 1982. How big would the future market be? What should the next generation of satellites look like? Also, should Inmarsat follow the Intelsat approach of buying and operating its own satellites, or should it continue with the lease approach? Both methods were authorized by its charter agreements. The "conservatives," led by Comsat, prevailed in the decision to select a "simple" global beam satellite and to seek proposals for both purchase and lease from industry. Inmarsat issued a request for proposals in 1983, following resolution of a number of contentious issues, including orbit locations, the feeder link frequency plan, and the inclusion of a 400-megahertz receiving antenna to support an emergency beacon system. After reviewing the only two proposals received, Inmarsat decided to purchase the satellites, although it later negotiated a financed lease.

The organization now had to face the management of a major space program, as did Intelsat, but with staff resources nowhere near the level of those available to Intelsat. Management oversight was provided at a dangerously low level, and major development problems arose both with the European bus and the American payload. The signatories poured in assistance and authorized a major staff increase, which put the program back on track. Finally, in late 1990, the first Inmarsat-2 was launched, more than seven years after the issuance of the request for proposals. As of today, all four satellites of the Inmarsat-2 series are performing well.

Even before the deployment of Inmarsat-2, work had begun on the next generation of satellites, Inmarsat-3. These satellites, deployed beginning in February 1996, constitute Inmarsat's first use of spot-beam technology. As a result, the size and cost of user terminals will decline. The procurement of launch services has produced a significant political breakthrough: the U.S. government has agreed to support the use of the Russian Proton launcher vehicle for one of the Inmarsat launches. That will be the first use of a Proton rocket with a U.S. commercial satellite.

In 1985, the Inmarsat Assembly adopted amendments to the agreements to extend Inmarsat's legal competence, although on a nonexclusive basis, to provide aeronautical services. These amendments went into force in 1989 after ratification by a sufficient number of signatory countries. Aviation was back in the picture after an absence of more than twenty years. Inmarsat proceeded to design a system capable of sharing the satellites originally procured only for maritime service. It coordinated its activities carefully with organizations, such as the International Civil Aviation Organization, the Airline Electronic Engineering Committee, and the Radio Technical Commission for Aeronautics. This careful international coordination has led to the global acceptance of Inmarsat as a standard system. Avionics for the Inmarsat aeronautical system have proven to be quite expensive, and Inmarsat is currently planning to introduce a lower cost aeronautical service to be offered in the spot-beam coverage of the Inmarsat-3 satellites.

As technology developments shrank the weight and size of the so-called "standard-A" ship terminal, considerable interest developed in producing portable terminals that could be packed in one or more suitcase-sized containers and taken to remote locations on land. Licensing issues are troublesome, but this kind of use has enjoyed explosive growth recently. The Inmarsat Assembly in 1989 adopted nonexclusive amendments to its agreements, very similar to the aeronautical amendments, to enable Inmarsat to enter land mobile markets. As of today, however, a sufficient number of countries have not ratified them.

The United States and Canada Make Their Own Plans

As the Inmarsat system was being established, parallel activities were under way to establish a mobile satellite system to serve the United States and Canada. In 1983, two U.S. companies, Mobilesat and Skylink, filed for experimental authority to launch and operate mobile satellite systems for U.S. services. NASA, with the Canadian Department of Communications, set up the MSAT-X experimental program to further explore and develop the technology. Geostar, a company that ultimately provided fleet management services via satellite communications in the late 1980s, filed its first application with the FCC to provide position determination services using satellites.

In 1984, the FCC established a deadline by which any entity interested in operating in the United States would be required to file an application. Twelve applicants filed, setting off a regulatory battle that took years to resolve. During the next several years, while

Inmarsat was expanding its customer base, the U.S. L-band mobile satellite communications industry consisted principally of legal briefs presented to the FCC.

However, interest in mobile satellite applications stimulated developments outside the L-band spectrum. In 1985, Irwin Jacobs and Andrew Viterbi launched Qualcomm to provide satellite-based mobile positioning and messaging services using spread-spectrum technology through Ku-band domestic satellites. Qualcomm currently is the largest operator of mobile satellite communications services in the world in terms of customer base, with more than 100,000 of its units on trucks and other vehicles in the United States, as well as more than 5,000 in Brazil, Japan, Eastern Europe, and Russia.

Meanwhile, the FCC ruled that the twelve original mobile satellite communications applicants had to form a single consortium, indicating its belief that neither the spectrum nor the market would support more than one licensee. More legal and corporate battles ensued, but finally, in 1988, a consortium, now known as the American Mobile Satellite Corporation (AMSC), was formed. Its principal shareholders are Hughes Communications, Singapore Telecom, AT&T Wireless Services (formerly McCaw), and Mtel Corporation.

AMSC received its FCC operating license in 1989. That license authorized the consortium to provide a full range of mobile voice and data services via satellite to customers on land, in the air, and on the sea within a service area consisting of the continental United States, Alaska, Hawaii, Puerto Rico, the U.S. Virgin Islands, and coastal waters up to 200 miles offshore. In addition to mobile services, AMSC was authorized to provide fixed voice and data services via satellite to locations within its service area that are not served by cellular or fixed telephony. The license authorized up to three geosynchronous satellites to provide these services.

On 7 April 1995, almost exactly thirty years after the launch of the world's first commercial satellite, Early Bird, AMSC launched its first satellite, AMSC-1, on a Lockheed Martin Atlas II rocket. It was the most powerful mobile satellite deployed to date. Service was inaugurated over North America. Early in 1996, Telesat Mobile Inc., the consortium's Canadian partner, launched an identical companion satellite. AMSC has an arrangement with its Canadian partner that will enable each satellite to provide backup to the services provided by the other satellite. The initiation of this service culminated nearly a decade of protracted entrepreneurial and regulatory activity.

New Low-Orbit Systems

In 1990, Motorola announced plans for Iridium, a global system of low-orbit satellites for mobile handheld communication. Subsequently, other proposals appeared, and Inmarsat had to shift its own thinking into high gear. It wished to enter the handheld market in competition with Iridium and the other proposed systems, but Inmarsat soon discovered that its own decision-making processes presented a severe handicap compared to those of the private ventures. In particular, Inmarsat owners were sharply divided over central issues, including the appropriateness of Inmarsat's entry into that market in the first place. Other owners saw the implementation of the land mobile program as a natural extension of Inmarsat's other systems, while still others wanted to take the venture out of Inmarsat entirely, although allowing for the possibility of using Inmarsat as a management contractor.

After years of debate, a compromise was struck. A separate affiliate organization was formed in which Inmarsat held a 15-percent investment share. This compromise meant that all Inmarsat signatories have an investment in the affiliate, whether or not they otherwise chose to participate. The balance of the initial investment opportunity, open initially only to Inmarsat signatories, was oversubscribed. Consequently, as when the original Inmarsat investment opportunity was offered, a pro rata reduction was introduced. Moreover, in late 1995, the announcement came that the first outsider, Hughes Aircraft, had joined the partnership. Hughes also was selected to build the space segment, a constellation of ten operational satellites in medium-altitude orbit. Inmarsat also must face competition from a host of regional system operators that will blanket most of the land areas of the globe with service provided from high-powered geostationary satellites to handheld units much like those envisioned to be used with low-orbit systems. In the future, the role of Inmarsat in global mobile satellite communications clearly will be diminished, but the work accomplished in developing the technology and markets has cleared the way for the coming world revolution in personal communications.

Chapter 22

Net Gain: The Use of Satellites at MCI

by Adam L. Gruen

From 1983 to 1987, MCI used satellites to enhance its national telephone network during a critical period in the corporation's history, even though MCI realized that using satellites for voice transmission was a bad idea and that other technical alternatives were available. In 1985, MCI knew that it was acquiring a money-losing operation called Satellite Business Systems. MCI's use of satellites during the 1980s was not at all a mistake, however, but rather a necessary part of greater and deliberate business schemes.

The Background

The MCI that exists today began in 1968 with a key decision by the U.S. Federal Communications Commission (FCC) affirming the right of telephone users to connect private communications equipment to the AT&T network. This was the so-called *Carterfone* decision of June 1968.[1] In August 1968, William G. McGowan established Microwave Communications of America, which changed its name to MCI three years later.[2]

A later decision in May 1971, called the *Specialized Common Carrier Services* decision, affirmed the right of corporations to offer only specialized services.[3] Carriers were only supposed to *connect* to the AT&T system; nobody had said anything about allowing anyone to *compete* with AT&T by offering basic phone service. This formula was a recipe for shenanigans. What exactly was a specialized service? Also, what was to prevent AT&T from gouging customers trying to connect to the system?

McGowan loved the potential of satellites for two reasons. First and most important, he saw them as a way to bypass the existing AT&T network. Second, he believed that satellites would be a cheap and effective way to reach remote areas without the necessity of building microwave towers every thirty or sixty miles (forty-eight or ninety-six kilometers) from one point to another. In October 1970, therefore, even before MCI had actually finished setting up its towers between Chicago and St. Louis, McGowan set up a subsidiary corporation called MCI Satellite Incorporated. Three months later, MCI and Lockheed Missiles & Space Company formed a joint venture called the MCI Lockheed Satellite Corporation.

---

1. Federal Communications Commission (FCC) Docket 16942, "Carterfone Device," *Decision*, 26 June 1968, 13 FCC 2d, 420.

2. Minutes, MICOM Board of Directors, Special Meeting of 27 August 1968, Minute Book A, MCI Corporate Archives, Washington, DC. MICOM changed its name to MCI Communications Corporation on 19 July 1971. Philip L. Cantelon, *The History of MCI, 1968–1988: The Early Years* (Dallas: Heritage Press, 1993), pp. 117, 570, fn. 10. *The History of MCI* is the property of MCI; the work is not available to the general public. All of the files and documents collected for the work are located in the MCI Corporate Archives, which is not open to members of the public.

3. FCC Docket 18920, "Specialized Common Carrier Services," *First Report and Order*, 3 June 1971, 29 FCC 2d, 872.

The plan was to build a $168 million satellite communications system. MCI Lockheed indeed was one of the first companies to request FCC authorization for domestic satellite-based communications.

The problem was that $168 million back then was really big money, the equivalent of about $400 million today. In 1972, MCI Lockheed needed cash and acquired a new partner called the Communications Satellite Corporation, known as Comsat. At that point, the enterprise became a three-headed joint venture called CML Satellite Corporation.[4] During the following two years, 1973 and 1974, however, MCI was in deep financial trouble. Undercapitalized to begin with, it had managed to survive by selling stock and borrowing money. To make matters worse, AT&T decided to pull the plug on interconnection, meaning that MCI could not use the existing network to offer its services. Ultimately, this was the kind of monopolistic behavior that was the basis for AT&T's defeat in the antitrust suits brought against it, but that was six to seven years in the future. In the meantime, all that MCI could see was that it had a lot of cash going out, but very little coming in. Therefore, MCI sold its third of CML Satellite Corporation to IBM in late 1974. Lockheed, which also had hit hard times, did the same. IBM and Comsat added a new partner, the Aetna Satellite Communications Company, and the entire enterprise was renamed Satellite Business Systems (SBS) in 1975.[5]

What Comes Around Goes Around

During the following seven years, MCI survived mostly by winning court decisions. There was a joke around town that MCI was actually a law firm with an antenna on its roof. The first victory came in 1977 and 1978, when MCI bypassed the FCC and won the right essentially to offer common carrier services.[6] The second major victory came in 1980, when MCI won a massive antitrust case against AT&T.[7] After exhausting the appeals process, AT&T agreed in 1982 to divest itself of most of its local holdings by 1 January 1984.[8] These victories meant that MCI could compete head-on with AT&T for voice, data, and video communications customers. MCI had a one-shot, roughly five-year chance at building a new nationwide coast-to-coast network in the period from 1979 to 1984 and grabbing as much residential market share as possible.

---

4.   Cantelon, *The History of MCI*, p. 341.

5.   *Telecommunications Reports* 51(26) (1 July 1985): 4.

6.   In the so-called *Execunet* decisions, the U.S. Court of Appeals of the District of Columbia overruled the FCC's decision to prohibit MCI from offering Execunet, a shared private-line service. See Cantelon, *The History of MCI*, pp. 250–56. In effect, the original ambiguity of the *Specialized Common Carrier Services* decision had finally come back to haunt the FCC. The federal court argued that if Execunet was a specialized service similar to a regular long-distance service, that was too bad. Once the FCC had come out in favor of competition and innovative services, it could not dictate the validity of specific offerings.

7.   The 1980 case of *MCI v. AT&T* began in 1974, when attorneys of Jenner and Block filed a suit for damages on behalf of MCI in the U.S. District Court for Northern Illinois on the grounds that AT&T had violated the Sherman Antitrust Act of 1887. On 13 June 1980, a twelve-member jury found AT&T guilty of abusing monopoly power in a relevant market, and it awarded MCI the astounding sum of $600 million in damages, which under federal law had to be trebled to $1.8 billion, making it the largest monetary award in U.S. history. However, this amount subsequently was reduced to $113 million on appeals. The real damage to AT&T was not monetary, but in public policy; having lost such a suit in civil court, there seemed to be no real hope of defending against pending antitrust action brought by the U.S. Department of Justice. See *Ibid.*, pp. 297–313.

8.   AT&T and the U.S. Department of Justice agreed to implement divestiture of twenty-two local operating companies—about two-thirds of AT&T's assets—on 24 August 1982. This agreement modified the Consent Decree of 1956 (known also as the Final Judgment). This agreement, which became known as the Modified Final Judgment, set the boundaries and conditions of divestiture that were to take effect on 1 January 1984. See *Ibid.*, pp. 326–27.

Building a coast-to-coast network was a daunting proposition. Three new kinds of technologies could be employed as part of a network: microwaves, satellites, and fiber optic cables. Microwave repeaters could carry only so much traffic before the quality of transmission suffered. For example, an analog FM radio could carry 2,100 circuits. Using a technique called single sideband radio, engineers could boost this to 5,400 circuits with some loss of transmission quality. Digital radios, although providing a much cleaner signal, could carry only 1,344 circuits. It was obvious that digital technology was better for transmitting data, but MCI at this point needed to build a nationwide network for voice as quickly and as cheaply as possible. Therefore, the old analog microwave technology, supplemented by single sideband radio, had to suffice.

To accommodate the growth in network capacity that surely would be needed following 1982, MCI had two alternatives: satellites and fiber optics. Each technology had its advocates within the corporation, and in the end, as with most institutions struggling with a choice between two options, MCI decided to pursue both.

McGowan and his chief of corporate development, H. Brian Thompson, favored satellites for two reasons. Ostensibly, the main argument in favor of satellites was an argument against fiber optics: fiber optics was an untried, untested technology unavailable in large quantities. Ignoring internal advice to the contrary, in 1979 McGowan stated that it would be generations before fiber optics became significant for long-distance point-to-point transmission. Perhaps the real reason he favored satellites was that they had sex appeal from a marketing perspective. Using satellites made MCI a space age company instead of one with just boring microwave towers.[9]

However, MCI's chief operating officer, V. Orville Wright, and the chief of transmission systems, Thomas Leming, contended that satellites were not appropriate for domestic voice transmission.[10] Satellites were appropriate for international voice, because it was the only obvious alternative to undersea cables. Also, satellites would be terrific for digital video and data because of its "bursty" nature, but in 1982 these markets were just opening up for MCI. Leming and Wright argued against using satellites for voice on the very simple basis that they caused a delay and an echo, which most customers would find unacceptably irritating.

Acknowledging the logic of that argument, but unwilling to abandon the allure of satellites altogether, Thompson pushed the notion that even if satellites were not ideal for voice, they offered network redundancy, which was critical to overall network operations. In other words, Thompson felt that customers would be willing to suffer the indignity of a delay or an echo rather than face the outright loss of service in the event of a partial network failure.

In the final analysis, it was McGowan's decision to make, and he decided in favor of satellites. In February 1983, MCI announced the largest purchase of satellite transponder capacity in telecommunications history: twenty-four transponders, carrying 48,000 circuits, were to be purchased for the Galaxy II and Galaxy III spacecraft under construction by Hughes Aircraft. MCI planned to spend between $200 and $300 million for the satellite portion of its network. In September 1983, and again in August 1984, two Delta launchers worked perfectly, and the two spacecraft began to operate.[11] Neither Leming nor Wright had given up on the belief that single-mode fiber optic cable was the pipeline of the future, however.

9.   McGowan quoted in F.I.R. Associates Inc., *Corporate Conference, MCI Communications Corporation*, 15 February 1979, p. 18, History Project Files, MCI Corporate Archives.
10.   Thomas Leming, interview with Philip Cantelon, 24 February 1988, San Francisco, CA, pp. 84–86, MCI Oral History Volumes, MCI Corporate History Archives.
11.   MCI Press Release, 3 February 1983; *Washington Times*, 4 February 1983, MCI Corporate Archives.

Bell Laboratories had invented—and AT&T already had deployed—multimodal optical cable. Single-mode optical cable theoretically was cheaper: it did not need to use as many repeaters. This could only be proven over fairly long distances. Leming reasoned that if MCI deployed single-mode cable over a long enough stretch of distance for a high-traffic-volume route—for example, from Washington, D.C., to New York City—then the economies of scale would bring the cost of the new technology down to the point of profitability.[12] This was a gamble, probably the biggest gamble MCI made in the 1980s. MCI's Washington-to-New York City fiber optical system was operational by March 1984, and by the end of that year, MCI proved that single-mode cable was enormously profitable.

Meanwhile, as Leming and Wright had predicted, in 1984 the number of customer complaints about delayed voice transmission skyrocketed. In 1985 and 1986, even McGowan had to acknowledge defeat. MCI eventually leased back some circuits to Hughes and rented out some transponders to other companies.[13] This brings us to the curiosity of the MCI decision to purchase SBS in June 1985, a full year after the decision to deploy fiber optic cable had proved itself.

Satellite Business Systems

Back in early 1984, MCI executives did not know that the fiber optic gamble would work. McGowan set up a corporate development group, code-named "Orbit," to study candidates in the satellite industry for possible acquisition. One of the candidates was SBS. During the eight years from 1975 to 1983, SBS produced no profits and demanded periodic infusions of cash. Orbit discovered that SBS in fact had lost exactly $120 million for the last three years in a row. By July 1984, as SBS was on its way to losing another $120 million, Comsat finally backed out, selling off its one-third interest to IBM and Aetna.[14]

IBM kept a 59-percent share of SBS, because the satellites' digital data capacity was useful to IBM internally. In East Fishkill, New York, IBM had a laboratory devoted to computer modeling of semiconductor interaction and surface physics interactions. Anyone in IBM designing a semiconductor chip communed with the Fishkill laboratory. This required a huge amount of bandwidth for data, which SBS transponders satisfied. IBM paid cheap bulk rates for the bandwidth, and SBS took the loss. Meanwhile, Aetna and Comsat lost big, in effect paying for two-thirds of what would have been IBM's costs. In 1984, then, IBM found itself stuck with a majority interest in what one journalist called "a dog it did not know how to make grow." According to Richard Liebhaber, then IBM's director of business policy and development, IBM realized it had three options: shut the whole thing down and write it off as a loss; continue to lose $120 million a year; or sell it to someone.[15]

If MCI knew that SBS was losing money, then why did MCI want to acquire SBS? The answer has to be understood within the context of the larger business picture at the time. The 1980s was the decade of the takeover, the corporate raider, and innovative financing tactics, such as the issuance of unsecured corporate bonds or raids on pension surpluses. McGowan and other MCI executives realized that MCI was vulnerable to a hostile takeover. What MCI needed, they realized, was an alliance with a company so big that no one would dare attempt a raid. McGowan called this "shark repellent."[16] An arrangement

12.   Leming interview, 24 February 1988, pp. 84–86.
13.   Cantelon, *The History of MCI*, p. 343.
14.   *Telecommunications Reports* 51(26) (1 July 1985): 4; see also *The Economist*, 29 June 1985, pp. 69–70.
15.   Richard Liebhaber, interview with Philip Cantelon, 1 and 13 April 1992, Washington, DC, p. 8, MCI Oral History Volumes, MCI Corporate Archives.
16.   Cantelon, *The History of MCI*, p. 434.

Figure 36

The first of three Satellite Business Systems (SBS) satellites designed solely to serve the communications needs of U.S. business and launched by NASA. MCI purchased an interest in Satellite Business Systems in 1975 in the hopes of acquiring needed carrier services. (Courtesy of NASA, photo no. 80-H-873)

with IBM was the perfect "shark repellent." By selling off a chunk of its stock to IBM and announcing the possibility of joint ventures and partnerships, MCI gained capital and, perhaps more important, instant credibility with Wall Street and potential business clients. In fact, two days after the alliance was announced, MCI's stock price jumped more than 30 percent.[17]

Meanwhile, IBM also gained. It purchased 16 percent of MCI stock. Many analysts expected Big Blue to purchase the rest of MCI within a few years. However, whether by partnership or acquisition, it seemed evident that IBM had gained an important toehold in the telecommunications industry, thereby hedging against the very real threat that AT&T posed as it moved into computers. Finally, a very specific part of the overall deal was that IBM would finally get to unload SBS on someone else.

By the time this deal was consummated in March 1986, MCI was well aware that satellites did not compare with fiber optics for either voice or data. The reason MCI wanted SBS was not for its transponders in space, but for its business customers on Earth. By this time, the battle for access to residential customers was waning, and MCI was beginning to chart out a new direction for its business policy during the rest of the 1980s. While it would continue to scrap for residential market share, MCI decided that its future lay with obtaining business and government contracts. Here, the rolodexes of the SBS sales representatives were golden. That was the theory behind the acquisition. As it turned out, MCI swallowed SBS whole, but it had trouble digesting the meal. Contacts were one thing, but business arrangements for new telecommunications services were another. Eventually, SBS was dismantled, and the unprofitable pieces were written off as losses. Some people stayed on with MCI; most left to find new jobs elsewhere.[18]

Conclusion

MCI got out of the satellite business as much as it possibly could after 1988. Only recently has the firm expressed interest in getting back into the game.[19] The reason why, again, has to do mostly with context. MCI is moving away from its traditional long-distance voice transmission business into the business of packaging, distributing, and owning digitized content. Its alliance with News Corporation was one step in this direction. MCI recently expressed an interest in purchasing rights to direct-broadcast satellite technology. The lesson in all of this is a simple one. Private industry will embrace unprofitable technologies because companies do not see them as unprofitable, even if they lose money. One has to be willing to expand one's definition of profitability from the narrow frame of return-on-investment to the bigger picture.

---

17.   *Ibid.*, p. 436.
18.   *Ibid.*, pp. 437–44.
19.   MCI wanted to bid in an FCC auction for licensing direct-broadcast satellite television service. On 29 September 1995, the Senate unanimously agreed to direct the FCC to proceed with an auction, and on 16 October 1995, the FCC announced that it would do so. "MCI Wins Round in Satellite TV License Battle," *Washington Post*, 30 September 1995, p. A14.

Appendix A
For Further Reading

Binod C. Agrawal and Kumkum Rai, *Women, Television, and Rural Development: An Evaluative Study of SITE in a Rajasthan Village* (Ahmedabad: Government of India, Space Applications Centre, Software Systems Group, Educational Resources Cell, 1980).

Joel Alper and Joseph Pelton, eds., *The INTELSAT Global Satellite System* (New York: American Institute of Aeronautics and Astronautics, 1984).

Asian Mass Communication Information and Research Centre (AMIC), ed., *Satellite Technology: The Communication Equaliser: An AMIC Compilation* (Singapore: AMIC, 1985).

Martin P. Brown, Jr., ed., *Compendium of Communication and Broadcast Satellites, 1958 to 1980* (New York: Institute of Electrical and Electronic Engineers (IEEE) Press, 1981), published for the IEEE Aerospace and Electronic Systems Society and distributed by John Wiley and Sons.

Radford Byerly, Jr., ed., *Space Policy Reconsidered* (Boulder, CO: Westview, 1989).

Alan D. Campen, ed., *The First Information War* (Fairfax, VA: AFCEA International Press, 1992).

Arthur C. Clarke, "Extra-Terrestrial Relays: Can Rocket Stations Give World-Wide Coverage?," *Wireless World* 51 (October 1945): 305–08.

Arthur C. Clarke, *The Exploration of Space* (New York: Harper & Row, 1952).

Arthur C. Clarke, *Voices from the Sky: Previews of the Coming Space Age* (New York: Harper & Row, 1965).

Linda R. Cohen and Roger G. Noll, eds., *The Technology Pork Barrel* (Washington, DC: Brookings Institution, 1991).

René Collette, "Space Communications in Europe: How Did We Make it Happen?," *History and Technology* 9 (1992): 83–93.

David Crowley and Paul Heyer, *Communication and History* (New York: Longmans, 1991).

Daniel Czitrom, *Media and the American Mind: From Morse to McLuhan* (Chapel Hill: University of North Carolina Press, 1982).

D.I. Dalgleish, *An Introduction to Satellite Communications* (London: Peter Peregrinus for the Institute of Electrical Engineers, 1989).

Lewis Anthony Dexter and David M. White, eds., *People, Society, and Mass Communications* (New York: Free Press, 1964)

Herbert S. Dordick, *Understanding Modern Telecommunications* (New York:, McGraw-Hill, 1986).

C.M. Drury, *A Domestic Satellite Communications System for Canada* (Ottawa: Queen's Printer, 1968).

Burton I. Edelson, Joseph N. Pelton, *et al.*, *Satellite Communications Systems and Technology: Europe, Japan, Russia* (Park Ridge, NJ: Noyes Data Corporation, 1994).

Donald C. Elder, *Out From Behind the Eight-Ball: A History of Project Echo*, AAS History Series, Vol. 16 (San Diego: American Astronautical Society, 1995).

Dennis R. Foote, Heather E. Hudson, and Edwin B. Parker, *Telemedicine in Alaska: The ATS-6 Biomedical Demonstration* (Palo Alto, CA: Stanford University Institute for Communication Research, 1976).

Arnold Frutkin, *International Cooperation in Space* (Englewood Cliffs, NJ: Prentice-Hall, 1965).

Jonathan F. Galloway, *The Politics and Technology of Satellite Communications* (Lexington, MA: D.C. Heath, 1972).

Louis A. Gebhard, *Evolution of Naval Radio-Electronics and Contributions of the Naval Research Laboratory*, Naval Research Laboratory Report 8300 (Washington, DC: Naval Research Laboratory, 1979), Chapter 11, "Satellite Electronics," pp. 395–413.

Theodore R. Hartz and Irvine Paghis, *Spacebound* (Ottawa: Government of Canada, Department of Communications, 1982).

Daniel R. Headrick, *The Invisible Weapon: Telecommunications and International Politics, 1851–1945* (New York: Oxford University Press, 1991).

W.J. Howell, Jr., ed., *World Broadcasting in the Age of the Satellite: Comparative Systems, Policies, and Issues in Mass Telecommunication* (Norwood, NJ: Ablex, 1986).

Heather E. Hudson, *Communications Satellites: Their Development and Impact* (New York: Free Press, 1990).

Heather E. Hudson, ed., *New Directions in Satellite Communications: Challenges for North and South* (Dedham, MA: Artech House, 1985).

Heather E. Hudson, *Three Case Studies on the Benefits of Telecommunications in Socio-Economic Development* (Geneva: International Telecommunications Union, 1983).

Heather E. Hudson, *When Telephones Reach the Village: The Role of Telecommunications in Rural Development* (Norwood, NJ: Ablex, 1984).

Heather E. Hudson, Douglas Goldschmidt, Edwin B. Parker, and Andrew P. Hardy, *The Role of Telecommunications in Socio-Economic Development: A Review of the Literature with Guidelines for Further Investigations* (Geneva: International Telecommunications Union, 1979).

Harold A. Innis, *The Bias of Communication* (Toronto: University of Toronto Press, 1991).

Intelsat, *Project SHARE* (Washington, DC: Intelsat, 1985).

Intelsat, *Project SHARE: A Final Report and Evaluation* (Washington, DC: Intelsat, 1988).

Leonard Jaffe, *Communications in Space* (New York: Holt, Rinehart and Winston, 1966).

Doris H. Jelly, *Canada: 25 Years in Space* (Montreal: Polyscience Publications, in cooperation with the National Museum of Science and Technology, 1988).

Thomas Karas, *The New High Ground: Strategies and Weapons of Space-Age War* (New York: Simon and Schuster, 1983).

Judith Tegger Kildow, *Intelsat: Policy-Maker's Dilemma* (Lexington, MA: Lexington Books, 1973).

Michael Kinsley, *Outer Space and Inner Sanctums: Government, Business, and Satellite Communication* (New York: John Wiley and Sons, 1976).

John Kirton, ed., *Canada, the United States, and Space* (Toronto: Canadian Institute for International Affairs, 1986).

John Krige, *The Launch of ELDO*, ESA HSR-7 (Noordwijk: European Space Agency (ESA), March 1993).

John Krige and Arturo Russo, *Europe in Space, 1960–1973*, ESA SP-1172 (Noordwijk: ESA, 1994).

Gerard Maral and M. Bousquet, *Satellite Communications Systems: Systems, Techniques, and Technology*, translated by J.C.C. Nelson, second edition (New York: John Wiley and Sons, 1993).

Harrie S.W. Massie and Malcolm Owen Robbins, *History of British Space Policy* (Cambridge: Cambridge University Press, 1986).

Walter McDougall, . . . *The Heavens and the Earth: A Political History of the Space Age* (New York: Basic Books, 1985).

Srinivas R. Melkote, *Communication for Development in the Third World: Theory and Practice* (New Delhi: Sage, 1991).

William Melody, Liorar Salter, and Paul Heyer, eds., *Culture, Communication and Dependency: The Tradition of H.A. Innis* (Norwood, NJ: Ablex, 1981).

Lloyd D. Musolf, ed., *Communications Satellites in Political Orbit* (San Francisco: Chandler, 1968).

Ralph M. Negrine, ed., *Satellite Broadcasting: The Politics and Implications of New Media* (New York: Routledge, 1988).

Ralph M. Negrine, ed., *Cable Television and the Future of Broadcasting* (London: Croom Helm, 1985).

Jacob Neufeld, *The Air Force in Space, 1970–1974* (Washington, DC: Office of Air Force History, August 1976).

Joseph N. Pelton, Marcel Perras, and Ashok Sinha, *Intelsat: The Global Telecommunications Network* (Honolulu: Pacific Telecommunications Conference, 1983).

Joseph N. Pelton and Marcellus Snow, eds., *Economic and Policy Problems in Space Communications* (New York: Praeger, 1977).

John R. Pierce, *The Beginnings of Satellite Communication* (San Francisco: San Francisco Press, 1968).

Jon T. Powell, *International Broadcasting by Satellite: Issues of Regulation, Barriers to Communication* (Westport, CT: Quorum Books, 1985).

David W.E. Rees, *Satellite Communications: The First Quarter Century of Service* (New York: John Wiley and Sons, 1990).

Arturo Russo, *The Early Development of the Telecommunications Satellite Programme in ESRO (1965–1971)*, ESA HSR-9 (Noordwijk: ESA, May 1993).

Arturo Russo, *ESRO's Telecommunications Programme and the OTS Project* (1970–1974), ESA HSR-13 (Noordwijk: ESA, 1994).

Herbert I. Schiller, *Mass Communications and American Empire* (New York: Augustus M. Kelley, 1970).

Lorenza Sebesta, *U.S.-European Cooperation in Space During the Sixties*, ESA HSR-14 (Noordwijk: ESA, 1994).

Delbert D. Smith, *Communication Via Satellite: A Vision in Retrospect* (Boston: A.W. Sijthoff, 1976).

Marcellus Snow, *The International Telecommunications Satellite Organization* (INTELSAT) (Baden-Baden: Nomos Verlagsgesellschaft, 1987).

Paul B. Stares, *Space and National Security* (Washington, DC: Brookings Institution, 1987).

Anthony Michael Tedeschi, *Live Via Satellite: The Story of COMSAT and the Technology that Changed World Communication* (Washington, DC: Acropolis Books, 1989).

Mark Williamson, *The Communications Satellite* (New York: A. Hilger, 1990).

Appendix B
Timeline of Selected Events in the Development
of Satellite Communications

Pre-1940s

1895	Tsiolkovsky describes a geosynchronous orbit.
1929	Noordung describes radio communications with a space station in a geosynchronous orbit using large antennas and solar power.

The 1940s

October 1945	Arthur C. Clarke's article, "Extra-Terrestrial Relays" in *Wireless World,* suggests global coverage with three satellites in a geosynchronous orbit.
10 January 1946	U.S. Army Signal Corps under John H. DeWitt, Jr., succeeds in bouncing radar waves off the Moon.
6 February 1946	Zoltán Bay reflects radar waves off the Moon.
12 May 1946	A RAND study proposes a synchronous communications relay system.
October 1946	Stanford University begins meteor radar studies; this is the start of meteor burst communications.

The 1950s

21 October 1951	The Naval Research Laboratory (NRL) carries out the first use of the Moon as a relay in a radio communications circuit.
8 November 1951	Researchers at the National Bureau of Standards Central Radio Propagation Laboratory (Sterling, Virginia) and Collins Radio (Cedar Rapids, Iowa) relay a telegraph message via the Moon.
24 July 1954	NRL achieves the first voice transmission via the Moon.
April 1955	John R. Pierce's article, "Orbital Radio Relays," is published in *Jet Propulsion.*
29 November 1955	NRL demonstrates transcontinental teleprinter communications from Washington, D.C., to San Diego via the Moon.
1956	The first transatlantic telephone cable (TAT-1) starts service between Britain and Canada.

23 January 1956   NRL achieves first transoceanic communications, from Washington, D.C., to Wahiawa, Oahu, Hawaii, via the Moon.

4 October 1957   The Soviet Union launches Sputnik I, the first artificial satellite, into orbit.

1958   The first teletype relay is accomplished by satellite (Courier 1B).

29 July 1958   The National Aeronautics and Space Administration (NASA) is created.

October 1958   A synchronous communications satellite project is proposed for the U.S. Department of Defense (DoD).

November 1958   NASA and DoD split the satellite communications program; NASA gets passive systems, while DoD has active systems.

18 December 1958   DoD launches Project SCORE, which brings the first broadcast from space of a voice message, with a delayed and real-time repeater.

1959   The second transatlantic telephone cable (TAT-2) begins.

1959   DoD's Communication Moon Relay system becomes operational between Washington, D.C., and Hawaii.

May 1959   Jodrell Bank begins lunar communications relay tests with Pye Telecommunications equipment.

6 June 1959   A lunar communications demonstration occurs at the opening of Canada's Prince Albert Radar Laboratory.

16 December 1959   A letter of agreement between NASA and Canada's Defence Research Board is signed for the Alouette satellite.

29 February 1960   Department of Defense combines synchronous satellite communications projects under Project Advent.

August 1960   NASA decides to pursue active satellite communications research; works out agreement with Department of Defense.

The 1960s

12 August 1960   Echo 1 is launched.

October 1960   AT&T requests a license from the Federal Communications Commission (FCC) for an experimental satellite.

October 1960   NASA and DoD reach agreement that NASA will leave synchronous work to DoD.

4 October 1960   DoD's Project Courier is launched, with a delayed repeater.

November 1960	NASA awards a contract for the Relay project's requirements to Space Technology Laboratories, a wholly owned subsidiary of Ramo-Wooldridge (later TRW).
December 1960	AT&T proposes a joint satellite communications effort to NASA.
30 December 1960	President Eisenhower gives NASA the lead role in satellite communications.
4 January 1961	NASA requests proposals for an experimental communications satellite.
March 1961	NASA's communications budget is increased by $10 million.
May 1961	RCA is selected over AT&T and Hughes for the Relay project.
23 June 1961	DoD approves NASA's Syncom project and the use of Advent ground stations.
24 July 1961	President Kennedy signs a policy statement on space communications.
28 July 1961	NASA signs a cooperative agreement with AT&T for Telstar.
11 August 1961	Hughes is selected for Syncom (a sole-source procurement).
21 October 1961	The launch of Project West Ford is unsuccessful.
May 1962	DoD's Project Advent is canceled.
10 July 1962	Telstar 1 is launched, resulting in the first transatlantic television via satellite.
31 August 1962	The Communications Satellite Act is signed, and the Communications Satellite Corporation (Comsat) is created.
29 September 1962	Alouette 1, Canada's top-side sounder and the first satellite by other than the Soviet Union or the United States, is launched.
4 October 1962	President Kennedy names the Comsat board of directors.
December 1962	The Conference of European Postal and Telecommunications Administrations forms a committee to study joining a U.S.-led global satellite communications system.
13 December 1962	Relay 1 is launched, the first communications satellite to transmit television worldwide.
1 February 1963	Comsat is incorporated.
7 May 1963	Telstar 2 is launched.

10 May 1963   Project West Ford (also known as "Project Needles") launches millions of hair-like copper wire dipole antennas into orbit, creating an artificial ionospheric communications relay.

26 July 1963   Syncom 2, the first geosynchronous communications satellite, is placed in orbit; Syncom 1 had failed during launch.

August 1963   An agreement is signed with NASA for Canadian participation in testing experimental communications satellites, including a commitment to build a ground station.

November 1963   The Advanced Technology Satellite (ATS) program initiated.

1964   The European Launcher Development Organization is established.

21 January 1964   Relay 2 is launched.

25 January 1964   Echo 2 is launched.

May 1964   The ATS project is approved at NASA headquarters.

August 1964   Intelsat is created (Interim Agreements).

19 July 1964   Syncom 3 is launched.

1965   The Initial Defense Satellite Communications System, which at first is called the Initial Defense Communications Satellite Program (IDCSP), begins; this will lead to the first operational military communications satellite.

11 February 1965   The first Lincoln Experimental Satellite (LES-1) is launched.

6 April 1965   Comsat's Early Bird (Intelsat I) is launched.

April 1965   Molniya 1 is launched; this is the first Soviet communications satellite.

6 May 1965   LES-2 is launched.

28 June 1965   Intelsat I begins routine operation between the United States and Europe; this is the beginning of commercial satellite communications.

21 December 1965   LES-3 and LES-4 are launched.

26 October 1966   Intelsat IIA is launched.

7 December 1966   NASA's ATS-1 is launched.

1967   The Intelsat II series begins; this is the first communications satellite capable of multiple-access transmissions.

| February 1967 | It is recommended that the prime Canadian space technology objective be its applications to domestic telecommunications and resource surveys. |

1 July 1967   LES-5 is launched.

8 November 1967   NASA's ATS-3 is launched.

1968   TACSAT, the first satellite to provide UHF mobile communications, is launched.

26 September 1968 LES-6 is launched.

January 1969   The first Intelsat III satellite begins service over the Atlantic Ocean.

30 January 1969   The first, International Satellites for Ionospheric Studies program satellite, ISIS 1, is launched.

20 July 1969   The landing of U.S. astronauts Armstrong and Aldrin on the Moon is relayed to Earth via Intelsat III satellites.

12 August 1969   NASA's ATS-5 is launched.

1 September 1969   Telesat Canada is established.

The 1970s

20 March 1970   NATO I is launched.

16 May 1970   The first aircraft-to-aircraft voice transmission occurs using LES-6.

31 March 1971   ISIS 2 is launched.

20 April 1971   The Communications Technology Satellite (CTS) agreement signed between Canada and the United States.

May 1971   The Intelsat Definitive Agreements are signed.

3 November 1971   The first two Defense Satellite Communications System (DSCS) military satellites are launched.

20 December 1971   The first Intelsat IV satellite is put into service.

10 November 1972   Anik A1, the first Canadian domestic geostationary communications satellite (C-band), is launched.

January 1973   NASA quits commercial satellite communications research and development until 1978; some projects remain until completion; and the ATS-G canceled.

20 April 1973	The Canadian Anik A2 satellite is launched.
July 1973	The Europeans decide to build their own launcher.
13 December 1973	DSCS III and DSCS IV are launched.
30 May 1974	NASA's ATS-6 is launched.
19 December 1974	The Franco-German geostationary communications satellite, Symphonie 1, is launched.
26 September 1975	Intelsat IVA (a modified Intelsat IV) is launched.
7 May 1975	Anik A3 is launched.
26 August 1975	Symphonie 2 is launched.
17 January 1976	The CTS (also known as Hermes)—the first high-powered, Ku-band satellite and the world's most powerful to date—is launched.
19 February 1976	Marisat I, the first communications satellite to provide commercial mobile satellite services, begins operation.
8 July 1976	The first Indonesia satellite, Palapa-A1, is brought into service.
14 March 1978	LES-8 and LES-9 are launched.
15 December 1978	Anik B, the first commercial hybrid satellite, is launched.
February 1979	France decides to create Télécom 1, a communications satellite system for domestic and overseas markets.
July 1979	ATS-6 is turned off.
16 July 1979	The Inmarsat Convention is entered into force.
November 1979	The CTS is turned off.

The 1980s

6 December 1980	The first Intelsat V satellite is launched.
1982	Inmarsat begins operation.
3 December 1982	The Advanced Communications Technology Satellite (ACTS) Mission Need Statement is signed by the NASA administrator.
4 August 1984	France's Télécom 1A is launched.
8 May 1985	France's Télécom 1B is launched.

20 November 1987   TV-SAT 1, another Franco-German satellite, is launched.

11 March 1988   France's Télécom 1C is launched.

28 October 1988   TDF-1, a Franco-German geostationary direct-broadcast television satellite, is launched.

6 August 1989   TV-SAT 2 is launched.

The 1990s

24 July 1990   TDF-2 is launched.

12 September 1993 NASA's ACTS is launched.

About the Authors

Jon Agar is a lecturer in the history of technology at the Centre for the History of Science, Technology and Medicine at Manchester University, Manchester, United Kingdom. He has created a computerized catalog of the Jodrell Bank Archives, and his essay on the construction of the Jodrell Bank big dish won the Singer Prize of the British Society for the History of Science.

José Altshuler is a scholar with the Centro de Estudios de Historia de la Ciencia y la Tecnologia, Havana, Cuba. He is the author of various articles and books on the history of science and technology, as well as a range of engineering science subjects. He has been both a professor of electrical engineering and a vice rector of the University of Havana. He also served as vice president of the Cuban Academy of Sciences and as president of the Cuban Space Commission, as well as president of the Cuban Society for the History of Science and Technology since 1993.

Bert C. Blevis joined the Canadian Defence Research Board in 1956. There he studied the Earth's upper atmosphere using radar reflections from the Moon and the aurora; then he investigated radio propagation in the non-ionized atmosphere, exploring the effects of atmospheric gases and precipitation on satellite communications. In 1969, Dr. Blevis joined the newly formed Canadian Department of Communications, where he was responsible for research on the applications of satellites to communications, broadcasting, and search and rescue. He also participated actively in international talks relating to satellite communications and its applications, as well as in negotiations concerning the use of the frequency spectrum and the geostationary orbit. His team, along with NASA, received an Emmy award for their work on the use of twelve-gigahertz satellites in broadcasting and the delivery of services to remote areas. He left government in 1987 as Director General, Space Technology and Applications, and has since worked in industry and as a private consultant. Over the span of his career in ionospheric research and space communications, he has written a large number of scientific and technical papers on a variety of subjects.

Andrew J. Butrica is a research historian and author of numerous articles and papers on the history of electricity and electrical engineering in the United States and France and the history of science and technology in nineteenth-century France. He is the author of *To See the Unseen: A History of Planetary Radar Astronomy*, published in the NASA History Series in 1996; a corporate history, *Out of Thin Air: A History of Air Products and Chemicals, Inc., 1940-1990*, published by Praeger in 1990; and co-editor of *The Papers of Thomas Edison: Vol. I: The Making of an Inventor, 1847-1873*, published by Johns Hopkins University Press in 1989.

Roberto Díaz-Martín is an active board member of the Cuban Society for the History of Science and Technology. He received an electrical engineering degree (telecommunications) from the University of Havana and a Ph.D. in electrical engineering (computer-assisted design) from the Institute of Radioengineering and Electronics of the Czechoslovak Academy of Sciences. Dr. Díaz-Martín served as a scientific deputy director with the Cuban Academy of Sciences and then as scientific secretary of the Cuban Space Commission before joining the Havana Center for Studies on the History of Science and Technology, where he has been a senior researcher since 1995. He has published several scientific papers in the fields of network theory and the computer-assisted design of electronic circuits and, more recently, in the history of science and technology, especially the history of international communications in Cuba.

Donald C. Elder teaches history at Eastern New Mexico University, Portales. He has published several articles on spaceflight and his book, *Out From the Behind the Eight-Ball: A History of Project Echo,* appeared in 1995 from Univelt, Inc., American Astronautical Society History Series. He is currently series editor for the AAAS American Astronautical Society History Series, beginning with volume 19. In addition, the University of Iowa Press published, in 1997, his book on the American Civil War as seen through the eyes of an Iowan, *"A Damned Iowa Greyhound": William Henry Harrison Clayton and the American Civil War.*

Franklin W. Floyd obtained degrees in electrical engineering from the Georgia Institute of Technology and the Massachusetts Institute of Technology (MIT), where he completed a doctoral thesis on x-ray astronomy. After serving in the U.S. Army Signal Corps, in 1963 he joined MIT's Lincoln Laboratory, where he contributed to the development of a variety of spacecraft electronics system. He was responsible for the attitude and orbit control systems of the eighth and ninth Lincoln Experimental Satellites (LES-8 and -9), as well as for the Lincoln Experimental Satellite Operations Center. Subsequently, he concentrated for several years on the development of UHF and millimeter-wave adaptive antennas for space communications applications. Between 1982 and 1985, he was project manager for the FLTSAT EHF Packages launched successfully in 1986 and 1989. After serving for three years as associate head of the Communications Division, in 1988 he left Lincoln Laboratory to join Stanford Telecommunications, Inc., as the founding vice president of the Microwave Division, which designed and manufactured wireless digital communications equipment and frequency synthesizers. He is a longtime member of the Institute of Electrical and Electronic Engineers (IEEE).

Jonathan F. Galloway teaches in the Department of Politics at Lake Forest College, Lake Forest, Illinois. He is the author of several publications, including *The Politics and Technology of Satellite Communications* (Lexington Books, 1972). He is also a member of the International Academy of Astronautics and the American Institute of Aeronautics and Astronautics (AIAA).

Adam L. Gruen is MCI's corporate historian and a member of the Network MCI Library. A graduate of Duke University, where he studied under Dr. Alex Roland, he enjoyed a Guggenheim Fellowship at the National Air and Space Museum and was formerly project director for NASA's Space Station History Project. A historian of technology with special expertise in the history of aerospace, networking systems, and modern structural and civil engineering, he also is author of the webwork vCity 1.0. Today, he lives with his wife, Beth Silver, and cat in Falls Church, Virginia.

Daniel R. Glover is a NASA engineer working for the Lewis Research Center's Satellite Networks and Architectures Branch in Cleveland. Since joining Lewis in 1982, most of his time has been spent working on spacecraft or space experiment design projects. Currently, he is working on Internet protocols over satellite. He received a bachelor of science in electrical engineering from The Ohio State University in 1980, a master of science in electrical engineering from the University of Toledo in 1986, and a Ph.D. from the University of Toledo in 1992. He is a licensed professional engineer in the State of Ohio. He is a member of the IEEE and the Technical Committee on History of the AIAA. He is the author of *Planet Quest,* published by the Space Telescope Science Institute's Exploration in Education program.

Daniel R. Headrick teaches history at Roosevelt University in Chicago. He is the author of several articles and books, including *Tools of Empire: Technology and European Imperialism in the Nineteenth Century* (Oxford University Press, 1981); *The Tentacle of Progress: Technology in the Age of Imperialism, 1850–1940* (Oxford University Press, 1988); and *The Invisible Weapon: Telecommunications and International Politics, 1851–1945* (Oxford University Press, 1991).

Edward J. Martin has spent the bulk of his forty-two-year career in the field of mobile communications, pioneering the use of satellites for the last thirty-three of those years. During the mid-1960s to early 1970s, he led Comsat's efforts to establish an aeronautical satellite system, and during the 1970s, he led the team that developed and operated Marisat, the world's first commercial mobile satellite system. In 1979, he formed Comsat's Inmarsat division and managed the transition from the U.S. to the international Marisat system. As the U.S. representative to the Inmarsat Council, he served as its first chair of the Finance Committee and as chair of the Council. Also at Comsat, he served on the Intelsat Board of Governors in various capacities, including vice president of Maritime Services. Since retiring from Comsat in 1988, he has been a private consultant to such clients as Hughes, Loral, GE Astrospace, Motorola/Iridium Inc., Honeywell, California Microwave, and Bell Atlantic International. He is currently president of MSUA (Mobile Satellite Users Association), an organization of industry and government members devoted to improving the mobile communications industry. A fellow of the IEEE, member-at-large of the IEEE Aerospace Policy Committee, and senior member of the AIAA, he is also a member of the Society of Satellite Professionals International.

Joseph N. Pelton currently serves as vice president of academic programs and dean of the International Space University in Strasbourg, France. During the previous seven years, he was the director of the Interdisciplinary Telecommunications Program at the University of Colorado at Boulder. The founding president of the Society of Satellite Professionals, a full member of the International Academy of Astronautics, and a member of Who's Who International, he held a number of management and executive positions with Intelsat for 22 years, including director of strategic policy and executive assistant to the director general. He currently chairs the 1997 NASA/National Science Foundation Panel on International Satellite Communications Technology and Systems. In addition, he is the author of fourteen books on telecommunications, the future of technology, and satellite communications, including *Global Talk: The Marriage of the Computer, World Communications, and Man* (Sijthoff & Noordhoff, 1981), for which he received the Eugene M. Emme Astronautical Literature Award from the American Astronautical Society. He also is editor of the *Journal of Space Communications* and originator of Project SHARE, for which he won the H. Rex Lee Award of the Public Service Satellite Consortium.

Arturo Russo is professor of the history of physics at the University of Palermo, Italy, and the author of several publications on the history of twentieth-century physics, including the history of quantum mechanics, cosmic-ray physics, high-energy physics, and space science. He is currently engaged in the history project of the European Space Agency (ESA). As part of that project, he published (with John Krige) the book *Europe in Space, 1960–1973* (ESA Publications Division, ESA SP-1172, September 1994), as well as several monographs on the history of the European space science and satellite telecommunications programs. Among those monographs, published by the ESA Publications Division, are: *The Definition of ESRO's First Scientific Satellite Programme, 1961–1966* (ESA HSR-2, October 1992); *Choosing ESRO's First Scientific Satellites* (ESA HSR-3, November 1992); *The Definition of a Scientific Policy: ESRO's Satellite Programme in 1969–1973* (ESA HSR-6, March 1993); *The Early Development of the Telecommunications Satellite Programme in ESRO, 1965–1971* (ESA HSR-9, May 1993); *ESRO's Telecommunications Programme and the OTS Project, 1970–1974* (ESA HSR-13, February 1994); and *The Scientific Programme Between ESRO and ESA: Choosing New Projects, 1973–1977* (ESA HSR-16, February 1995).

Lorenza Sebesta has worked with the European Space Agency (ESA) History Project for the last five years at the European University Institute in Florence, Italy. Currently, she holds an ESA-Jean Monnet Chair in European Political Integration at Faculty of Political Sciences, Course of International and Diplomatic Sciences in Bologna, Campus of Forlì. She has been writing extensively on security in the 1950s (*L'Europa indifesa,* Firenze: Ponte alle Graize, 1991) and on topics related to U.S.-European cooperation in space. She is currently working on two book projects: one with John Krige and Arturo Russo on the history of ESA and one with John M. Logsdon on U.S.-European relations in space.

Brian Shoesmith is a senior lecturer in media studies in the Department of Media Studies and director of the Centre for Asian Communication, Media and Cultural Studies at Edith Cowan University, Mt. Lawley, in the Province of Western Australia. He is the co-editor of *The Moving Image: Film and Television in Western Australia, 1896–1985* (History and Film Association of Australia, 1985), as well as several journal articles, and he is currently writing a book for John Libbey & Co., London, on the social and political consequences of satellite television in Asia from 1971 to the present.

David N. Spires teaches history at the University of Colorado at Boulder. As a career Air Force officer, he served on the faculty of the Air Force Academy, Colorado Springs; in intelligence assignments in Vietnam, Europe, and Turkey; and as staff historian at the headquarters of the U.S. Air Forces in Europe. His publications include articles and presentations on the German army and military space issues, as well as books on the pre-Hitler Germany Army, U.S./Greek military relations, and strategic defense issues. His forthcoming book is *Air Power for Patton's Army: The XIX Tactical Air Command in the Second World War.* He has just completed another book titled *Beyond Horizons: A Half Century of Air Force Space Leadership.*

Raman Srinivasan is a graduate student in the history and philosophy of science program at the University of Pennsylvania. His dissertation topic is a study of the technology and politics of satellite communications in India.

Rick W. Sturdevant, a historian with the U.S. Air Force, works in the Office of History at Headquarters Air Force Space Command, Peterson Air Force Base, Colorado. He received his Ph.D. from the University of California at Santa Barbara and has been a public historian since 1983. He has published articles in several journals and magazines, as well as "The United States Air Force Organizes for Space: The Operational Quest, 1943–1993," in *Organizing for the Use of Space: Historical Perspectives on a Persistent Issue* (Univelt, Inc., American Astronautical Society History Series, 1995).

David K. van Keuren is a historian with the Naval Research Laboratory. He has a Ph.D. in the history of science from the University of Pennsylvania (1982), was an Andrew W. Mellon Fellow with the Library of the American Philosophical Society (1984–85), a research consultant with the Center for the History of Chemistry (1985–86), and a summer scholar with the Center for Advanced Studies in the Behavioral Sciences (1986). He is editor of *The Estate of Social Knowledge,* published by the Johns Hopkins University Press; has published several essays on the history of nineteenth- and twentieth-century British and American science; and is currently conducting research on the history of Project Mohole, which reflects his interest in science-government relations during the twentieth century.

Craig B. Waff is a graduate of the Johns Hopkins University program in the history of science and managing editor of physical sciences for Collier's Encyclopedia. He has written variously on eighteenth-century mathematical and planetary astronomy, and he wrote contract histories of the Galileo project and the Deep Space Network for NASA's Jet Propulsion Laboratory.

William W. Ward, a senior member of the IEEE, obtained a bachelor of science in electrical engineering from Texas A&M University and master of science and Ph.D. degrees from the California Institute of Technology in electrical engineering. He served in the U.S. Army Signal Corps in the Pacific arena during World War II. He joined the staff of MIT's Lincoln Laboratory in 1952, where he worked on a assortment of radar and communications systems engineering problems. His radar work included airborne early-warning and ground-based surveillance radars, space tracking and range instrumentation for NASA's Project Mercury, and ballistic missile testing. Beginning in 1965, he worked on space communications systems—mainly the development of systems that served the diverse needs of both military and civil users by means of reliable satellite links. He helped design, build, test, and operate in orbit LES-5, -6, -8, and -9, as well as two EHF Packages launched on FLTSAT satellites. Since retiring from his position as Lincoln Laboratory manager of satellite operations in 1994, he continues to tinker with a few old satellites that refuse to die, and he lectures, writes, consults, and raises vegetables in the summertime.

David J. Whalen is the general manager of engineering (chief engineer) of Asia Satellite Telecommunications Co., Ltd. (AsiaSat), in Hong Kong. He recently completed a Ph.D. degree in public policy (science and technology) from George Washington University, where he wrote a dissertation on the origins of satellite communications technology. He has worked in the communications satellite field since 1974. Dr. Whalen's original degrees were in astronomy, and he later earned an MBA. His major interest is in the interplay of politics and economics in technological innovation.

Nigel Wright is a graduate student in the PREST (Policy Research in Engineering, Science and Technology) department of the University of Manchester, England. His dissertation addresses British and Commonwealth telecommunications in the postwar period to 1965, specifically focusing on the introduction of the transoceanic coaxial telephone cable and communications satellites into intercontinental communications networks—looking at these developments within the wider context of British foreign and economic policy in the postwar years.

Zhu Yilin is secretary-general of the Science and Technology Commission of the Chinese Academy of Space Technology. He has participated extensively in the Chinese satellite communications program.

Glossary of Acronyms

AAS	American Astronautical Society
ABC	American Broadcasting Company
ACTS	Advanced Communications Technology Satellite (NASA)
AFB	Air Force Base
AFCRC	Air Force Cambridge Research Center
AFSPACECOM	Air Force Space Command
AHCIET	Asociación Hispanoamericana de Centros de Investigación y Empresas de Telecomunicaciones
AIAA	American Institute of Aeronautics and Astronautics
AMIC	Asian Mass Communication Research and Information Centre
AMSC	American Mobile Satellite Corporation (satellite)
ARPA	Advanced Research Projects Agency (DoD)
ASC	American Satellite Company (also the acronym for the company's satellite)
AT&T	American Telephone and Telegraph
ATA	Air Transport Association
ATC	Air Training Command
ATS	Applications Technology Satellite (NASA)
Az-El	azimuth-elevation (type of antenna mounting)
BBC	British Broadcasting Company
BBCWST	BBC World Service Television
BMEWS	Ballistic Missile Early Warning System
Caltech	California Institute of Technology
CCTV	Central China Television
CEPT	European Conference of Postal and Telecommunication Administrations (translated from French acronym)
CERN	European Organization for Nuclear Research (English translation)
CETS	European Conference on Satellite Communications (translated from French acronym)
Comsat	Communications Satellite Corporation
CNES	Centre National d'Études Spatiales (French space agency)
CNN	Cable News Network
CNNI	Cable News Network International
CSIRO	Commonwealth Scientific and Industrial Research Organisation
CTS	Communications Technology Satellite (NASA-Canadian program)
DCA	Defense Communications Agency (DoD)
DCS	Defense Communications System (DoD)
DEC	declination
DEW	Distance Early Warning (Line)
DFH	Dong Fang Hong (Chinese satellite)
DoD	(U.S.) Department of Defense
DSCS	Defense Satellite Communications System (DoD)
DSIR	Department of Scientific and Industrial Research (Manchester University)
EBU	European Broadcasting Union
ECS	European Communication Satellite (ESA series) or Environmental Communication Satellite (China)

EHF	extremely high frequency
ELDO	European Launcher Development Organization
ESA	European Space Agency
ESC	European Space Conference
ESRO	European Space Research Organization
Euratom	European Atomic Energy Community
EUTELSAT	European Telecommunications Satellite (Organization)
FAA	Federal Aviation Administration (DOT)
FCC	Federal Communications Commission
FEP	FLTSAT EHF Package
GAO	Government Accounting Office
GEOS	Geodynamics Experimental Ocean Satellite
GNP	gross national product
HA	hour angle
H.M.S.O.	His *or* Her Majesty's Stationer's Office (British government printing office)
IDCSP	Initial Defense Communications Satellite Program (DoD)
IDSCS	Initial Defense Satellite Communications System Program (DoD; formerly IDCSP)
IEEE	Institute of Electric and Electronic Engineers
IGIA	Interagency Group on International Aviation
IGY	International Geophysical Year
INCOSPAR	Indian National Committee for Outer Space Research
Inmarsat	International Mobile (formerly Maritime) Satellite (Organization)
INSAT	Indian National Satellite
Intelsat	International Telecommunications Satellite (Organization)
IRE	Institute of Radio Engineers
ISIS	International Satellites for Ionospheric Studies (NASA-Canadian program)
ITT	International Telephone and Telegraph
ITU	International Telecommunications Union
JPL	Jet Propulsion Laboratory (NASA)
KDD	Kokusai, Denshin, Denwa (Japanese company)
LES	Lincoln Experimental Satellite
LET	Lincoln Experimental Terminal
MBB	Messerschmit-Bolkow-Blohm (German company)
MILSATCOM	Military Satellite Communications (DoD)
Milstar	Military, Strategic, Tactical and Relay (Air Force satellite program)
MIT	Massachusetts Institute of Technology
MLF	Multilateral Force
MSAT	Mobile communications satellite (Canadian civilian spacecraft)
MUSAT	Mobile UHF Satellite (never implemented Canadian proposal)
NACA	National Advisory Committee for Aeronautics
NASA	National Aeronautics and Space Administration
NATO	North Atlantic Treaty Organization
NRL	Naval Research Laboratory
NRO	National Reconnaissance Office (DoD)
NSAM	National Security Action Memorandum
OCS	Operational Communication Satellite (China)
OECD	Organization for Economic Cooperation and Development

OTS	Orbiting Test Satellite (ESRO)
PAMOR	Passive Moon Relay
PAS	Perigee-Apogee System
RCA	Radio Corporation of America
RCTI	Rajawah Citra Televisi Indonesia
RP	Research Publication (NASA)
RRSS	Returnable Remote Sensing Satellite (China)
RTG	radioisotope thermoelectric generator
SAMSO	Space and Missile Systems Organization (U.S. Air Force)
SBS	Satellite Business Systems (also the acronym for the company's satellite)
SCAMP	Single-Channel Advanced Milstar Portable (terminal)
SCORE	Signal Communication by Orbiting Relay Equipment (DoD satellite)
SCOTT	Single-Channel Objective Tactical Terminal
SÉREB	Société pour l'Étude et la Realisation d'Engins Balistiques
SHARE	Satellites for Health and Rural Education in English
SHF	superhigh frequency
SITE	Satellite Instructional Television Experiment (India)
SMU	Southern Methodist University
SP	Special Publication (NASA)
Star	Satellite Television Asia Region
STEP	Symphonie Telecommunications Experimental Project
Syncom	Synchronous Communications (satellite)
TACSAT	Tactical Satellite (U.S. Army)
TAT	transatlantic telephone (cable)
TDRSS	Tracking and Data Relay Satellite System
TIROS	Television and Infrared Operational Satellite
TRW	Thompson-Ramo-Wooldridge, Inc.
TTS	Technical Test Satellite (China)
UHF	ultrahigh frequency
URSI	Union Radioscientifique Internationale
USIA	U.S. Information Agency
USSPACECOM	U.S. Space Command (DoD)
USSR	Union of Soviet Socialist Republics
VHF	very high frequency

Index

The NASA History Series

Reference Works, NASA SP-4000

Grimwood, James M. *Project Mercury: A Chronology* (NASA SP-4001, 1963)

Grimwood, James M., and Hacker, Barton C., with Vorzimmer, Peter J. *Project Gemini Technology and Operations: A Chronology* (NASA SP-4002, 1969)

Link, Mae Mills. *Space Medicine in Project Mercury* (NASA SP-4003, 1965)

Astronautics and Aeronautics, 1963: Chronology of Science, Technology, and Policy (NASA SP-4004, 1964)

Astronautics and Aeronautics, 1964: Chronology of Science, Technology, and Policy (NASA SP-4005, 1965)

Astronautics and Aeronautics, 1965: Chronology of Science, Technology, and Policy (NASA SP-4006, 1966)

Astronautics and Aeronautics, 1966: Chronology of Science, Technology, and Policy (NASA SP-4007, 1967)

Astronautics and Aeronautics, 1967: Chronology of Science, Technology, and Policy (NASA SP-4008, 1968)

Ertel, Ivan D., and Morse, Mary Louise. *The Apollo Spacecraft: A Chronology, Volume I, Through November 7, 1962* (NASA SP-4009, 1969)

Morse, Mary Louise, and Bays, Jean Kernahan. *The Apollo Spacecraft: A Chronology, Volume II, November 8, 1962–September 30, 1964* (NASA SP-4009, 1973)

Brooks, Courtney G., and Ertel, Ivan D. *The Apollo Spacecraft: A Chronology, Volume III, October 1, 1964–January 20, 1966* (NASA SP-4009, 1973)

Ertel, Ivan D., and Newkirk, Roland W., with Brooks, Courtney G. *The Apollo Spacecraft: A Chronology, Volume IV, January 21, 1966–July 13, 1974* (NASA SP-4009, 1978)

Astronautics and Aeronautics, 1968: Chronology of Science, Technology, and Policy (NASA SP-4010, 1969)

Newkirk, Roland W., and Ertel, Ivan D., with Brooks, Courtney G. *Skylab: A Chronology* (NASA SP-4011, 1977)

Van Nimmen, Jane, and Bruno, Leonard C., with Rosholt, Robert L. *NASA Historical Data Book, Vol. I: NASA Resources, 1958–1968* (NASA SP-4012, 1976, rep. ed. 1988)

Ezell, Linda Neuman. *NASA Historical Data Book, Vol II: Programs and Projects, 1958–1968* (NASA SP-4012, 1988)

Ezell, Linda Neuman. *NASA Historical Data Book, Vol. III: Programs and Projects, 1969–1978* (NASA SP-4012, 1988)

Astronautics and Aeronautics, 1969: Chronology of Science, Technology, and Policy (NASA SP-4014, 1970)

Astronautics and Aeronautics, 1970: Chronology of Science, Technology, and Policy (NASA SP-4015, 1972)

Astronautics and Aeronautics, 1971: Chronology of Science, Technology, and Policy (NASA SP-4016, 1972)

Astronautics and Aeronautics, 1972: Chronology of Science, Technology, and Policy (NASA SP-4017, 1974)

Astronautics and Aeronautics, 1973: Chronology of Science, Technology, and Policy (NASA SP-4018, 1975)

Astronautics and Aeronautics, 1974: Chronology of Science, Technology, and Policy (NASA SP-4019, 1977)

Astronautics and Aeronautics, 1975: Chronology of Science, Technology, and Policy (NASA SP-4020, 1979)

Astronautics and Aeronautics, 1976: Chronology of Science, Technology, and Policy (NASA SP-4021, 1984)

Astronautics and Aeronautics, 1977: Chronology of Science, Technology, and Policy (NASA SP-4022, 1986)

Astronautics and Aeronautics, 1978: Chronology of Science, Technology, and Policy (NASA SP-4023, 1986)

Astronautics and Aeronautics, 1979–1984: Chronology of Science, Technology, and Policy (NASA SP-4024, 1988)

Astronautics and Aeronautics, 1985: Chronology of Science, Technology, and Policy (NASA SP-4025, 1990)

Gawdiak, Ihor Y. Compiler. *NASA Historical Data Book, Vol. IV: NASA Resources, 1969–1978* (NASA SP-4012, 1994)

Noordung, Hermann. *The Problem of Space Travel: The Rocket Motor.* In Ernst Stuhlinger and J.D. Hunley, with Jennifer Garland, editors (NASA SP-4026, 1995)

Management Histories, NASA SP-4100

Rosholt, Robert L. *An Administrative History of NASA, 1958–1963* (NASA SP-4101, 1966)

Levine, Arnold S. *Managing NASA in the Apollo Era* (NASA SP-4102, 1982)

Roland, Alex. *Model Research: The National Advisory Committee for Aeronautics, 1915–1958* (NASA SP-4103, 1985)

Fries, Sylvia D. *NASA Engineers and the Age of Apollo* (NASA SP-4104, 1992)

Glennan, T. Keith. *The Birth of NASA: The Diary of T. Keith Glennan*, edited by J.D. Hunley (NASA SP-4105, 1993)

Seamans, Robert C., Jr. *Aiming at Targets: The Autobiography of Robert C. Seamans, Jr.* (NASA SP-4106, 1996)

Project Histories, NASA SP-4200

Swenson, Loyd S., Jr., Grimwood, James M., and Alexander, Charles C. *This New Ocean: A History of Project Mercury* (NASA SP-4201, 1966)

Green, Constance McL., and Lomask, Milton. *Vanguard: A History* (NASA SP-4202, 1970; rep. ed. Smithsonian Institution Press, 1971)

Hacker, Barton C., and Grimwood, James M. *On Shoulders of Titans: A History of Project Gemini* (NASA SP-4203, 1977)

Benson, Charles D. and Faherty, William Barnaby. *Moonport: A History of Apollo Launch Facilities and Operations* (NASA SP-4204, 1978)

Brooks, Courtney G., Grimwood, James M., and Swenson, Loyd S., Jr. *Chariots for Apollo: A History of Manned Lunar Spacecraft* (NASA SP-4205, 1979)

Bilstein, Roger E. *Stages to Saturn: A Technological History of the Apollo/Saturn Launch Vehicles* (NASA SP-4206, 1980)

Compton, W. David, and Benson, Charles D. *Living and Working in Space: A History of Skylab* (NASA SP-4208, 1983)

Ezell, Edward Clinton, and Ezell, Linda Neuman. *The Partnership: A History of the Apollo-Soyuz Test Project* (NASA SP-4209, 1978)

Hall, R. Cargill. *Lunar Impact: A History of Project Ranger* (NASA SP-4210, 1977)

Newell, Homer E. *Beyond the Atmosphere: Early Years of Space Science* (NASA SP-4211, 1980)

Ezell, Edward Clinton, and Ezell, Linda Neuman. *On Mars: Exploration of the Red Planet, 1958–1978* (NASA SP-4212, 1984)

Pitts, John A. *The Human Factor: Biomedicine in the Manned Space Program to 1980* (NASA SP-4213, 1985)

Compton, W. David. *Where No Man Has Gone Before: A History of Apollo Lunar Exploration Missions* (NASA SP-4214, 1989)

Naugle, John E. *First Among Equals: The Selection of NASA Space Science Experiments* (NASA SP-4215, 1991)

Wallace, Lane E. *Airborne Trailblazer: Two Decades with NASA Langley's Boeing 737 Flying Laboratory* (NASA SP-4216, 1994)

Butrica, Andrew J. Editor. *Beyond the Ionosphere: Fifty Years of Space Communication* (NASA SP-4217, 1997)

Butrica, Andrew J. *To See the Unseen: A History of Planetary Radar Astronomy* (NASA SP-4218, 1996)

Center Histories, NASA SP-4300

Rosenthal, Alfred. *Venture into Space: Early Years of Goddard Space Flight Center* (NASA SP-4301, 1985)

Hartman, Edwin, P. *Adventures in Research: A History of Ames Research Center, 1940–1965* (NASA SP-4302, 1970)

Hallion, Richard P. *On the Frontier: Flight Research at Dryden, 1946–1981* (NASA SP-4303, 1984)

Muenger, Elizabeth A. *Searching the Horizon: A History of Ames Research Center, 1940–1976* (NASA SP-4304, 1985)

Hansen, James R. *Engineer in Charge: A History of the Langley Aeronautical Laboratory, 1917–1958* (NASA SP-4305, 1987)

Dawson, Virginia P. *Engines and Innovation: Lewis Laboratory and American Propulsion Technology* (NASA SP-4306, 1991)

Dethloff, Henry C. *"Suddenly Tomorrow Came . . .": A History of the Johnson Space Center, 1957–1990* (NASA SP-4307, 1993)

Hansen, James R. *Spaceflight Revolution: NASA Langley Research Center From Sputnik to Apollo* (NASA SP-4308, 1995)

General Histories, NASA SP-4400

Corliss, William R. *NASA Sounding Rockets, 1958–1968: A Historical Summary* (NASA SP-4401, 1971)

Wells, Helen T., Whiteley, Susan H., and Karegeannes, Carrie. *Origins of NASA Names* (NASA SP-4402, 1976)

Anderson, Frank W., Jr. *Orders of Magnitude: A History of NACA and NASA, 1915–1980* (NASA SP-4403, 1981)

Sloop, John L. *Liquid Hydrogen as a Propulsion Fuel, 1945–1959* (NASA SP-4404, 1978)

Roland, Alex. *A Spacefaring People: Perspectives on Early Spaceflight* (NASA SP-4405, 1985)

Bilstein, Roger E. *Orders of Magnitude: A History of the NACA and NASA, 1915–1990* (NASA SP-4406, 1989)

Logsdon, John M., with Lear, Linda J., Warren-Findley, Jannelle, Williamson, Ray A., and Day, Dwayne A., eds. *Exploring the Unknown: Selected Documents in the History of the U.S. Civil Space Program, Volume I: Organizing for Exploration* (NASA SP-4407, 1995)

Logsdon, John M., with Day, Dwayne A., and Launius, Roger D., eds. *Exploring the Unknown: Selected Documents in the History of the U.S. Civil Space Program, Volume II: External Relationships* (NASA SP-4407, 1996)

NEW IN THE NASA HISTORY SERIES

Exploring the Unknown: Selected Documents in the History of the U.S. Civil Space Program. John M. Logsdon, general editor. *Volume I: Organizing for Exploration* and *Volume II: External Relationships.* NASA SP-4407, 1995-1996.

The first two volumes of this projected five-volume documentary history have already become essential references on the history and development of the U.S. civil space program. Each volume covers specific issues in space program development and includes more than 100 key documents—many published for the first time. Each is introduced by a headnote providing context, bibliographical details, and background information necessary to understand the document. These are organized into major sections, each beginning with an introductory essay that keys the documents to major events in the history of space exploration. This is an important new resources for those who seek to understand the development of the American effort in space.

To See the Unseen: A History of Planetary Radar Astronomy. By Andrew J. Butrica. NASA SP-4218, 1996.

To See the Unseen is an indispensable study of a little-known but important field in space science. The past 50 years have brought forward a unique capability to expand scientific knowledge of the Solar System through the use of radar to conduct planetary astronomy. Scientists can aim a carefully controlled radio signal at a planet (or some other Solar System target, such as a planetary satellite, asteroid, or a ring system), detect its echo, and analyze the information that the echo carries. Andrew Butrica has written a comprehensive and illuminating history that is quite rigorous and systematic in its methodology. *To See the Unseen* explores the development of the radar astronomy specialty within the context of the larger community of scientists.

Aiming at Targets: The Autobiography of Robert C. Seamans, Jr. By Robert C. Seamans, Jr. NASA SP-4106, 1996.

How did the Apollo astronauts ever get to set foot on the Moon? Who was really involved in leading the world's most complex technological feat? The autobiography of former NASA Deputy Administrator Seamans tells a fascinating part of this story. He worked closely with James E. Webb and Hugh L. Dryden in a highly successful "management triad" for much of the Apollo program during the 1960s. After leaving NASA, he continued his distinguished federal career as Secretary of the Air Force and head of the Energy Research and Development Agency in the 1970s. This book is highly valuable reading for those interested in aerospace, government, and military history and may be useful for today's managers, who will learn how at least one person handled previously novel challenges in science, technology, and public policy.